Building Large Language Models from Scratch

Design, Train, and Deploy LLMs with PyTorch

Dilyan Grigorov

Apress®

Building Large Language Models from Scratch: Design, Train, and Deploy LLMs with PyTorch

Dilyan Grigorov
Floor 7,
Varna, Bulgaria

ISBN-13 (pbk): 979-8-8688-2296-4 ISBN-13 (electronic): 979-8-8688-2297-1
https://doi.org/10.1007/979-8-8688-2297-1

Managing Director, Apress Media LLC: Welmoed Spahr
Acquisitions Editor: Celestin Suresh John
Coordinating Editor: Gryffin Winkler

Cover image designed by Giovanni_cg on unsplash.com

Distributed to the book trade worldwide by Springer Science+Business Media New York, 1 New York Plaza, New York, NY 10004. Phone 1-800-SPRINGER, fax (201) 348-4505, e-mail orders-ny@springer-sbm.com, or visit www.springeronline.com. Apress Media, LLC is a Delaware LLC and the sole member (owner) is Springer Science + Business Media Finance Inc (SSBM Finance Inc). SSBM Finance Inc is a **Delaware** corporation.

For information on translations, please e-mail booktranslations@springernature.com; for reprint, paperback, or audio rights, please e-mail bookpermissions@springernature.com.

Apress titles may be purchased in bulk for academic, corporate, or promotional use. eBook versions and licenses are also available for most titles. For more information, reference our Print and eBook Bulk Sales web page at http://www.apress.com/bulk-sales.

Any source code or other supplementary material referenced by the author in this book is available to readers on GitHub. For more detailed information, please visit https://www.apress.com/gp/services/source-code.

If disposing of this product, please recycle the paper

*To the **University of Bath**, where ideas flourish and boundaries are meant to be pushed. Within these walls, I found not just a place to study but a home where curiosity is celebrated and ambitious dreams are nurtured into reality.*

*To **Dr. Harish Tayyar Madabushi**, my supervisor, mentor, and guide through the intricate world of language models. Your unwavering belief in my potential gave me the courage to tackle challenges I once thought insurmountable.*

To my son and my family!
With profound gratitude and warmest regards.

Table of Contents

About the Author

Dilyan Grigorov is a software developer with a passion for Python software development, generative deep learning and machine learning, data structures, and algorithms. He is an advocate for open source and the Python language itself. He has 17 years of industry experience programming in Python and has spent 5 of those years researching and testing generative AI solutions. Dilyan graduated with excellence from Stanford's AI specialization, where he was taught by world-renowned researchers including Andrew Ng, Fei-Fei Li, Percy Liang, and Christopher Manning. He is currently a PhD candidate at the University of Bath, UK, a prestigious research-intensive institution ranked among the top universities in the UK and internationally recognized for its excellence in computer science and artificial intelligence research.

His passion for AI stems from his background as an SEO specialist dealing with search engine algorithms daily. He enjoys engaging with the software community, often giving talks at local meetups and larger conferences. In his spare time, he enjoys reading books, hiking in the mountains, taking long walks, playing with his son, and playing the piano.

About the Technical Reviewer

Tuhin Sharma is Sr. Principal Data Scientist at Red Hat in the Data and AI group. Prior to that, he worked at Hypersonix as an AI architect. He also cofounded and has been CEO of Binaize, a website conversion intelligence product for e-commerce SMBs. Previously, he was part of IBM Watson, where he worked on NLP and ML projects, a few of which were featured on Star Sports and CNN-IBN. He received a master's degree from IIT Roorkee and a bachelor's degree from IIEST Shibpur in computer science. He loves to code and collaborate on open-source projects. He is one of the top 25 contributors to pandas. He has four research papers and five patents in the fields of AI and NLP. He is a reviewer of the IEEE MASS conference and Springer Nature and Packt publications in the AI track. He writes deep learning articles for O'Reilly in collaboration with the AWS MXNET team. He is a regular speaker at prominent AI conferences like O'Reilly Strata and AI, PyCon, PyData, ODSC, GIDS, Devconf, etc.

Introduction

Large Language Models have transformed how we interact with technology, powering everything from chatbots to code generators. Yet for many developers, these models remain mysterious black boxes—impressive in their capabilities but opaque in their construction.

This book demystifies that process. Rather than treating LLMs as abstract concepts or relying solely on pre-built libraries, we'll build one from the ground up. You'll implement each component yourself: tokenization, embeddings, attention mechanisms, and transformer architectures. Along the way, you'll gain an intuitive understanding of why these models work and how to adapt them for your own applications.

Whether you're a software engineer looking to deepen your AI expertise or a researcher wanting to experiment with novel architectures, this hands-on approach will give you the foundation to not just use LLMs, but to truly understand and innovate with them.

Let's begin building!

CHAPTER 1

What Is a Large Language Model? Getting Started with Libraries and Environment Setup for Building an LLM from Scratch

Getting Started with the Foundations of Language Modeling

In the rapidly evolving field of artificial intelligence, **large language models (LLMs)** have emerged as foundational tools in natural language processing (NLP). These models are at the core of systems capable of translating languages, writing essays, generating code, and even composing poetry—all from a simple prompt. The transformative potential of LLMs has reshaped how we understand the intersection of language, computation, and machine learning. However, while many interact with these systems daily, few understand how they are built from the ground up.

D. Grigorov, *Building Large Language Models from Scratch*, https://doi.org/10.1007/979-8-8688-2297-1_1

This book aims to bridge that gap by guiding readers through the process of building an LLM from scratch. The objective is not merely to teach how to use pretrained models, but to **understand their architecture, components, and underlying mathematics**, thereby empowering learners to customize, innovate, and contribute to the future of NLP.

To anchor our learning experience, this course offers a **comprehensive, beginner-friendly pathway** to constructing LLMs without requiring an advanced mathematical background. Inspired by Andrej Karpathy's influential GPT-from-scratch lectures, the curriculum takes a bottom-up approach, starting from simple data operations and progressing toward full transformer-based architecture.

Throughout the course of the book, learners engage with **hands-on Python coding** in **Jupyter notebooks**, using **PyTorch** as the primary deep learning framework. Importantly, all development is conducted **locally**, without reliance on proprietary cloud platforms or expensive GPU clusters. This ensures accessibility while maintaining the technical rigor necessary to build real-world models.

The curriculum systematically explores key components of modern language models. Early topics include **data handling**, such as working with large datasets and implementing memory-efficient strategies. **Tokenization**—the process of converting raw text into usable numerical representations—is introduced with an emphasis on character-level models. From there, learners are guided through the implementation of **neural network fundamentals**, including tensors, activation functions, embeddings, and forward passes.

A substantial portion of the course is dedicated to **PyTorch's modeling capabilities**, emphasizing GPU acceleration via **CUDA** and best practices in structuring model pipelines. The course also introduces key optimization techniques, including **gradient descent**, **AdamW optimization**, **loss function design**, and **training-validation splits** to evaluate model generalization.

As learners progress, the curriculum moves from **simple bigram models** to **transformer-based architectures**, culminating in a decoder-only model structure similar to GPT. Concepts such as **multihead self-attention**, **positional encodings**, **residual connections**, and **dropout regularization** are presented in a clear, incremental fashion. The course concludes with practical applications of model training, **evaluation modes**, **hyperparameter tuning**, and **serialization** for deployment and fine-tuning.

Moreover, attention is given to the ethical, technical, and operational challenges in deploying LLMs at scale. Topics like **data bias**, **overfitting**, **activation functions** (ReLU, sigmoid, tanh), and **model efficiency** are discussed to foster a nuanced understanding of both the power and limitations of LLMs. Tools such as **Hugging Face Transformers** are introduced to connect custom-built models with the broader ecosystem of modern NLP tools.

In the chapter that follows, we will take the core insights from this course and expand them into a structured, readable, and academically sound guide to building a language model from scratch. By grounding our learning in code, mathematics, and practical examples, readers will come away with not just theoretical knowledge but the ability to actively create and experiment with their own language models.

What Is a Large Language Model (LLM)?

In recent years, large language models (LLMs) have emerged as one of the most influential developments in the fields of artificial intelligence (AI) and natural language processing (NLP). These models underpin a growing number of applications ranging from automated translation and conversational agents to content generation, code synthesis, and complex reasoning tasks.

Despite their increasing ubiquity, the conceptual and technical underpinnings of LLMs are often misunderstood or oversimplified. This chapter seeks to provide a rigorous yet accessible introduction to the principles, architecture, and significance of large language models.

At its core, a language model is a probabilistic system designed to predict the next element in a sequence of text, given its preceding context. Traditionally, these models operated at the level of words or characters, estimating the likelihood of sequences based on statistical co-occurrence patterns.

The journey to modern LLMs began with early statistical language models in the 1980s and 1990s, such as n-gram models, which relied on simple co-occurrence probabilities. The introduction of neural networks in the 2000s, particularly recurrent neural networks (RNNs) and long short-term memory (LSTM) models, marked a significant leap in capturing sequential dependencies.

However, it was the advent of the transformer architecture in 2017 that catalyzed the explosive growth of LLMs, enabling unprecedented scalability and performance. This evolution reflects not only technological advancements but also a shift in AI research toward leveraging massive datasets and computational power.

Modern LLMs, however, extend this paradigm dramatically in scale and complexity. A "large" language model typically refers to one that has been trained on a vast corpora—encompassing billions of words—and comprises hundreds of millions to hundreds of billions of parameters, which are the tunable elements that the model learns during training.

The principal mechanism through which contemporary LLMs function is the **transformer architecture**, first introduced by Vaswani et al. - in their "Attention is All You Need" paper (2017) in the landmark paper *"Attention is All You Need."* This architecture eschews the recurrent structures used in earlier models in favor of **self-attention mechanisms**, which allow the model to weigh the relevance of different words in a sequence regardless of their position.

The real power of the transformer architecture lies in its **self-attention mechanism**, which allows the model to dynamically weigh the relevance of each token in a sequence, regardless of its position.

A defining feature of these models is their use of **tokenization**, the process of converting text into a series of numerical units (tokens) that the model can process. These tokens may represent characters, subwords, or whole words, depending on the model's design. Each token is mapped to a high-dimensional vector through a process known as **embedding**, which provides the model with a continuous representation of discrete inputs.

These embeddings are then processed through multiple transformer layers, each comprising attention and feedforward sub-layers, to produce an output distribution over the vocabulary. The model selects the next token based on this distribution, thereby enabling it to generate text sequentially.

For each token, self-attention computes Query, Key, and Value vectors, using their dot products to assign attention scores that determine how much focus each token receives. Multihead attention extends this by capturing diverse relationships (e.g., syntactic, semantic) in parallel, while stacked transformer layers refine these representations.

This mechanism enables LLMs to model long-range dependencies and contextual nuances, making them adept at tasks like translation, summarization, and dialogue. However, self-attention's quadratic complexity poses computational challenges, prompting innovations like sparse attention and efficient approximations, which are critical for scaling LLMs to process longer sequences or run on resource-constrained devices. These advancements, coupled with attention's role in multimodal models, underscore its centrality to the capabilities and ongoing evolution of LLMs.

The versatility of LLMs has led to their adoption across diverse domains. For instance, in healthcare, LLMs assist in summarizing medical records and generating preliminary diagnostic reports. In education, they power intelligent tutoring systems and automated essay grading. In creative industries, tools like AI-driven scriptwriting assistants and music lyric generators rely on LLMs to produce human-like outputs.

Additionally, LLMs are integral to virtual assistants (e.g., Siri, Alexa) and customer service chatbots, handling millions of interactions daily. These applications underscore the transformative potential of LLMs while also raising questions about their reliability and ethical deployment in sensitive contexts.

One of the central training tasks for LLMs is **causal language modeling**, also known as autoregressive modeling. In this framework, the model learns to predict the next token in a sequence by minimizing the difference between its predictions and the actual next tokens observed in the training data.

Over the course of training, often requiring thousands of GPU hours, the model learns not just syntactic rules but also semantic relationships, factual knowledge, and pragmatic patterns embedded in human language.

While LLMs are pretrained on vast datasets to achieve general language proficiency, they are often fine-tuned for specific tasks to enhance their performance in targeted applications. Fine-tuning involves additional training on smaller, task-specific datasets, allowing the model to adapt its general knowledge to domains like legal document analysis, sentiment classification, or question-answering.

Despite their remarkable capabilities, LLMs are not without limitations. Notably, these models do not possess **true understanding** of the language they process. Instead, they operate by detecting and exploiting statistical regularities, which can sometimes lead to **hallucinated content**, factual inaccuracies, or inappropriate outputs.

The Attention Mechanism—Cornerstone of Modern Large Language Models (LLMs)

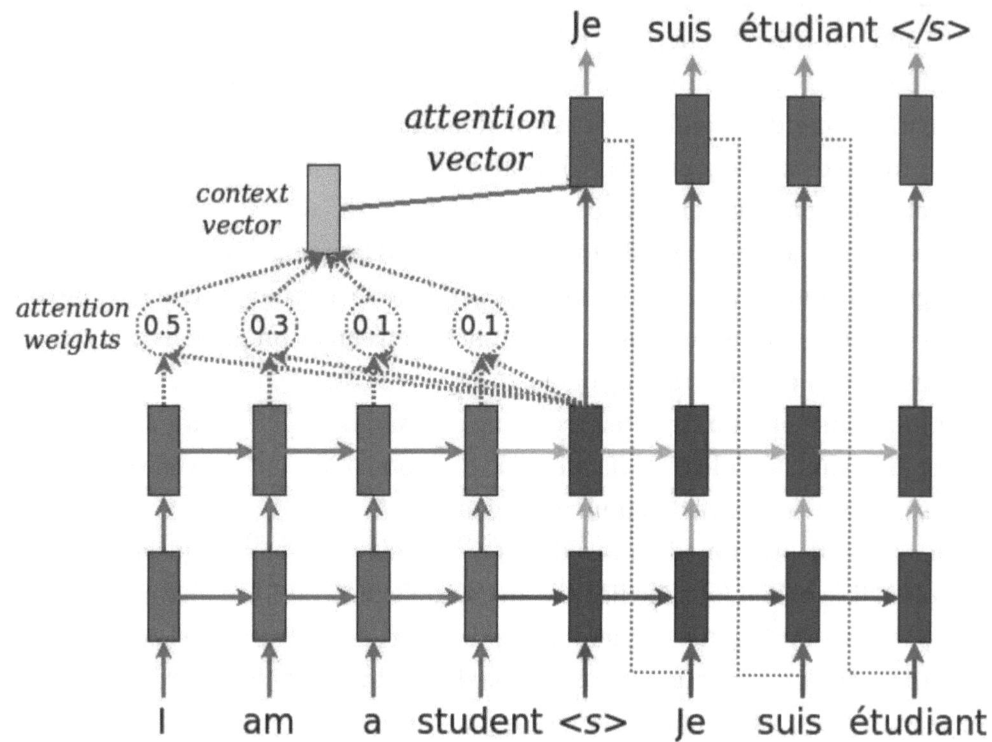

Figure 1-1. *Attention Mechanism, Source: SyncedReview.com*

The attention mechanism (Figure 1-1) enables their remarkable ability to process and generate human-like text. Since the introduction of the transformer architecture in the 2017 paper "Attention Is All You Need" by Vaswani et al., attention mechanisms have redefined natural language processing (NLP) by addressing limitations of earlier models like recurrent neural networks (RNNs). Below, I'll expand on the role of attention mechanisms in LLMs, their technical underpinnings, variants, and significance as of May 2025, tailored to complement the existing text on LLMs you provided.

Attention mechanisms are the heart of modern LLMs, enabling them to process complex linguistic relationships with unprecedented efficiency and flexibility. Their ability to focus on relevant tokens, model long-range dependencies, and adapt to diverse tasks has made transformers the backbone of NLP. As of May 2025, ongoing innovations in sparse, efficient, and multimodal attention continue to push the boundaries of what LLMs can achieve while addressing challenges like computational cost and bias. By integrating a focused discussion of attention into your text, you can provide readers with a deeper appreciation of the technical ingenuity driving LLMs' success.

What Are Attention Mechanisms?

At its core, an attention mechanism allows a model to focus on specific parts of an input sequence when processing or generating text, rather than treating all parts equally. In the context of LLMs, attention determines how much "weight" or importance each token (e.g., a word or subword) in a sequence should have when computing the representation of another token. This dynamic weighting enables LLMs to capture complex relationships, such as long-range dependencies, syntactic structures, and semantic associations, across vast stretches of text.

Unlike RNNs, which process sequences sequentially and struggle with long-term dependencies due to vanishing gradients, attention mechanisms operate in parallel, considering all tokens simultaneously. This parallelism not only improves computational efficiency but also allows LLMs to model relationships between distant tokens—critical for understanding context in lengthy documents or conversations.

Detailed Explanation About How Self-Attention Works in Transformers

The transformer architecture, which underpins virtually all modern LLMs (e.g., GPT, BERT, LLaMA), relies on a specific type of attention called **self-attention**. Here's a detailed breakdown of how it functions:

1. **Token Embeddings and Positional Encoding:**

    - Input text is tokenized into numerical units (tokens), each mapped to a high-dimensional vector via an embedding layer. These embeddings capture semantic and syntactic properties.

    - Since transformers process tokens in parallel (unlike RNNs, which rely on sequence order), positional encodings are added to embeddings to indicate the position of each token in the sequence.

2. **Self-Attention Mechanism:**

    - For each token, the model computes three vectors: **Query (Q)**, **Key (K)**, and **Value (V)**, derived from the token's embedding via learned linear transformations.

    - The attention score between two tokens is calculated as the dot product of the query vector of one token and the key vector of another, scaled by the square root of the dimension of the key vector (to stabilize gradients). This score measures how relevant one token is to another.

    - Scores are normalized using a softmax function to create attention weights, which determine how much focus each token should receive.

    - The output for each token is a weighted sum of the Value vectors, where weights come from the attention scores. This produces a new representation for each token that incorporates contextual information from all other tokens.

3. **Multihead Attention:**

    - To capture diverse relationships (e.g., syntactic, semantic, or positional), transformers use **multihead attention**, where self-attention is computed multiple times (heads) in parallel, each with different learned projections of Q, K, and V.

    - The outputs of all heads are concatenated and transformed to produce the final attention output, allowing the model to attend to different aspects of the input simultaneously.

4. **Layer Stacking:**

    - A transformer layer combines multihead attention with a feedforward neural network, normalization, and residual connections to refine token representations.

    - LLMs stack multiple transformer layers (often dozens or hundreds), enabling increasingly abstract and contextually rich representations as information flows through the network.

This self-attention process is what allows LLMs to weigh the relevance of every token in a sequence, regardless of its position, making them exceptionally adept at tasks like translation, summarization, and dialogue generation.

Why Attention Is Central to LLMs

Attention mechanisms are the linchpin of LLMs for several reasons:

1. **Long-Range Dependency Modeling:**
    - Attention enables LLMs to capture relationships between tokens that are far apart in a sequence, such as connecting a pronoun to its referent in a long paragraph. For example, in the sentence "The scientist, after years of research, discovered a new species," attention ensures the model links "scientist" to "discovered" despite intervening words.
2. **Contextual Understanding:**
    - By assigning different weights to tokens based on context, attention allows LLMs to generate coherent and contextually relevant text. For instance, in a question-answering task, the model can focus on the specific words in the question and relevant parts of a document, ignoring irrelevant details.
3. **Scalability:**
    - Attention's parallel processing makes it computationally efficient compared to RNNs, enabling LLMs to scale to billions of parameters and process vast datasets. This scalability underpins the "scaling laws" mentioned in your text, where larger models and datasets yield better performance.
4. **Flexibility Across Tasks:**
    - Attention's ability to dynamically focus on relevant tokens makes transformers versatile for diverse NLP tasks, from autoregressive text generation (e.g., GPT) to bidirectional context modeling (e.g., BERT) and even multimodal tasks (e.g., image-captioning with CLIP).

Variants and Innovations in Attention (As of May 2025)

Since 2017, attention mechanisms have evolved to address computational bottlenecks, improve efficiency, and extend functionality. Here are key advancements relevant to LLMs in 2025:

1. **Sparse Attention:**
    - Standard self-attention computes scores for all token pairs, resulting in quadratic complexity ($O(n^2)$ for sequence length n), which is computationally expensive for long sequences.
    - Sparse attention variants, like those in **Longformer** or **BigBird**, reduce complexity to O(n) or O(n log n) by limiting attention to a subset of tokens (e.g., nearby tokens or a fixed window). These are critical for processing documents with thousands of tokens, common in legal or scientific applications.
2. **Efficient Attention:**
    - Models like **Linformer** and **Performer** use low-rank approximations or kernel-based methods to approximate full attention, reducing memory and compute requirements.
    - These innovations enable LLMs to run on resource-constrained devices, democratizing access to advanced NLP tools.
3. **Hierarchical Attention:**
    - For very long sequences, hierarchical attention processes text at multiple levels (e.g., sentences, then paragraphs), as seen in models like **HiFormer**. This approach improves efficiency while maintaining contextual understanding.
4. **Cross-modal Attention:**
    - In multimodal LLMs (e.g., DALL·E 3, CLIP), attention mechanisms integrate text and other data types, such as images or audio. Cross-modal attention allows the model to align textual descriptions with visual features, enabling tasks like image generation or visual question answering.

5. **Memory-Augmented Attention:**
    - Recent research explores attention mechanisms that incorporate external memory banks, allowing LLMs to store and retrieve information from past contexts or knowledge bases. This is particularly useful for tasks requiring factual consistency or long-term memory, such as extended dialogues.

These advancements reflect ongoing efforts to make attention mechanisms more efficient, scalable, and adaptable, ensuring LLMs remain at the forefront of AI innovation.

Challenges and Limitations of Attention

Despite their power, attention mechanisms introduce challenges that impact LLMs:

1. **Computational Cost:**
    - Full self-attention's quadratic complexity limits its applicability to very long sequences, even with modern GPUs. Sparse and efficient attention variants mitigate this but may sacrifice some performance.
2. **Overfitting to Patterns:**
    - Attention can overly focus on frequent patterns in training data, leading to biases or hallucinations. For example, an LLM might prioritize common phrases over rare but correct ones, producing plausible but inaccurate outputs.
3. **Interpretability:**
    - Attention weights are often interpreted as indicators of token importance, but they don't fully explain model decisions. Recent research (as of 2025) aims to improve interpretability by analyzing attention flows, but this remains an open challenge.

4. **Bias Amplification:**

    - Since attention learns from training data, it can amplify biases present in the corpus. For instance, if certain demographics are underrepresented, attention may assign lower weights to related tokens, leading to biased outputs.

These limitations underscore the need for ongoing research into robust, fair, and efficient attention mechanisms.

Practical Significance in LLMs

Attention mechanisms are not just a technical innovation—they are the engine driving LLMs' real-world impact. For example:

- In **conversational AI** (e.g., Grok, accessible on grok.com or X apps), attention ensures the model tracks dialogue context, responding coherently to user queries.
- In **translation**, attention aligns words across languages, capturing nuanced relationships (e.g., idiomatic expressions).
- In **content generation**, attention enables LLMs to maintain narrative coherence over long passages, as seen in AI-assisted writing tools.

Moreover, attention's role in multimodal models expands its impact beyond text. For instance, in 2025, LLMs with cross-modal attention power applications like automated video editing, where the model aligns textual scripts with visual scenes.

The Tools We Will Use to Build a Large Language Model from Scratch

Building a large language model (LLM) from scratch requires a set of accessible yet powerful tools. This course adopts a local-first approach, allowing learners to work entirely on their own machines without the need for cloud infrastructure. Below is an overview of the key technologies used:

- **Python**: A widely used programming language in AI and data science. Python's clear syntax and extensive libraries make it ideal for rapid development and experimentation in machine learning.

- **Jupyter Notebook**: An interactive coding environment that supports real-time execution, inline visualizations, and explanatory text. It enables seamless experimentation with model components, code walkthroughs, and documentation.
- **PyTorch**: A dynamic deep learning framework that simplifies tensor operations, neural network construction, and model training. PyTorch is used throughout the course to implement and optimize the LLM architecture.
- **CUDA**: NVIDIA's GPU computing platform that allows PyTorch to offload computations to the GPU. This greatly accelerates matrix operations and training times, especially for larger models.
- **NumPy**: A core library for numerical computing in Python. It provides support for efficient array and matrix manipulation and is often used alongside PyTorch for auxiliary calculations and preprocessing tasks.
- **Matplotlib**: A visualization library used to generate plots and graphs. It is essential for tracking training performance, visualizing loss curves, and understanding model behavior.
- ***The Wizard of Oz***: A public domain literary text from Project Gutenberg. Used as the primary dataset, it provides real-world language data for character-level modeling and training demonstrations.

Together, these tools create a hands-on environment for learning the inner workings of LLMs—from data preparation to neural architecture design and model evaluation.

Why Python for Building Large Language Models?

Figure 1-2. *Python Logo, Source: Python Software Foundation*

Readable, Simple, and Accessible for All Skill Levels

Python's clean, intuitive syntax makes it easy to express complex ideas with minimal code. This greatly speeds up prototyping and debugging while also lowering the barrier for newcomers. Its gentle learning curve allows researchers, engineers, and domain experts to collaborate effectively when building LLMs.

Extensive Machine Learning Ecosystem and Strong Community Support

Python's rich collection of ML and NLP libraries—such as PyTorch, TensorFlow, NumPy, scikit-learn, and Hugging Face Transformers—provides ready-to-use tools for model components, training loops, and data processing. Combined with a massive academic and industry community, developers have access to abundant tutorials, documentation, and best practices.

Integrated Tools for Visualization, Experimentation, and Full ML Pipelines

Python (Figure 1-2) works seamlessly with Jupyter Notebook, Matplotlib, Seaborn, and experiment-tracking tools, making it easy to visualize training metrics, debug intermediate outputs, and document experiments. This integration supports a full end-to-end workflow—from preprocessing and tokenization to training, evaluation, and deployment—without switching languages.

High Performance Through GPU Acceleration, Scalability, and Low-Level Interoperability

Frameworks like PyTorch offer built-in CUDA and GPU support, enabling efficient training of large models that require massive computation. Python's ecosystem also includes distributed computing tools (Ray, Dask, DeepSpeed, DistributedDataParallel) for scaling across multiple GPUs or clusters. When needed, Python can interface with C++ or Rust to optimize performance-critical components.

Production-Ready Deployment and Access to Pretrained Models for Transfer Learning

Python enables a smooth transition from research to production with deployment frameworks such as FastAPI, Flask, ONNX, and TorchScript. The Hugging Face ecosystem provides thousands of pretrained LLMs and datasets, allowing developers to fine-tune models efficiently rather than training from scratch. Together, these capabilities make Python a powerful and practical choice for modern LLM development.

Why Jupyter Notebook for Building Large Language Models?

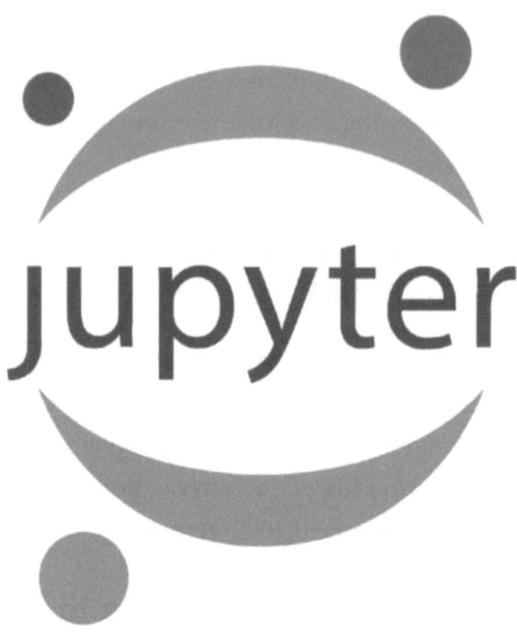

Figure 1-3. *Jupyter Logo*

Jupyter Notebook is a powerful, interactive development environment widely adopted in data science, education, and machine learning research. It plays a central role in this course on large language model (LLM) development due to its unique ability to support **incremental, visual, and literate programming**—all critical for exploring complex neural architectures like transformers.

Interactive, Iterative Coding with Immediate Feedback

Jupyter's cell-based execution model allows users to run and modify small pieces of code independently, making it ideal for experimenting with model components, inspecting tensors, and refining logic without restarting an entire script. This interactive workflow dramatically accelerates learning, debugging, and model exploration.

Unified Environment for Code, Documentation, and Visualization

By combining executable code, Markdown explanations, and visual outputs in one place, Jupyter supports literate programming and transparent experimentation. This format is perfect for demonstrating concepts like attention mechanisms, loss curves, or gradient behavior while keeping explanations and results together in a single, readable document.

Deep Debugging Transparency and Real-Time Inspection

Jupyter (Figure 1-3) makes it easy to observe model behavior at any stage of execution—printing intermediate results, checking tensor shapes, visualizing weights or gradients, and validating assumptions step by step. This granularity is especially valuable when building LLMs from scratch, where debugging often requires inspecting internal model states as they evolve.

Optimized for Learning, Collaboration, and Reproducible Research

Widely used in education and research, Jupyter notebooks are easy to share, version, and reproduce. They support collaborative workflows through tools like Git, nbdime, and JupyterHub, making them ideal for teaching, group projects, and experiment tracking. Their structure ensures that all code, outputs, and documentation are preserved together for transparent and repeatable experimentation.

Seamless Integration with Scientific Libraries and Scalable Environments

Jupyter works natively with Python's scientific stack—NumPy, PyTorch, Matplotlib, Pandas—and also supports advanced visualization tools like Plotly and Bokeh. It can run locally with minimal setup or scale to cloud and multiuser environments through Colab, SageMaker, and JupyterHub. This versatility enables everything from simple local experiments to large-scale LLM training with GPU or TPU acceleration.

Why PyTorch for Building Large Language Models?

PyTorch has solidified its position as a leading deep learning framework for natural language processing (NLP), particularly for developing large language models (LLMs). Created by Meta AI (formerly Facebook AI Research), PyTorch strikes an optimal balance between flexibility, performance, and usability, making it the go-to choice for researchers prototyping novel architectures and engineers deploying production-scale models. Its intuitive design, robust ecosystem, and alignment with modern AI workflows make it uniquely suited for the complex demands of LLM development.

Dynamic, Flexible Execution for Fast Experimentation and Debugging

PyTorch's dynamic computational graph (eager execution) lets developers build and modify models at runtime, enabling immediate inspection of tensors, gradients, and intermediate outputs. This flexibility is invaluable for LLM experimentation, where researchers frequently adjust attention layers, positional encodings, or normalization strategies. Because execution follows standard Python control flow, debugging is intuitive—one can step through code, print tensors, and test custom components without rebuilding an entire graph.

Pythonic Design and Deep Integration with the Scientific Ecosystem

PyTorch behaves naturally within Python, using familiar classes, loops, and conditionals rather than a domain-specific language. This lowers the learning curve and makes complex LLM codebases easier to understand and maintain. It also integrates seamlessly with Python's scientific stack—NumPy, Pandas, SciPy—making data manipulation, tokenization pipelines, and preprocessing straightforward. This alignment ensures accessibility for beginners while providing the power needed for advanced research.

High-Performance GPU Acceleration and Parallel Computing at Scale

PyTorch provides robust support for GPU acceleration through simple APIs like .to("cuda"), enabling fast matrix operations essential for transformer models. It extends beyond single-GPU training with strong **parallel computing** capabilities:

- **Data parallelism** via DataParallel and DistributedDataParallel
- **Model and pipeline parallelism** for extremely large LLMs
- **Distributed training** through torch.distributed, DeepSpeed, and Horovod
- **Cluster-scale scaling** across multiple nodes and hardware accelerators

These capabilities allow LLMs with billions of parameters to be trained efficiently across multi-GPU rigs or cloud clusters, significantly reducing training time.

Modular Architecture and a Rich, Research-Friendly Ecosystem

The torch.nn module offers clean, modular building blocks—embeddings, linear layers, attention modules, normalization layers—that make transformer architectures easy to implement and extend. PyTorch Lightning abstracts away boilerplate training logic, while Hugging Face Transformers provides thousands of pretrained models for fine-tuning. Visualization tools, experiment trackers, and open-source LLM codebases further support rapid experimentation and reproducible research. This ecosystem allows developers to move quickly from ideas to fully functional LLM implementations.

Production-Ready Deployment, Interoperability, and Future-Proof Development

With TorchScript, ONNX export, and integrations with optimization engines like TensorRT, PyTorch enables smooth deployment to servers, cloud environments, edge devices, or high-performance inference systems. It interoperates with C++ via libtorch, supports custom kernels, and integrates with Rust and other languages for performance-critical components. Ongoing advancements—such as torch.compile, better support for new hardware (Hopper, ROCm, TPUs), and improvements to distributed training—ensure that PyTorch remains aligned with future deep learning and LLM trends.

Why CUDA for Building Large Language Models?

Figure 1-4. *Nvidia CUDA, Source: nvidia.com*

Before we continue, it's important to note that the first models are intentionally kept CPU-friendly so they can run locally without specialized hardware. CUDA is introduced only when we move to larger experiments, because while CUDA tools provide major speed and efficiency advantages, they also come with significantly higher costs.

Large language models (LLMs), such as GPT, BERT, and LLaMA, require immense computational resources to train and deploy due to their billions of parameters and complex architectures. CUDA, NVIDIA's (Figure 1-4) parallel computing platform and programming model, has become the cornerstone of modern deep learning, particularly for LLMs. By enabling highly efficient computation on NVIDIA GPUs, CUDA provides the performance, scalability, and flexibility needed to tackle the unique challenges of LLM development. Below, we explore the key reasons why CUDA is indispensable for building LLMs.

Massive Parallelism for Compute-Intensive Operations

LLMs rely on operations like matrix multiplications, tensor contractions, and attention computations, which are inherently parallelizable. CUDA excels at parallel processing by leveraging thousands of GPU cores to execute these operations simultaneously. Key benefits include

- **High Throughput**: A single NVIDIA GPU, such as the A100, can have thousands of CUDA cores (e.g., 6,912 in the A100 40GB), enabling parallel execution of millions of mathematical operations per cycle.
- **Efficient Matrix Operations**: CUDA optimizes matrix multiplications, which dominate transformer-based LLMs (e.g., in attention layers and feed-forward networks), achieving significant speedups over CPU-based computation.
- **Scalability for Large Models**: The parallel nature of CUDA allows LLMs with billions of parameters to be trained efficiently by distributing computations across GPU cores.

For example, computing the attention mechanism in a transformer, which involves large matrix operations like `Q * K^T` (query-key dot product), can be accelerated by orders of magnitude on CUDA-enabled GPUs compared to CPUs.

Seamless Integration with Deep Learning Frameworks

CUDA is tightly integrated with popular deep learning frameworks like PyTorch, TensorFlow, and JAX, which are widely used for LLM development. This integration simplifies GPU-accelerated workflows:

- **High-Level Abstractions**: Frameworks provide simple commands like PyTorch's `.to('cuda')` or TensorFlow's `tf.device('/GPU:0')` to move tensors and models to GPUs, abstracting away low-level CUDA programming.
- **Optimized Libraries**: CUDA powers NVIDIA's cuDNN (CUDA Deep Neural Network library) and cuBLAS (CUDA Basic Linear Algebra Subroutines), which provide highly optimized implementations of deep learning primitives (e.g., convolutions, matrix multiplications) used in LLMs.
- **Broad Compatibility**: CUDA supports a wide range of NVIDIA GPUs, from consumer-grade RTX cards to data-center-grade A100 and H100 GPUs, ensuring flexibility across development environments.

This integration allows developers to focus on model design and experimentation while CUDA handles the underlying performance optimization.

Support for Mixed Precision Training

Training LLMs is memory-intensive, often requiring tens or hundreds of gigabytes of GPU memory. CUDA's support for mixed precision training addresses this challenge by combining 16-bit (half-precision) and 32-bit (full-precision) floating-point computations:

- **Memory Efficiency**: Using half-precision (FP16) or bfloat16 reduces memory usage, allowing larger models or batch sizes to fit on a single GPU. For example, a 1.5B-parameter model that requires 3GB in FP32 can be trained with ~1.5GB in FP16.
- **Performance Boost**: CUDA's Tensor Cores, available in NVIDIA GPUs like the Volta, Ampere, and Hopper architectures, are optimized for mixed precision, delivering up to 10x higher throughput for matrix operations compared to FP32.
- **Stability with Automatic Scaling**: CUDA's `cuda.amp` (Automatic Mixed Precision) in PyTorch or similar tools in TensorFlow automatically manages precision to prevent numerical instability, ensuring training accuracy.

Mixed precision training, enabled by CUDA, is critical for scaling LLM training to billion-parameter models while reducing costs and training time.

Optimized Libraries for Deep Learning

CUDA powers a suite of NVIDIA libraries that are tailored for deep learning and LLM development, enhancing performance and developer productivity:

- **cuDNN**: This library provides optimized implementations of neural network operations, such as attention mechanisms, layer normalization, and activation functions, which are core components of transformers.
- **cuBLAS**: Offers high-performance linear algebra routines, accelerating matrix and vector operations that dominate LLM computations.

- **NCCL (NVIDIA Collective Communications Library)**: Enables efficient multi-GPU and multinode communication, critical for distributed training of LLMs across clusters.
- **CUTLASS**: A CUDA-based library for custom matrix multiplication kernels, allowing developers to optimize performance for specific LLM architectures.

These libraries ensure that LLMs can be trained and deployed with maximum efficiency, leveraging GPU hardware to its full potential.

Scalability for Distributed Training

Training LLMs often requires distributing computations across multiple GPUs or nodes due to their size and data requirements. CUDA provides robust support for distributed training:

- **Data Parallelism**: CUDA integrates with frameworks like PyTorch's `torch.nn.DataParallel` or `torch.distributed` to split mini-batches across GPUs, speeding up training.
- **Model Parallelism**: For models too large for a single GPU, CUDA supports splitting layers or parameters across devices, facilitated by libraries like DeepSpeed or Megatron-LM.
- **Cluster Scalability**: CUDA's compatibility with NCCL and MPI (Message Passing Interface) enables efficient communication in multinode setups, such as training a 1T-parameter model on a supercomputing cluster.
- **Advanced Frameworks**: CUDA underpins tools like DeepSpeed's ZeRO (Zero Redundancy Optimizer) and NVIDIA's Megatron-LM, which optimize memory and communication for distributed LLM training.

This scalability makes CUDA essential for training state-of-the-art LLMs, which often require hundreds of GPUs and petabytes of data.

Support for Advanced Hardware Features

CUDA leverages NVIDIA's cutting-edge GPU architectures (e.g., Volta, Ampere, Hopper), unlocking advanced hardware features for LLM development:

- **Tensor Cores**: These specialized cores accelerate matrix operations, delivering up to 1,250 TFLOPS of FP16 performance on an H100 GPU. Tensor Cores are particularly effective for transformer computations, such as attention and feed-forward layers.
- **Transformer Engine**: NVIDIA's Transformer Engine, built on CUDA, optimizes transformer-specific operations, including FP8 precision for ultra-efficient training and inference.
- **NVLink and GPU Direct**: High-speed interconnects like NVLink and GPU Direct reduce data transfer bottlenecks in multi-GPU setups, improving training efficiency for LLMs.
- **HBM3 Memory**: CUDA supports GPUs with high-bandwidth memory (e.g., 141GB on the H100), enabling faster data access for large models and datasets.

These hardware advancements, accessible through CUDA, ensure that LLMs can be trained and deployed with unprecedented performance.

Flexibility for Custom Kernels and Optimizations

While CUDA's libraries provide optimized primitives, LLM development often requires custom computations, such as novel attention mechanisms or specialized tokenization schemes. CUDA's programmability allows developers to write custom kernels:

- **CUDA C++ and Python APIs**: Developers can write low-level CUDA C++ kernels or use Python libraries like CuPy for custom GPU-accelerated computations.
- **CUTLASS and Triton**: These tools simplify the creation of custom matrix multiplication and attention kernels, enabling performance optimizations tailored to specific LLM architectures.

- **Integration with PyTorch/JAX**: Custom CUDA kernels can be seamlessly integrated into high-level frameworks, allowing researchers to experiment with innovative designs without sacrificing performance.

This flexibility is critical for research, where new LLM architectures (e.g., sparse transformers, mixture-of-experts) often require custom implementations.

Rapid Prototyping and Experimentation

CUDA's integration with frameworks like PyTorch enables rapid prototyping, which is essential for LLM research:

- **Interactive Workflows**: In PyTorch, developers can move tensors to CUDA devices and execute operations interactively, inspecting outputs or gradients in real time. This is particularly useful for debugging attention layers or testing new loss functions.
- **Experiment Scalability**: CUDA allows researchers to scale experiments from a single GPU (e.g., RTX 4090 for prototyping) to multi-GPU clusters (e.g., DGX A100 systems) without changing code.
- **Support for Small-Scale Testing**: CUDA's compatibility with consumer GPUs makes it accessible for students and independent researchers to prototype LLMs on affordable hardware before scaling to data-center GPUs.

This balance of accessibility and scalability accelerates the iterative process of designing and testing LLM architectures.

Support for Inference Optimization

Deploying LLMs for real-time applications, such as chatbots or text generation, requires low-latency inference. CUDA enables several optimization techniques:

- **Quantization**: CUDA supports 8-bit (INT8) and 4-bit quantization, reducing model size and inference latency while maintaining accuracy. Libraries like `bitsandbytes` leverage CUDA for efficient quantized inference.

- **Graph Optimization**: CUDA integrates with NVIDIA's TensorRT, which optimizes computational graphs for inference, fusing layers and reducing memory overhead.
- **Batching and Streaming**: CUDA's parallel processing enables efficient batching of inference requests, improving throughput for production systems.
- **FP8 Inference**: The Hopper architecture's FP8 precision, supported by CUDA, delivers high-performance inference for LLMs with minimal accuracy loss.

These optimizations ensure that LLMs can be deployed in resource-constrained environments, such as edge devices or cloud APIs.

Extensive Ecosystem and Community Support

CUDA benefits from a robust ecosystem and community, which enhances its utility for LLM development:

- **NVIDIA NGC**: The NVIDIA GPU Cloud provides prebuilt containers, pretrained models, and tutorials optimized for CUDA, streamlining LLM development workflows.
- **Documentation and Tutorials**: NVIDIA's CUDA Toolkit includes comprehensive documentation, while forums like the NVIDIA Developer Forums offer community support for troubleshooting.
- **Open-Source Contributions**: Libraries like Hugging Face Transformers, DeepSpeed, and Megatron-LM are optimized for CUDA, providing reusable codebases for LLM training and inference.
- **Educational Resources**: Courses from NVIDIA's Deep Learning Institute and platforms like Coursera offer CUDA-specific training for deep learning, making it accessible to learners.

This ecosystem ensures that developers can find tools, resources, and support to maximize CUDA's potential for LLMs.

Energy Efficiency and Cost Savings

Training LLMs is resource-intensive, with significant energy and financial costs. CUDA's efficiency helps mitigate these challenges:

- **Optimized Performance**: CUDA's ability to leverage Tensor Cores and mixed precision reduces training time, lowering energy consumption and cloud computing costs.
- **Hardware Versatility**: CUDA's compatibility with a range of GPUs (from RTX to H100) allows developers to choose cost-effective hardware for their needs, such as using consumer GPUs for prototyping.
- **Cloud Integration**: CUDA's support for cloud platforms like AWS, Google Cloud, and Azure enables access to on-demand GPUs, reducing the need for expensive on-premises hardware.

For example, training a 175B-parameter model like GPT-3 can cost millions on CPUs but is significantly cheaper on CUDA-enabled GPUs due to faster training and lower energy use.

Future-Readiness and Innovation

CUDA is continuously updated to support emerging trends in LLM development:

- **New Architectures**: CUDA supports NVIDIA's latest GPUs (e.g., Hopper, Blackwell), ensuring compatibility with future hardware advancements.
- **AI Accelerators**: CUDA's integration with NVIDIA's Grace CPU and upcoming AI accelerators positions it for next-generation LLM workflows.
- **Sparse Computing**: CUDA supports sparse attention and pruning techniques, which reduce computational costs for LLMs without sacrificing performance.
- **Quantum and Federated Learning**: CUDA's flexibility makes it adaptable to emerging paradigms, such as quantum machine learning or federated LLM training.

This forward-looking approach ensures that CUDA remains a cornerstone of LLM innovation.

Accessibility for Diverse Developers

CUDA's versatility makes it accessible to a wide range of developers:

- **Consumer-Grade GPUs**: Students and independent researchers can use affordable GPUs like the NVIDIA RTX 3060 or 4090 for small-scale LLM experiments.
- **Cloud Accessibility**: CUDA's integration with cloud providers allows developers to rent high-end GPUs (e.g., A100, H100) for large-scale training without upfront hardware costs.
- **Educational Support**: NVIDIA's CUDA Toolkit is free to download, and resources like the CUDA Programming Guide lower the barrier to entry for learning GPU programming.

This inclusivity fosters a diverse community of LLM developers, from hobbyists to enterprise teams.

Support for Research and Custom Architectures

CUDA's programmability empowers researchers to explore novel LLM designs:

- **Custom Attention Mechanisms**: Researchers can implement sparse or performant attention variants (e.g., FlashAttention) using CUDA kernels to optimize memory and speed.
- **Experimental Models**: CUDA supports architectures like mixture-of-experts, retrieval-augmented generation (RAG), or multimodal LLMs, which require custom compute patterns.
- **Open-Source Innovation**: Many state-of-the-art LLMs, such as Meta's LLaMA or xAI's Grok, rely on CUDA for training and fine-tuning, with codebases shared publicly to inspire further research.

This research-friendly environment makes CUDA a preferred choice for pushing the boundaries of LLM capabilities.

CUDA's unparalleled performance, scalability, and flexibility make it the backbone of large language model development. By leveraging NVIDIA GPUs' massive parallelism, optimized libraries, and advanced hardware features, CUDA enables researchers and engineers to train and deploy LLMs with unprecedented efficiency. Its seamless integration with deep learning frameworks, support for mixed precision, and robust ecosystem ensure that developers can focus on innovation rather than low-level optimization.

Whether you're prototyping a transformer on a single GPU, training a trillion-parameter model on a supercomputing cluster, or deploying a chatbot in production, CUDA provides the tools and performance to succeed. As LLMs continue to evolve, CUDA's ongoing advancements and community support position it as the foundation for the future of AI.

Why NumPy and Matplotlib for Building Large Language Models?

Figure 1-5. *NumPy Logo, Source: numpy.org*

Large language models (LLMs), such as GPT, BERT, and LLaMA, rely on complex workflows that involve data preprocessing, numerical computations, and result analysis. While deep learning frameworks like PyTorch or TensorFlow handle model training, NumPy (Figure 1-5) and Matplotlib are indispensable for foundational tasks in the LLM pipeline.

NumPy provides high-performance numerical computing capabilities, and Matplotlib enables powerful data visualization, together forming a robust toolkit for researchers, engineers, and students working on LLMs. Below, we explore the key reasons why NumPy and Matplotlib are critical for building LLMs.

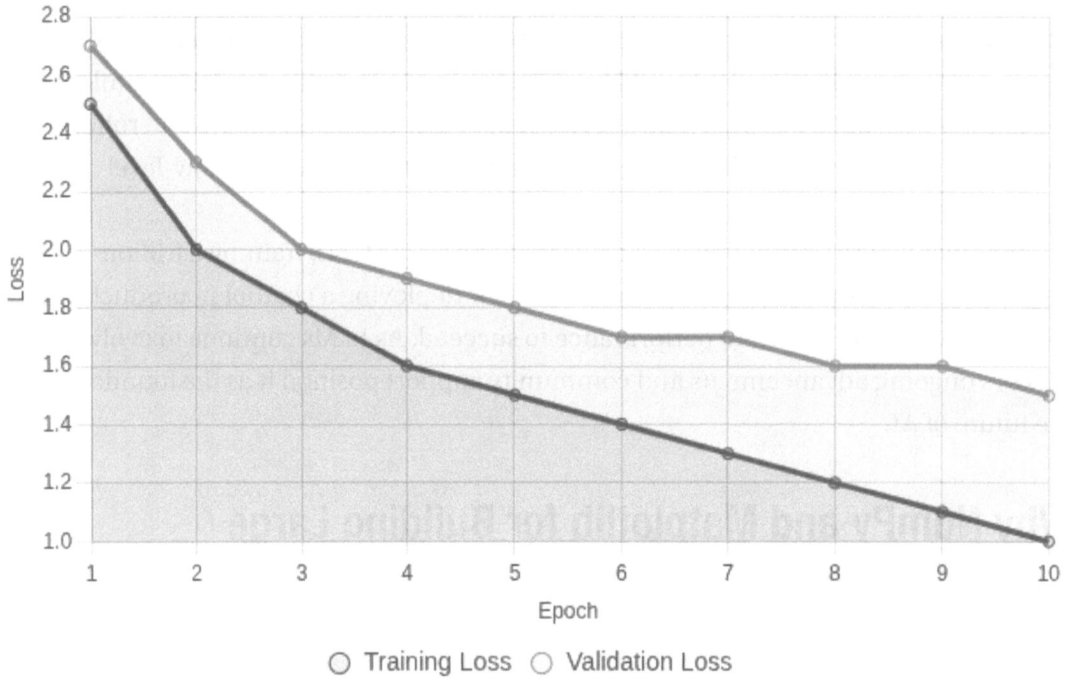

Figure 1-6. *Training vs. Validation Loss Visualisation*

NumPy: Efficient Numerical Computations for Data Processing

LLM development begins with data preprocessing, such as tokenization, embedding creation, and dataset manipulation, which require efficient numerical operations. NumPy, a fundamental Python library for numerical computing, excels in these tasks due to its optimized array-based computations.

- **High-Performance Arrays**: NumPy's `ndarray` (N-dimensional array) is a fast, memory-efficient data structure for handling large datasets. For example, converting text data into numerical token IDs or embeddings involves manipulating large matrices, which NumPy processes orders of magnitude faster than native Python lists.
- **Vectorized Operations**: NumPy supports vectorized operations, allowing element-wise computations (e.g., scaling, normalization) without explicit loops. This is critical for tasks like normalizing word embeddings or computing similarity matrices for input data.

- **Broad Functionality**: NumPy provides functions for linear algebra (e.g., `np.dot`, `np.linalg`), random sampling (e.g., `np.random`), and statistical operations (e.g., `np.mean`, `np.std`), which are used in data preprocessing, feature engineering, and model initialization.
- **Interoperability with Deep Learning Frameworks**: NumPy arrays are compatible with PyTorch (`torch.from_numpy`) and TensorFlow (`tf.convert_to_tensor`), enabling seamless data transfer between preprocessing pipelines and model training. For instance, tokenized text data prepared in NumPy can be directly converted to tensors for training.

Example: When preparing a dataset for an LLM, NumPy can efficiently convert a vocabulary of 50,000 tokens into a one-hot encoded matrix or compute cosine similarities between word embeddings, tasks that would be prohibitively slow in pure Python.

NumPy: Memory Efficiency for Large-Scale Data

LLMs often process massive datasets, such as billions of tokens from web corpora or books. NumPy's memory-efficient design is crucial for handling these datasets:

- **Compact Data Representation**: NumPy arrays use contiguous memory blocks and fixed data types (e.g., `float32`, `int64`), reducing memory overhead compared to Python lists. This allows large datasets, like embedding matrices or token ID arrays, to fit in memory.
- **In-Place Operations**: NumPy supports in-place modifications (e.g., `arr += 1`) to minimize memory allocation, which is essential when manipulating large matrices during preprocessing.
- **Slicing and Indexing**: NumPy's advanced indexing (e.g., `arr[:, 1:10]`) enables efficient subsetting of data, such as selecting specific tokens or features from a dataset without copying memory.

NumPy: Foundation for Preprocessing and Feature Engineering

LLM pipelines involve complex preprocessing steps, such as tokenization, padding, and embedding lookup, where NumPy shines:

- **Tokenization and Encoding**: NumPy can convert tokenized text into numerical arrays (e.g., token IDs or one-hot encodings) for input to LLMs. For example, a sentence like "I love AI" can be mapped to `[12, 45, 78]` using a NumPy array.
- **Padding and Truncation**: NumPy's array manipulation functions (e.g., `np.pad`, `np.zeros`) simplify padding sequences to a fixed length or truncating long sequences, ensuring uniform input sizes for batch processing.
- **Embedding Preprocessing**: NumPy is used to preprocess or initialize embedding matrices, such as loading pretrained GloVe or Word2Vec embeddings and normalizing them for model input.

Example: For a transformer model, NumPy can create a batch of padded token sequences with shape `(batch_size, max_seq_len)` and compute positional encodings using trigonometric functions (`np.sin`, `np.cos`), which are then fed into the model.

NumPy: Support for Custom Algorithms and Research

LLM research often involves custom algorithms, such as novel attention mechanisms or data augmentation techniques, where NumPy's flexibility is invaluable:

- **Prototyping Algorithms**: NumPy's intuitive syntax allows researchers to prototype algorithms, such as computing attention scores or implementing custom loss functions, before porting them to GPU-accelerated frameworks.
- **Data Augmentation**: NumPy supports random sampling and transformations (e.g., `np.random.permutation`) for augmenting text data, such as shuffling sentences or injecting noise for robustness training.
- **Numerical Stability**: NumPy's precise control over data types and operations helps researchers avoid issues like numerical overflow when experimenting with new LLM architectures.

Matplotlib: Visualization for Model Understanding

Training and evaluating LLMs require visualizing metrics, model outputs, and internal states to understand performance and diagnose issues (Figure 1-6). Matplotlib, a versatile Python plotting library, is the go-to tool for creating these visualizations:

- **Training Metrics**: Matplotlib can plot loss curves, perplexity, or accuracy over epochs, helping developers monitor training progress. For example, a plot of training vs. validation loss can reveal overfitting.
- **Attention Visualizations**: Matplotlib is used to create heatmaps of attention weights in transformer models, showing how tokens attend to each other (e.g., visualizing self-attention in BERT).
- **Embedding Analysis**: Matplotlib can plot 2D projections of high-dimensional embeddings (e.g., using t-SNE or PCA), helping researchers understand semantic relationships learned by the model.

Matplotlib: Debugging and Interpretability

Debugging LLMs involves inspecting intermediate outputs, such as gradients, activations, or token embeddings. Matplotlib enhances interpretability through visual tools:

- **Gradient Analysis**: Plotting gradient magnitudes (e.g., using `plt.hist`) helps diagnose issues like vanishing or exploding gradients in deep transformer models.
- **Activation Distributions**: Matplotlib can visualize the distribution of layer activations (e.g., using histograms or box plots), revealing whether normalization layers are functioning correctly.
- **Token-Level Insights**: Bar plots or line graphs can show token probabilities or confidence scores during inference, aiding in error analysis.

Example: To debug a transformer's attention mechanism, Matplotlib can create a heatmap of attention scores for a single input sentence, highlighting whether the model attends to relevant tokens (e.g., nouns or verbs).

Matplotlib: Customizable and Publication-Ready Visuals

Matplotlib's flexibility allows developers to create tailored visualizations for LLM projects, from quick exploratory plots to publication-quality figures:

- **Customization**: Matplotlib supports extensive customization of colors, fonts, labels, and layouts, enabling clear and professional visualizations. For example, a researcher can create a multipanel plot comparing loss curves for different learning rates.
- **Interactive Plots**: Matplotlib's integration with Jupyter Notebook allows interactive exploration of plots, such as zooming into loss spikes or hovering over data points.
- **Export Options**: Plots can be saved in formats like PNG, PDF, or SVG, making them suitable for reports, presentations, or academic papers.

NumPy and Matplotlib: Seamless Integration with Python Ecosystem

Both NumPy and Matplotlib integrate seamlessly with Python's scientific computing stack, enhancing their utility in LLM development:

- **NumPy as a Backbone**: NumPy serves as the foundation for libraries like Pandas (data manipulation), SciPy (scientific computing), and scikit-learn (machine learning), which are used in LLM preprocessing and evaluation.
- **Matplotlib with Other Tools**: Matplotlib works with libraries like Seaborn for enhanced statistical visualizations or Plotly for interactive plots, expanding its capabilities for LLM analysis.
- **Jupyter Notebook Synergy**: In Jupyter, NumPy handles data preparation, and Matplotlib generates inline plots, creating a cohesive workflow for experimenting with LLMs.

Example: In a Jupyter Notebook, a developer can use NumPy to compute token frequency distributions for a dataset and Matplotlib to plot a histogram of these frequencies, all within a single cell.

NumPy and Matplotlib: Accessibility for Beginners

NumPy and Matplotlib are beginner-friendly, making them ideal for students and newcomers to LLM development:

- **Gentle Learning Curve**: NumPy's syntax is intuitive (e.g., `np.array([1, 2, 3])`), and Matplotlib's plotting functions (e.g., `plt.plot`) are straightforward, allowing learners to focus on LLM concepts.
- **Extensive Documentation**: Both libraries have comprehensive documentation, tutorials, and community resources (e.g., Stack Overflow, NumPy User Guide), supporting self-paced learning.
- **Open-Source and Free**: NumPy and Matplotlib are freely available, ensuring accessibility for independent researchers and students without institutional resources.

NumPy: Support for Dataset Creation and Evaluation

LLM development involves creating and evaluating datasets, where NumPy's array operations are critical:

- **Dataset Creation**: NumPy can generate synthetic datasets for testing LLMs (e.g., random token sequences) or preprocess real-world datasets by filtering, sorting, or grouping data.
- **Evaluation Metrics**: NumPy supports computations for metrics like BLEU, ROUGE, or perplexity, which are used to evaluate LLM performance. For example, computing cosine similarity between predicted and target embeddings is a one-liner with `np.dot`.
- **Batch Processing**: NumPy's array slicing enables efficient batch creation, ensuring that large datasets are processed in manageable chunks during training or evaluation.

Example: To evaluate a text generation model, NumPy can compute the average perplexity across a test set by exponentiating the cross-entropy loss for each sample.

Matplotlib: Support for Experiment Tracking

LLM development involves running multiple experiments (e.g., varying hyperparameters or architectures), and Matplotlib helps track and compare results:

- **Hyperparameter Analysis**: Matplotlib can plot performance metrics (e.g., validation loss) for different learning rates, batch sizes, or dropout rates, aiding in hyperparameter tuning.
- **Model Comparison**: Side-by-side plots or subplots can compare the performance of different LLMs (e.g., transformer vs. LSTM) or training strategies (e.g., fine-tuning vs. pretraining).
- **Time-Series Analysis**: Matplotlib can visualize training dynamics, such as how loss evolves over time, helping identify optimal training durations.

Example: A researcher can use Matplotlib to create a subplot grid comparing the training loss of four LLMs with different attention mechanisms, highlighting which performs best.

NumPy and Matplotlib: Support for Research and Innovation

LLM research often involves novel data processing or visualization techniques, where NumPy and Matplotlib provide flexibility:

- **NumPy for Custom Pipelines**: Researchers can use NumPy to implement custom preprocessing, such as hierarchical tokenization or data augmentation with noise injection.
- **Matplotlib for Novel Visualizations**: Matplotlib's low-level API allows researchers to create bespoke plots, such as 3D visualizations of embedding spaces or animated attention flows.
- **Open-Source Contributions**: Many open-source LLM projects (e.g., Hugging Face datasets) use NumPy for data handling and Matplotlib for visualization, enabling researchers to build on existing tools.

NumPy and Matplotlib: Lightweight and Local-First

Unlike cloud-based or heavy IDE tools, NumPy and Matplotlib are lightweight and run locally, aligning with accessible LLM development:

- **Minimal Setup**: Both libraries can be installed with a single `pip install` command and run on standard hardware, making them ideal for students or researchers with limited resources.
- **Offline Workflows**: NumPy and Matplotlib enable data processing and visualization without internet access, ensuring uninterrupted work in low-connectivity environments.
- **Cross-Platform Compatibility**: They work on Windows, macOS, and Linux, providing flexibility across development setups.

NumPy and Matplotlib: Community and Ecosystem Support

Both libraries benefit from vibrant communities and extensive ecosystems, enhancing their utility for LLM development:

- **Community Resources**: NumPy and Matplotlib have active communities on GitHub, Stack Overflow, and forums like Reddit, providing solutions to common issues.
- **Tutorials and Courses**: Resources like SciPy's documentation, Matplotlib's gallery, and online courses (e.g., Coursera, fast.ai) offer practical guides for using these tools in machine learning.
- **Integration with Advanced Tools**: NumPy underpins libraries like Hugging Face's datasets, and Matplotlib complements tools like Weights & Biases for experiment tracking.

NumPy and Matplotlib (Figure 1-7) are indispensable for building large language models, providing the numerical horsepower and visual clarity needed for data processing, analysis, and debugging. NumPy's efficient array operations and memory management enable scalable preprocessing and feature engineering, while Matplotlib's versatile plotting capabilities bring transparency to model performance and behavior. Together, they form a lightweight, accessible, and powerful toolkit that integrates seamlessly with Python's scientific ecosystem, supporting both beginner-friendly learning and cutting-edge research.

Whether you're tokenizing a massive corpus, visualizing attention heatmaps, or comparing hyperparameter experiments, NumPy and Matplotlib empower you to tackle the challenges of LLM development with confidence. As LLMs continue to advance, these libraries remain foundational tools for transforming raw data into insights and models into breakthroughs.

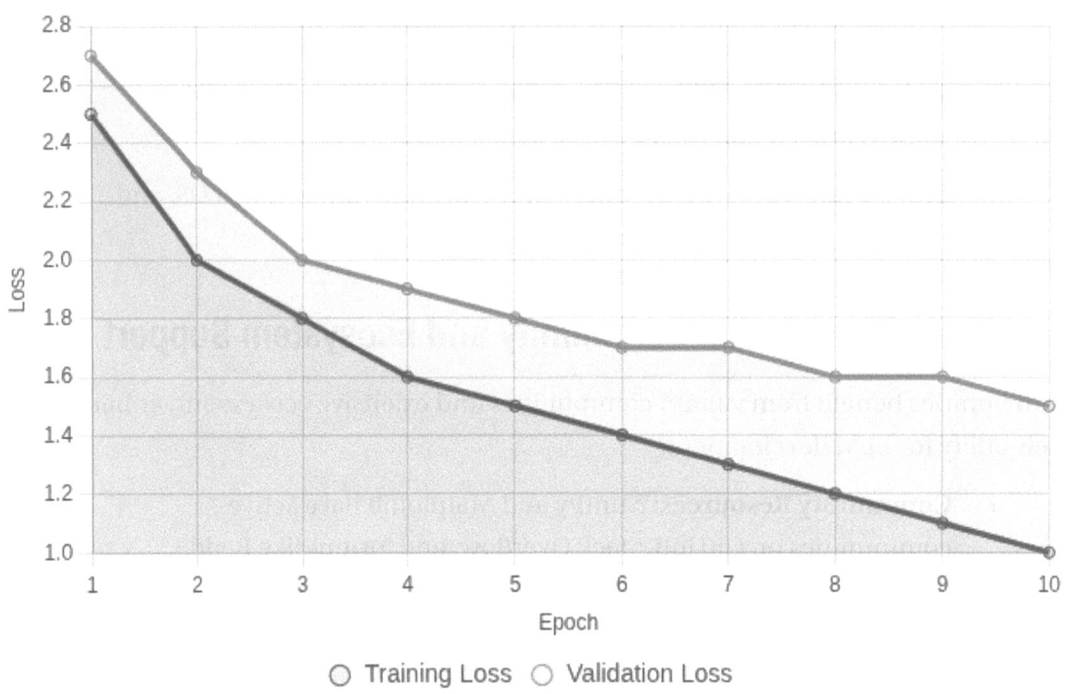

Figure 1-7. *Matplotlib Visualization Example*

TinyStories, Shakespeare, *The Wizard of Oz*, and OpenWebText

Four Key Datasets for Building and Understanding LLMs

Large language models depend on the data they are trained on, and different datasets serve different purposes across the development pipeline. Some are small and clean, perfect for teaching concepts or training tiny models. Others are large and diverse, intended for training full-scale transformers capable of handling real-world language.

TinyStories, **Shakespeare**, ***The Wizard of Oz***, and **OpenWebText** together illustrate this range, from "toy" datasets ideal for learning to massive corpora used in state-of-the-art research.

TinyStories: Learning LLM Fundamentals Through Simple Narratives

TinyStories is a synthetic dataset made of short children's stories written in very simple, clean language. Its purpose is not realism but approachability: it allows even tiny transformer models—sometimes with only a few million parameters—to learn coherent grammar and narrative flow. Because the text is deliberately constrained, students and researchers can train full language models on consumer hardware, often within minutes. The outputs are easy to read, easy to debug, and easy to understand, which makes TinyStories one of the best datasets for teaching how transformers work or testing new model architectures in a controlled environment. However, its simplicity is also its limitation: a model trained only on TinyStories will not generalize outside the narrow world of childlike storytelling.

Shakespeare: Stylized, Poetic, and Ideal for Demonstrations

The Shakespeare corpus, usually taken from the public-domain plays and sonnets, is one of the most commonly used datasets in introductory machine learning tutorials. Its small size and distinctive style make it perfect for character-level models, text-generation demos, or experiments in stylistic fine-tuning. Because Shakespeare's writing is so recognizable, it's easy to judge whether a model "gets" the style. The language, however, is highly archaic and not representative of modern English, so like TinyStories, it is best suited for learning and experimentation rather than large-scale, general-purpose training.

The Wizard of Oz: A Compact, Coherent Narrative for Prototyping

L. Frank Baum's *The Wonderful Wizard of Oz* serves a similar role but offers a more modern and coherent narrative structure than Shakespeare. It is still small—only around 40,000 words—but its storytelling is straightforward, descriptive, and cleanly edited. This makes it excellent for prototyping tokenizer setups, testing preprocessing pipelines, or running educational experiments in language modeling. Because the entire text fits easily into memory, developers often use it for debugging: errors are easier to spot, and training runs complete quickly. Like the other small datasets, though, its narrow style and limited scope prevent it from being useful for training robust LLMs on its own.

OpenWebText: Large-Scale, Diverse, and Suitable for Real LLM Training

At the opposite end of the spectrum is OpenWebText, a massive web-scraped dataset of online articles, posts, essays, and discussions. It was created to mimic the dataset used to train GPT-2 and contains a wide variety of writing styles, topics, and linguistic patterns. Unlike the smaller literary datasets, OpenWebText provides the breadth and diversity necessary for real large-scale pretraining. It reflects modern language, including slang, technical jargon, and everyday conversation. Because it is so large—tens of gigabytes uncompressed—it demands extensive preprocessing, deduplication, and filtering. Tools like PyLZMA are often used here, since the dataset is typically distributed in compressed formats such as .xz to save space. For serious LLM research, OpenWebText is one of the most accessible and widely used corpora.

How These Datasets Work Together

These datasets naturally complement each other across the development workflow. TinyStories, Shakespeare, and *The Wizard of Oz* are perfect for early experiments: they run fast, produce interpretable results, and make debugging straightforward. They help beginners understand tokenization, embeddings, attention mechanisms, and model training—all without needing a GPU cluster. Meanwhile, OpenWebText is what developers turn to when they're ready to train larger models or replicate the behavior of modern LLMs. It provides the scale, complexity, and linguistic variety that small literary or synthetic datasets simply cannot offer.

Ethical and Practical Considerations

The smaller datasets—TinyStories, Shakespeare, and *The Wizard of Oz*—are public domain or synthetic, making them easy to use without licensing concerns. Their simplicity is both a strength and a limitation, providing clean input at the cost of diversity. OpenWebText brings its own complications: web-scraped material includes noise, bias, and sometimes sensitive content, and processing such a large dataset requires substantial computing power. Still, with proper filtering and documentation, it remains a cornerstone of open-source LLM development.

Why PyLZMA for Building Large Language Models?

Large language models (LLMs) require huge datasets and generate large artifacts, making efficient data storage and transfer essential. PyLZMA—a Python interface to the LZMA compression algorithm—is not a core training or modeling tool but plays a valuable supporting role in LLM workflows by providing high-ratio, lossless compression.

Why PyLZMA Matters for LLM Development

1. **Compressing Massive Text Corpora**

   LZMA achieves much higher compression ratios than ZIP/GZIP (often reducing text to 10–20% of its size). This lowers storage costs, speeds up dataset transfers, and enables large corpora (e.g., OpenWebText) to fit on limited local or cloud storage.

2. **Working with Compressed Dataset Formats**

   Many datasets (Common Crawl, Wikipedia) ship as .xz/.lzma. PyLZMA enables direct, on-the-fly decompression inside Python pipelines, integrating cleanly with tokenizers and preprocessing scripts.

3. **Compressing Checkpoints and Artifacts**

   LLM checkpoints can be several GB each. PyLZMA reduces their size dramatically while preserving integrity, making storage, versioning, and sharing more practical.

4. **Optimizing Storage in Constrained Environments**

   Useful for academic labs, small teams, or edge devices where disk space is limited. Compresses datasets, model weights, and intermediate files to fit within resource constraints or reduce cloud costs.

5. **Seamless Integration with Python Workflows**

   Works naturally alongside NumPy, Pandas, and Hugging Face libraries. Easy to script for automated compression/decompression, including in Jupyter notebooks.

6. **Support for Custom Formats**

   Can compress NumPy arrays, synthetic datasets, and other custom artifacts better than many default formats. Offers configurable compression levels for different use cases.

7. **Useful for Open-Source and Collaborative Research**

   Matches common distribution formats for public datasets, supports reproducibility, and is referenced in LLM tutorials (e.g., freeCodeCamp, LLM-from-scratch projects).

8. **Helpful in Preprocessing, Deduplication, and Cleaning**

   Large intermediate files from deduplication or filtering pipelines can be compressed to keep preprocessing scalable and disk usage manageable.

9. **Cross-Platform Compatibility**

   Works on Windows, macOS, and Linux and with legacy .lzma/.xz archives often found in research.

10. **Data Integrity and Security Benefits**

    Provides checksums and reliable lossless compression, helping ensure datasets and checkpoints are not corrupted.

11. **Installation Caveats**

    PyLZMA can be difficult to install on some systems (e.g., macOS ARM). Workarounds include conda or falling back to Python's built-in lzma module.

12. **How It Compares to Alternatives**

    LZMA compresses much better than GZIP/BZIP2 (though slower), making it ideal for large static datasets but less suitable for real-time or frequently updated data.

Although PyLZMA isn't required for training LLMs, it is highly useful for managing the enormous datasets and artifacts involved in LLM development. Its strengths—high compression ratio, Python integration, and compatibility with standard dataset formats—make it a practical tool for scalable, efficient LLM workflows.

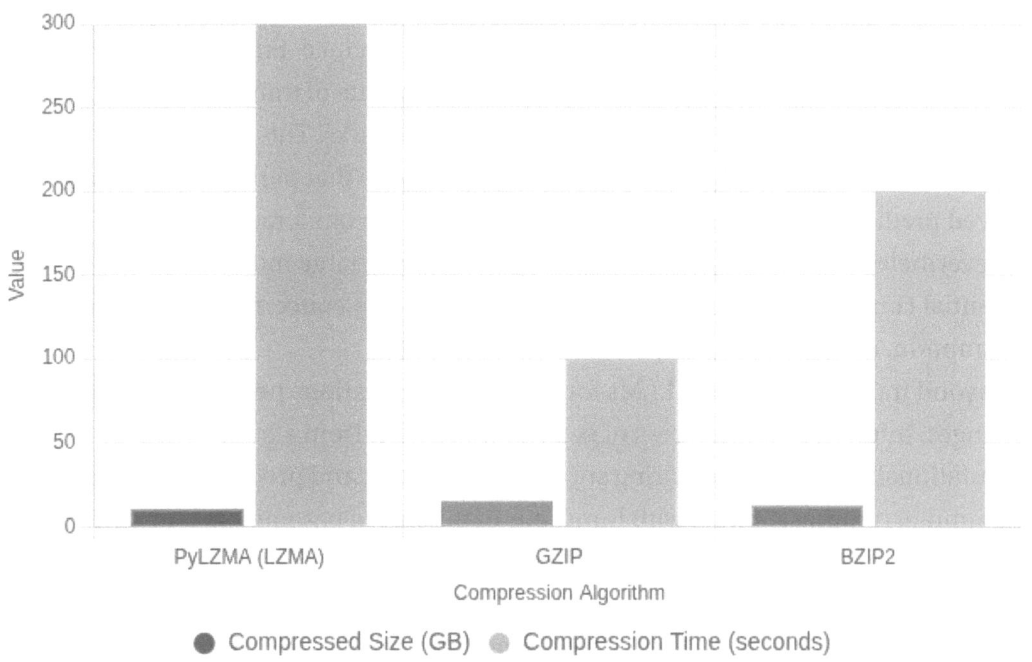

Figure 1-8. *Comparison Between PyLZMA and Other Compression Methods*

These references (Figure 1-8) underscore PyLZMA's niche but established role in LLM data workflows, particularly for researchers building models from scratch.

PyLZMA is a useful supporting tool in LLM development, helping manage the large datasets and artifacts that training produces. Its high compression ratio and compatibility with common formats (.lzma, .xz) reduce storage demands, speed up transfers, and simplify preprocessing workflows. Though not central to model training like CUDA or PyTorch, it enables researchers—especially those with limited resources—to efficiently compress datasets, checkpoints, and intermediate files. Despite slower compression and occasional installation issues, PyLZMA remains a practical solution for handling the heavy data needs of modern LLM pipelines and collaborative research.

Advanced Concepts

Techniques such as reinforcement learning from human feedback (RLHF) are also used to align an LLM model's outputs with human preferences, improving their usability in conversational systems. This adaptability makes LLMs highly versatile, though fine-tuning requires careful design to avoid overfitting or amplifying biases present in the task-specific data.

The scale of LLMs plays a pivotal role in their performance. Empirical studies have shown that increasing the size of the model and the volume of training data leads to consistent improvements across a range of NLP benchmarks. This phenomenon, often referred to as the "scaling laws" of deep learning, suggests that performance gains can be achieved predictably by enlarging models and datasets in tandem.

Nevertheless, this scaling is not without cost: training large models requires substantial computational resources and raises significant concerns about energy consumption, environmental impact, and accessibility.

Beyond training, deploying LLMs for real-time applications poses significant challenges. Inference—the process of generating outputs from a trained model—is computationally intensive, requiring substantial memory and processing power. For example, running a model with hundreds of billions of parameters may demand specialized hardware like high-end GPUs or TPUs, leading to high operational costs.

Additionally, latency is a concern for applications requiring rapid responses, such as chatbots or real-time translation. Techniques like model quantization, pruning, and distillation are often employed to create smaller, more efficient versions of LLMs without sacrificing too much performance. These trade-offs highlight the engineering complexities of making LLMs accessible and practical for widespread use.

Addressing the ethical challenges of LLMs requires a multifaceted approach. Techniques like debiasing datasets, fairness-aware fine-tuning, and interpretability tools aim to reduce harmful outputs, though no solution is foolproof. Governance frameworks, such as AI ethics boards and regulatory policies, are increasingly advocated to ensure responsible development and deployment. Moreover, transparency in model design and documentation, such as model cards, helps users understand an LLM's capabilities and limitations.

Public engagement and interdisciplinary collaboration—spanning technologists, ethicists, and policymakers—are also critical to aligning LLMs with societal values and mitigating risks like misinformation or job displacement. These efforts remain an active area of research and policy development, reflecting the profound societal stakes of LLM proliferation.

Moreover, since LLMs learn from large-scale human-authored texts, they can inadvertently encode and reproduce social biases, stereotypes, and other harmful content. As such, the development and deployment of LLMs necessitate careful consideration of ethical, societal, and safety implications.

While LLMs traditionally focus on text, recent advancements have led to multimodal models that integrate text with other data types, such as images, audio, or even video. For example, models like DALL·E and CLIP (developed by OpenAI) combine text and visual processing to generate images from textual descriptions or understand visual content. These multimodal LLMs expand the scope of applications, enabling tasks like automated image captioning, visual question answering, and even creative content generation across media. However, they also introduce new challenges, such as the need for diverse multimodal datasets and increased computational complexity.

Language models represent a convergence of massive data, powerful algorithms, and scalable computing infrastructure. They are capable of generating coherent and contextually relevant text, performing complex reasoning tasks, and adapting to a wide variety of applications with minimal task-specific tuning. Yet, their operation is grounded in statistical inference rather than comprehension, and their outputs must be interpreted with appropriate caution. As we delve deeper into building an LLM from scratch, understanding these foundational principles will be essential to both appreciating the model's capabilities and recognizing its limitations.

The Shakespeare, TinyStories, *Wizard of Oz*, and *OpenWebText* datasets play distinct but vital roles in building large language models, catering to different needs in the development pipeline. *The Wizard of Oz* offers a small, high-quality, and accessible dataset for prototyping, education, and fine-tuning, enabling researchers and students to experiment with LLM concepts on minimal hardware. OpenWebText, with its massive scale and diverse content, is a cornerstone for pretraining modern LLMs, providing the breadth needed for general-purpose language understanding.

Together, they complement each other, supporting both rapid iteration and production-scale training while integrating seamlessly with Python-based tools like PyLZMA, NumPy, Matplotlib, PyTorch, and CUDA. Despite challenges like *The Wizard of Oz*'s limited diversity or OpenWebText's noise and ethical concerns, their strategic use empowers developers to create robust, versatile LLMs. Whether you're a beginner tokenizing a classic novel or a researcher pretraining a transformer on web-scale data, these datasets provide the foundation for advancing AI innovation.

In Summary

As we've seen, building a large language model from scratch begins with a solid foundation—understanding the tools, libraries, and environment that enable such an ambitious endeavor. From selecting Python and PyTorch for their accessibility and flexibility to leveraging Jupyter Notebooks, CUDA, and efficient data management tools like NumPy, Matplotlib, and PyLZMA, we've laid the groundwork for hands-on, scalable LLM development. These components not only power modern NLP workflows but also set the stage for the deeper theoretical and computational principles that lie ahead.

In the next chapter, we'll transition from basic array operations to tensors—the fundamental data structure for neural networks. We'll also introduce key linear algebra concepts that underpin neural computation, explore how to structure datasets for training and validation, and delve into the bigram model—our first step into building a functioning language model. These topics mark the beginning of our journey into the mathematics and mechanics of model design, preparing us to move from setup to actual language modeling.

CHAPTER 2

Foundational Concepts in LLM Development

Large Language Models Common Architecture

In the age of artificial intelligence, Large Language Models (LLMs) have become the invisible architects of our digital conversations. From crafting human-like prose to answering complex questions, translating languages, or even composing poetry, these models power technologies that feel almost magical. But behind this magic lies a carefully constructed framework of **computational architectures**—ingenious designs that enable machines to understand and generate human language with unprecedented fluency. This book embarks on a journey to unravel these architectures, exploring the blueprints that have transformed AI into a cornerstone of modern innovation.

At the heart of every LLM is a structural marvel, most notably the **transformer**, a revolutionary architecture introduced in 2017 that redefined how machines process language. With its ability to weigh the importance of words in a sentence, capture intricate contextual relationships, and scale to billions—or even trillions—of parameters, the transformer has become the gold standard for LLMs like GPT, BERT, and beyond.

Yet, the story of LLMs is not limited to transformers alone. **From the sequential processing of Recurrent Neural Networks (RNNs) to the efficiency-driven Mixture of Experts (MoE)** and the emerging promise of **State Space Models (SSMs)**, the evolution of LLM architectures reflects a relentless pursuit of performance, scalability, and efficiency.

D. Grigorov, *Building Large Language Models from Scratch*, https://doi.org/10.1007/979-8-8688-2297-1_2

This chapter is designed to demystify these architectures, offering a clear and comprehensive guide to their inner workings. We will explore how transformers leverage self-attention to understand context, why decoder-only models excel at text generation, and how innovations like sparse attention or MoE are pushing the boundaries of what LLMs can achieve.

We will also delve into the trade-offs—computational costs, memory demands, and the challenges of modeling long-range dependencies—that shape the design of these systems. Whether you are a student of AI, a practitioner building the next generation of language models, or simply curious about the technology shaping our world, this book will equip you with the knowledge to understand the engines driving LLMs.

As we stand at the cusp of new architectural paradigms, from efficient transformers to novel approaches like Mamba, the field of LLMs is evolving at a breakneck pace. This chapter not only lays the foundation for understanding today's architectures but also prepares you to anticipate the innovations of tomorrow. Join us as we decode and code together the structures that enable machines to speak, think, and create—unveiling the architectural genius behind the words that shape our future.

Transformer Architecture: Conceptual and Mathematical Deep Dive

The transformer architecture, introduced in the 2017 paper "Attention is All You Need" by Vaswani et al., is a landmark in machine learning, particularly for sequence-to-sequence tasks like natural language processing (NLP), computer vision, and beyond. Unlike its predecessors, such as Recurrent Neural Networks (RNNs) and Long Short-Term Memory (LSTM) networks, the transformer eliminates recurrence entirely, relying on attention mechanisms to model relationships between tokens in a sequence.

This design enables parallel processing, significantly speeding up training and improving performance on tasks like machine translation, text summarization, and question answering. Its flexibility and scalability have made it the foundation for models like BERT, GPT, and T5, which dominate modern AI applications.

The transformer architecture has transformed the landscape of machine learning by offering a powerful, scalable, and flexible framework for modeling sequences. Its reliance on attention mechanisms, parallel processing, and adaptability has made it the go-to architecture for a wide range of applications, from NLP to vision and beyond. Despite its computational demands, ongoing research continues to improve its efficiency, ensuring its dominance in AI for years to come.

Unlike recurrent neural networks (RNNs) or long short-term memory (LSTM) networks, which process sequences sequentially and struggle with long-range dependencies, transformers rely entirely on attention mechanisms. This design enables parallel computation, captures relationships between distant tokens, and scales effectively with large datasets and models.

Below, we provide a comprehensive explanation of the mathematical foundations of the transformer, covering input embeddings, positional encodings, attention mechanisms, feed-forward networks, residual connections, layer normalization, output layers, loss functions, and encoder-decoder interactions. All mathematical expressions are presented with clear explanations to ensure understanding.

Input Embedding

The transformer begins by converting discrete tokens (e.g., words, subwords, or characters) into continuous vector representations suitable for neural network processing. This step transforms sparse, high-dimensional inputs into dense, low-dimensional vectors that capture semantic meaning.

Consider a vocabulary of size V (e.g., 50,000 words or subwords, as in typical NLP models). Each token is represented by an index, and the embedding layer maps these indices to dense vectors of dimension d_{model} (commonly 512 or 768). The embedding matrix is defined as

$$W_e \in \mathbb{R}^{V \times d_{\text{model}}}$$

For a token with index i (e.g., the index of "cat" in the vocabulary), its embedding is a row of the matrix:

$$Emb(i) = W_e[i] \in \mathbb{R}^{d_{\text{model}}}$$

This vector is learned during training, allowing the model to assign similar embeddings to semantically related tokens *(e.g., "cat" and "dog" may have closer vectors than "cat" and "table").*

For a sequence of n tokens with indices $x = [x_1, x_2, \ldots, x_n]$ (e.g., the sentence ["the", "cat", "runs"]), the embedding layer produces a matrix:

$$X = [\text{Emb}(x_1), \text{Emb}(x_2), \ldots, \text{Emb}(x_n)] \text{ in } R^{n \times d_{\text{model}}}$$

Each row of X is the embedding vector of a token, forming the initial representation of the sequence.

The embedding layer reduces the dimensionality of tokens from V (potentially tens of thousands) to d_{model}, making computations feasible while preserving semantic information. The learned embeddings capture relationships between tokens based on their co-occurrence in the training data, enabling the model to generalize to unseen sequences.

Positional Encoding

Since transformers process all tokens in a sequence simultaneously, unlike RNNs, which process tokens sequentially, they lack an inherent sense of token order. Positional encodings are added to the embeddings to provide information about each token's position in the sequence.

Note *PE(pos + k) can be represented as a linear function of PE(pos), which is why the sinusoidal approach works.*

Sinusoidal positional encodings (used in the original transformer) assign each position in the sequence a vector built from sine and cosine waves of different frequencies. At first glance this seems arbitrary, but it was chosen for a very specific reason: ***a shift in position can be expressed using linear operations on the encoding****, meaning that moving from position pos to position pos + k does not require recomputing everything from scratch. The representation "moves" in a predictable, algebraically smooth way.*

To understand why this matters, it helps to recall the fundamental challenge. Transformers have no inherent sense of order, because self-attention is permutation-invariant. Positional encodings are therefore injected to give the model a notion of sequence position. However, not all ways of encoding position are equal: some treat each position independently, while others—like the sinusoidal method—treat positions in a structured, mathematically meaningful way.

Sine and cosine functions have a very special property:

$$\sin(a+b) = \sin(a)\cos(b) + \cos(a)\sin(b)$$

$$\cos(a+b) = \cos(a)\cos(b) - \sin(a)\sin(b)$$

These identities mean that "shifting the input" (moving from a to a + b) results in a ***linear transformation*** *of the pair. That is, the encoding at position* ***pos + k*** *can be written as a matrix acting on the encoding at* ***pos****.*

This holds independently for each frequency used in the positional encoding. Because the full positional encoding vector is just a stack of many such sine/cosine pairs, the entire vector inherits this property.

The original transformer uses fixed **sinusoidal functions to encode positions**, defined for position pos (from 1 to n) and dimension i (from 0 to $d_{\text{model}} - 1$):

$$PE(pos, 2i) = \sin\left(\frac{pos}{10000^{2i/d_{\text{model}}}}\right)$$

$$PE(pos, 2i+1) = \cos\left(\frac{pos}{10000^{2i/d_{\text{model}}}}\right)$$

These equations assign a unique vector to each position. The denominator $10000^{2i/d_{\text{model}}}$ creates a range of frequencies, with lower dimensions (small i) capturing low-frequency (global) positional information and higher dimensions capturing high-frequency (local) information.

For a sequence of length n, the positional encodings form a matrix:

$$PE \in \mathbb{R}^{n \times d_{\text{model}}}$$

where each row $PE(pos, :)$ is the positional encoding for position pos.

The input to the transformer is the sum of the embedding and positional encoding matrices:

$$X_{\text{input}} = X + PE \in \mathbb{R}^{n \times d_{\text{model}}}$$

Some modern transformers learn positional encodings as parameters, optimized during training:

$$PE_{\text{learned}} \in \mathbb{R}^{n_{\max} \times d_{\text{model}}}$$

where $n_{\max}$ is the maximum sequence length the model can handle.

Positional encodings ensure the model distinguishes between sequences where order matters, such as "The cat chased the dog" versus "The dog chased the cat." The sinusoidal functions produce smooth, periodic patterns that allow the model to generalize to longer sequences than seen during training. The scaling factor $10000^{2i/d_{\text{model}}}$ ensures that positional differences are meaningful across dimensions, enabling the attention mechanism to weigh positional relationships effectively.

Attention Mechanism

The attention mechanism is the heart of the transformer, enabling it to model relationships between tokens by focusing on relevant parts of the input sequence. The transformer uses scaled dot-product attention and multi-head attention to capture diverse dependencies.

Scaled Dot-Product Attention

For an input matrix $X \in \mathbb{R}^{n \times d_{\text{model}}}$ (e.g., the embedded input sequence with positional encodings), the attention mechanism computes three vectors for each token: queries, keys, and values. These are obtained via linear transformations:

$$Q = XW_Q \in \mathbb{R}^{n \times d_k}$$

$$K = XW_K \in \mathbb{R}^{n \times d_k}$$

$$V = XW_V \in \mathbb{R}^{n \times d_v}$$

where $W_Q, W_K \in \mathbb{R}^{d_{\text{model}} \times d_k}$, $W_V \in \mathbb{R}^{d_{\text{model}} \times d_v}$, and typically $d_k = d_v = d_{\text{model}}/h$, with h being the number of attention heads (e.g., 8 or 12).

The attention scores are computed as a scaled dot product:

$$\text{Scores} = \frac{QK^T}{\sqrt{d_k}} \in \mathbb{R}^{n \times n}$$

Each element $[QK^T]_{ij}$ represents the similarity between the query of token i and the key of token j. The scaling factor $\sqrt{d_k}$ prevents large values in high-dimensional settings, which could push the softmax into regions with small gradients, hindering training.

The scores are normalized using the softmax function to form attention weights:

$$A = \text{softmax}\left(\frac{QK^T}{\sqrt{d_k}}\right) \in \mathbb{R}^{n \times n}$$

where

$$\text{softmax}(S)_{ij} = \frac{\exp(S_{ij})}{\sum_{k=1}^{n} \exp(S_{ik})}$$

This ensures that the weights for each token sum to 1, representing a probability distribution over the input tokens.

The attention output is a weighted sum of the values:

$$\text{Attention}(Q, K, V) = AV \in \mathbb{R}^{n \times d_v}$$

Each row of the output is a weighted combination of the value vectors, where the weights reflect the relevance of other tokens.

The dot product measures how aligned a token's query is with other tokens' keys, allowing the model to focus on relevant tokens (e.g., in "The cat chased the dog," the word "chased" might attend strongly to "cat" and "dog"). The softmax ensures that attention is distributed in a normalized way, making the output interpretable as a weighted average of relevant information.

Multihead Attention

To capture different types of relationships (e.g., syntactic, semantic, or positional), the transformer uses multiple attention heads. The queries, keys, and values are split into h heads, each with reduced dimensionality $d_k = d_v = d_{\text{model}}/h$. For head i:

$$Q_i = XW_{Q_i}, \quad K_i = XW_{K_i}, \quad V_i = XW_{V_i}$$

$$\text{head}_i = \text{Attention}(Q_i, K_i, V_i) = \text{softmax}\left(\frac{Q_i K_i^T}{\sqrt{d_k}}\right) V_i$$

The outputs of all heads are concatenated and transformed:

$$\text{MultiHead}(Q, K, V) = \text{Concat}(\text{head}_1, \text{head}_2, \ldots, \text{head}_h)W_O \in \mathbb{R}^{n \times d_{\text{model}}}$$

where $W_O \in \mathbb{R}^{h \cdot d_v \times d_{\text{model}}}$

Multihead attention allows the model to attend to different aspects of the input simultaneously. For example, one head might focus on nearby tokens (local context), while another captures long-range dependencies (e.g., connecting a pronoun to its referent). The concatenation and linear transformation combine these perspectives into a unified representation.

Masked Attention (Decoder)

In the transformer decoder, masked self-attention ensures that a token only attends to previous positions in the output sequence, which is critical for autoregressive tasks such as language generation. This is achieved by applying a mask matrix to the attention scores, preventing the model from accessing future tokens during prediction.

A mask matrix $(M \in \mathbb{R}^{n \times n})$ is defined, where:

$$M_{ij} = \begin{cases} 0 & \text{if } i \geq j \\ -\infty & \text{otherwise} \end{cases}$$

The attention computation incorporates this mask to restrict attention to previous and current positions:

$$Attention(Q, K, V) = \text{softmax}\left(\frac{QK^T}{\sqrt{d_k}} + M\right)V$$

Here, $(Q, K, V \in \mathbb{R}^{n \times d_k})$ are the query, key, and value matrices, and (d_k) is the dimension of each attention head. The $(-\infty)$ values in (M) force the softmax to assign zero weight to future tokens (positions where $(i < j)$), ensuring that the attention mechanism only considers past and current tokens when computing the output for position (i).

The $-\infty$ values force the softmax to assign zero weight to future tokens, ensuring predictions rely only on past and current tokens.

Masking is essential for tasks like machine translation, where the decoder generates the target sequence one token at a time without access to future tokens. This mimics the autoregressive nature of language generation, ensuring the model learns to predict tokens in sequence.

Feed-Forward Neural Network

Each encoder and decoder layer includes a position-wise feed-forward network (FFN), applied independently to each token's representation to introduce non-linearity and enhance expressiveness. For a token's vector $x \in \mathbb{R}^{d_{\text{model}}}$

$$\text{FFN}(x) = \max(0, xW_1 + b_1)W_2 + b_2$$

where:

- $W_1 \in \mathbb{R}^{d_{\text{model}} \times d_{\text{ff}}}$, $b_1 \in \mathbb{R}^{d_{\text{ff}}}$ project the input to a higher-dimensional space (d_{ff} is typically 2048 or 4096).
- $\max(0, \cdot)$https://www.codecogs.com/eqnedit.php?latex=%20%5Cmax(0%2C%20%5Ccdot)%20#0 is the ReLU activation function.
- $W_2 \in \mathbb{R}^{d_{\text{ff}} \times d_{\text{model}}}$, $b_2 \in \mathbb{R}^{d_{\text{model}}}$ project back to the original dimension.

Some modern transformers (e.g., BERT, GPT) use the Gaussian Error Linear Unit (GELU) instead of ReLU:

$$\text{GELU}(x) = x \cdot \Phi(x) \approx 0.5x\left(1 + \tanh\left(\sqrt{\frac{2}{\pi}}(x + 0.044715x^3)\right)\right)$$

where $\Phi(x)$ is the cumulative distribution function of the standard normal distribution.

The FFN acts as a per-token transformation, allowing the model to refine each token's representation independently. The expansion to d_{ff} increases the model's capacity to learn complex patterns, while the non-linear activation enables the capture of non-linear relationships. Applying the FFN position-wise ensures computational efficiency, as it can be parallelized across tokens.

Residual Connections and Layer Normalization

Each sub-layer (attention and FFN) in the encoder and decoder includes a residual connection followed by layer normalization to stabilize training and improve gradient flow. For a sub-layer function $F(x)$ (e.g., multi-head attention or FFN):

$$y = x + F(x) \in \mathbb{R}^{n \times d_{\text{model}}}$$

This adds the input directly to the sub-layer's output, allowing the model to learn identity functions and mitigate vanishing gradients.

Layer normalization normalizes the output across the d_{model} dimensions for each token:

$$\text{LayerNorm}(y) = \frac{y - \mu}{\sqrt{\sigma^2 + \epsilon}} \cdot \gamma + \beta$$

where

- $\mu = \frac{1}{d_{\text{model}}} \sum_{i=1}^{d_{\text{model}}} y_i$ is the mean across the feature dimension.
- $\sigma^2 = \frac{1}{d_{\text{model}}} \sum_{i=1}^{d_{\text{model}}} (y_i - \mu)^2$ is the variance.
- ϵ is a small constant (e.g., 10^{-5}) for numerical stability.
- $\gamma, \beta \in \mathbb{R}^{d_{\text{model}}}$ are learned scaling and shifting parameters.

Residual connections ensure that the model can retain information from earlier layers, making it easier to train deep networks by preventing gradient degradation. Layer normalization standardizes the activations, reducing internal covariate shift and stabilizing training. This combination allows transformers to stack many layers (e.g., 12 or 24) without suffering from optimization issues.

Output Layer

The decoder's final output is used to predict the next token in the sequence. For the decoder's output matrix $Z \in \mathbb{R}^{n \times d_{\text{model}}}$:

$$\text{Logits} = ZW_{\text{vocab}} + b_{\text{vocab}} \in \mathbb{R}^{n \times V}$$

where $W_{\text{vocab}} \in \mathbb{R}^{d_{\text{model}} \times V}$, $b_{\text{vocab}} \in \mathbb{R}^V$ map the output to the vocabulary size. The logits are converted to probabilities using the softmax function:

$$P(y_i|Z) = \text{softmax}(\text{Logits})_i = \frac{\exp(\text{Logits}_i)}{\sum_{j=1}^{V} \exp(\text{Logits}_j)} \in \mathbb{R}^V$$

This gives the probability distribution over all possible next tokens for each position.

The linear layer projects the decoder's output to the vocabulary space, and the softmax ensures a valid probability distribution. During inference, the token with the highest probability is typically selected (greedy decoding), though techniques like beam search or sampling can be used for better results.

Loss Function

During training, the transformer minimizes the cross-entropy loss to align predicted probabilities with the true target sequence. For a target sequence $y = [y_1, y_2, \ldots, y_n]$ and predicted probabilities $P(y_i)$:

$$\mathcal{L} = -\frac{1}{n}\sum_{i=1}^{n} \log P(y_i)$$

This measures the negative log-likelihood of the true tokens under the model's predictions.

The **cross-entropy loss** encourages the model to assign high probabilities to the correct tokens. During training, the decoder uses teacher forcing, where the true target sequence is fed as input, allowing the model to learn from ground-truth tokens rather than its own predictions.

Encoder-Decoder Interaction

In sequence-to-sequence tasks like machine translation, the encoder processes the input sequence (e.g., a sentence in the source language) to produce a contextualized representation:

$$H \in \mathbb{R}^{n \times d_{\text{model}}}$$

In the decoder's encoder-decoder attention layer, the decoder's queries attend to the encoder's keys and values:

$$Q_{\text{decoder}} = X_{\text{decoder}} W_{Q_{\text{decoder}}} \in \mathbb{R}^{n \times d_k}$$

$$K_{\text{encoder}} = H W_{K_{\text{encoder}}}, \quad V_{\text{encoder}} = H W_{V_{\text{encoder}}} \in \mathbb{R}^{n \times d_v}$$

$$\text{Attention}(Q_{\text{decoder}}, K_{\text{encoder}}, V_{\text{encoder}}) = \text{softmax}\left(\frac{Q_{\text{decoder}} K_{\text{encoder}}^T}{\sqrt{d_k}}\right) V_{\text{encoder}}$$

This mechanism allows the decoder to focus on relevant parts of the input sequence when generating the output. For example, in translating "El gato corre" to "The cat runs," the decoder attends to "gato" when generating "cat," ensuring alignment between source and target sequences.

Complexity Analysis

The computational complexity of the transformer is driven by its attention and feed-forward components:

Time Complexity: Self-attention involves computing QK^T ($O(n^2 \cdot d_k)$) and multiplying with V ($O(n^2 \cdot d_v)$), leading to $O(n^2 \cdot d_{\text{model}})$ per layer. The FFN is $O(n \cdot d_{\text{model}} \cdot d_{\text{ff}})$, as it processes each token independently. For L layers, the total complexity is

$$O(L \cdot (n^2 \cdot d_{\text{model}} + n \cdot d_{\text{model}} \cdot d_{\text{ff}}))$$

Space Complexity: Storing Q, K, V and attention scores requires $O(n^2 + n \cdot d_{\text{model}})$ space, with additional memory for intermediate activations.

The quadratic complexity of self-attention ($O(n^2)$ is a major bottleneck for long sequences, prompting research into efficient variants like Sparse transformers or Linformer, which reduce this to $O(n \log n)$ or $O(n)$. The FFN's linear complexity per token makes it less demanding, but the high dimensionality of d_{ff} contributes significantly to the overall computational cost.

Advantages of the Transformer

- **Parallelization**: By processing all tokens simultaneously, transformers avoid the sequential bottlenecks of RNNs, enabling faster training and inference on modern hardware like GPUs and TPUs.
- **Long-Range Dependencies**: The attention mechanism allows the model to directly connect any two tokens in a sequence, regardless of their distance, overcoming the vanishing gradient problem that limits RNNs and LSTMs.

- **Scalability**: Transformers scale effectively with more data, larger models, and increased computational resources, as evidenced by their success in massive models like GPT-3 (175 billion parameters) and LLaMA.
- **Flexibility**: The architecture is highly adaptable, supporting tasks beyond NLP, such as image processing (vision transformers), speech recognition, and even scientific applications like protein structure prediction (AlphaFold).

Limitations

- **Computational Complexity**: The self-attention mechanism has a quadratic complexity with respect to sequence length ($O(n^2)$), making it computationally expensive for very long sequences. Variants like Performer or Linformer address this by approximating attention.
- **Memory Requirements**: Large transformer models require significant memory, limiting their deployment on resource-constrained devices.
- **Lack of Explicit Sequential Modeling**: While positional encodings provide sequence information, they are less natural than the sequential processing of RNNs for tasks where order is critical (e.g., time-series modeling).
- **Training Data Dependence**: Transformers require large amounts of data to achieve optimal performance, and their generalization can be limited in low-data regimes.

Applications

The transformer architecture underpins most state-of-the-art models in NLP:

- **BERT (Bidirectional Encoder Representations from Transformers)**: Uses only the encoder for tasks like text classification, question answering, and named entity recognition by leveraging bidirectional context.
- **GPT (Generative Pretrained Transformer)**: Uses only the decoder for autoregressive tasks like text generation and dialogue systems.
- **T5 (Text-to-Text Transfer Transformer)**: Frames all NLP tasks as text-to-text, enabling a unified approach to classification, translation, and summarization. Beyond NLP, transformers are applied in:
- **Computer Vision**: Vision transformers (ViT) treat images as sequences of patches, achieving performance comparable to convolutional neural networks.
- **Audio Processing**: Models like Wav2Vec use transformers for speech recognition and audio classification.
- **Scientific Applications**: AlphaFold by DeepMind uses transformers to predict protein structures with unprecedented accuracy.

Variants and Improvements

Since its introduction, the transformer has inspired numerous optimizations:

- **Efficient Attention**: Models like Sparse Transformers, Longformer, and BigBird reduce the quadratic complexity of attention for longer sequences.
- **Pretraining and Fine-Tuning**: Techniques like pretraining on large corpora (e.g., Wikipedia, Common Crawl) and fine-tuning on specific tasks have become standard, as seen in BERT and RoBERTa.
- **Compact Models**: DistilBERT and TinyBERT reduce model size for faster inference on resource-constrained devices.
- **Multimodal Transformers**: Models like CLIP and DALL-E integrate text and images, extending transformers to multimodal tasks.

The Comprehensive Mathematics Behind Training Large Language Models

Explaining the Purpose and Scope

Large language models (LLMs), such as GPT, BERT, and their successors, have transformed natural language processing by enabling machines to understand and generate human-like text. These models, built on the transformer architecture, rely on intricate mathematical frameworks to learn from vast datasets, capturing complex patterns in language.

Training an LLM involves optimizing billions of parameters to model the probability distribution over sequences of tokens, a task that requires sophisticated techniques from probability, linear algebra, and optimization. This information provides a comprehensive exploration of the mathematical principles underlying LLM training, covering the objective function, optimization algorithms, regularization methods, scaling laws, and advanced techniques like fine-tuning and reinforcement learning from human feedback (RLHF). Each section is designed to offer a deep understanding of the concepts, with detailed explanations and mathematical formulations to illuminate the processes involved.

Defining the Goal of Language Modeling

The primary objective of training an LLM is to model the probability distribution over sequences of tokens (e.g., words, subwords, or characters). This is typically framed as a language modeling task, where the model learns to predict the next token in a sequence given the preceding tokens (for autoregressive models like GPT) or to predict masked tokens given their context (for bidirectional models like BERT). The mathematical foundation of this process lies in probability theory, specifically in maximizing the likelihood of observed data.

Next-Token Prediction

Modeling Sequential Probability

In autoregressive LLMs, the goal is to model the joint probability of a sequence of tokens $(w_1, w_2, \ldots, w_n)$. Assuming the sequence is generated sequentially, the joint probability can be decomposed using the chain rule of probability:

$$P(w_1, w_2, \ldots, w_n) = \prod_{i=1}^{n} P(w_i | w_1, w_2, \ldots, w_{i-1})$$

Here, $(P(w_i | w_1, \ldots, w_{i-1}))$ represents the conditional probability of the (i)-th token given all previous tokens. The model learns to approximate this distribution by predicting each token based on its context.

Defining the Loss Function

To train the model, we minimize the negative log-likelihood (NLL), which corresponds to the cross-entropy loss between the predicted and true token distributions:

$$L = -\frac{1}{n} \sum_{i=1}^{n} \log P(w_i | w_1, \ldots, w_{i-1})$$

This loss measures how well the model's predicted probabilities align with the true tokens. A lower loss indicates that the model assigns higher probabilities to the correct tokens.

Softmax for Probability Distribution

The model outputs a probability distribution over the vocabulary (V) (which may contain tens of thousands of tokens) for each token position. This is achieved using the softmax function applied to the output logits $(z_i \in \mathbb{R}^{|V|})$:

$$P(w_i | w_1, \ldots, w_{i-1}) = \frac{\exp(z_i[w_i])}{\sum_{w \in V} \exp(z_i[w])}$$

Here, $(z_i[w_i])$ is the logit corresponding to the true token (w_i), and the denominator normalizes the probabilities across all possible tokens in the vocabulary. Computing the softmax for large vocabularies is computationally intensive due to the summation over $(|V|)$. To address this, techniques like sampled softmax (sampling a subset of negative tokens) or \textit{hierarchical softmax} (organizing the vocabulary into a tree) are often used during training to reduce computational costs.

Practical Considerations

In practice, the dataset consists of billions of tokens (e.g., from web corpora, books, or code), and the model processes sequences in fixed-length windows (e.g., 512 or 2048 tokens). The loss is computed over these sequences, and the model learns to capture long-range dependencies through the transformer's attention mechanism.

Masked Language Modeling

Bidirectional Context Modeling

For bidirectional models like BERT, training employs masked language modeling (MLM). In MLM, a percentage of tokens (e.g., 15%) in the input sequence are masked (replaced with a special [MASK] token, a random token, or left unchanged), and the model predicts the original tokens based on the surrounding context. The objective is to maximize the likelihood of the masked tokens:

$$L_{\text{MLM}} = -\sum_{i \in \text{masked}} \log P(w_i | w_{\text{context}})$$

where (w_{context}) includes all non-masked tokens in the sequence. This approach allows the model to learn bidirectional dependencies, making it suitable for tasks like text classification or question answering, where understanding the full context is crucial.

Comparison with Next-Token Prediction

Unlike next-token prediction, which is unidirectional (left-to-right), MLM leverages both left and right contexts, enabling the model to develop a richer understanding of token relationships. However, this comes at the cost of requiring a different inference strategy, as the model is not trained to generate sequences autoregressively.

Optimization: Gradient-Based Learning

Training an LLM involves optimizing billions of parameters (θ) (including weights, biases, and embeddings) to minimize the loss function $(\mathcal{L})$. This is a high-dimensional, non-convex optimization problem, addressed using gradient-based methods from numerical optimization.

Gradient Descent

Iterative Parameter Updates

The core optimization technique is gradient descent, which iteratively updates the parameters in the direction that reduces the loss:

$$\theta \leftarrow \theta - \eta \nabla_\theta \mathcal{L}$$

Here, \(\eta \) is the learning rate (e.g., \(10^{-4} \) to \(10^{-6} \)), a hyperparameter that controls the step size, and \(\nabla_\theta \mathcal{L} \) is the gradient of the loss with respect to the parameters, indicating the direction of steepest ascent. The negative gradient points toward the direction of loss reduction.

Stochastic Gradient Descent

Computing the gradient over the entire dataset (full-batch gradient descent) is infeasible due to the massive size of training corpora (e.g., terabytes of text). Instead, stochastic gradient descent (SGD) or mini-batch gradient descent is used, where the gradient is approximated over a small batch of (B) sequences:

$$\nabla_\theta \mathcal{L} \approx \frac{1}{B} \sum_{i=1}^{B} \nabla_\theta \mathcal{L}_i$$

Typical batch sizes range from 512 to 4096 sequences, depending on hardware and model size. Mini-batch SGD balances computational efficiency with gradient accuracy, enabling stable convergence.

Adam Optimizer

Adaptive Optimization

The Adam optimizer, a variant of SGD, is widely used for LLMs due to its efficiency in handling high-dimensional parameter spaces and non-convex loss landscapes. Adam combines the benefits of momentum (to accelerate gradients in consistent directions) and adaptive learning rates (to adjust step sizes per parameter). The update rules are:

$$m_t \& = \beta_1 m_{t-1} + (1-\beta_1)\nabla_\theta \mathcal{L}$$

$$v_t \& = \beta_2 v_{t-1} + (1-\beta_2)(\nabla_\theta \mathcal{L})^2$$

$$\hat{m}_t \& = \frac{m_t}{1-\beta_1^t}, \hat{v}_t = \frac{v_t}{1-\beta_2^t}$$

$$\theta_t \& = \theta_{t-1} - \eta \frac{\hat{m}_t}{\sqrt{\hat{v}_t}+\epsilon}$$

Here:

- (m_t): First moment (mean of gradients), incorporating momentum with decay rate $\beta_1 \approx 0.9$.
- (v_t): Second moment (uncentered variance of gradients), with decay rate $\beta_2 \approx 0.999$.
- $(\hat{m}_t, \hat{v}_t)$: Bias-corrected moments to account for initialization at $t=0$.
- $(\epsilon \approx 10^{-8})$: Small constant to prevent division by zero.
- (η): Learning rate, typically scheduled to decay over time (e.g., linear or cosine decay).

Why Adam Works for LLMs

Adam's adaptive learning rates (via $\sqrt{\hat{v}_t}$) allow it to handle sparse gradients and varying scales of parameters in LLMs, which often have billions of parameters with diverse roles (e.g., attention weights vs. embeddings). This adaptability ensures faster convergence compared to vanilla SGD, especially in the early stages of training.

Backpropagation

Computing Gradients

Gradients are computed using backpropagation, which applies the chain rule to propagate errors backward through the model's layers. For a transformer with (L) layers, the loss depends on the output of the final layer (z_L), and the gradient for a parameter (θ_l) in layer (l) is

$$\nabla_{\theta_l}\mathcal{L} = \frac{\partial \mathcal{L}}{\partial z_L} \cdot \frac{\partial z_L}{\partial z_{L-1}} \cdots \frac{\partial z_{l+1}}{\partial z_l} \cdot \frac{\partial z_l}{\partial \theta_l}$$

Each $(\frac{\partial z_{i+1}}{\partial z_i})$ represents the Jacobian of the layer's transformation (e.g., attention or feed-forward). This process is computationally intensive due to the depth of transformers (e.g., 12–96 layers) and the large sequence lengths (e.g., 2048 tokens).

Role of Automatic Differentiation

Modern frameworks like PyTorch and TensorFlow use automatic differentiation to compute gradients efficiently, avoiding manual derivation of complex layer interactions. Backpropagation scales to billions of parameters by leveraging optimized matrix operations on GPUs or TPUs.

Regularization Techniques

Preventing Overfitting

Given the massive parameter count of LLMs (e.g., 175 billion for GPT-3), overfitting is a significant risk, especially when training on finite datasets. Regularization techniques modify the training process to improve generalization and stability.

Dropout

Randomly Deactivating Neurons

Dropout randomly sets a fraction (p) (e.g., 0.1) of activations to zero during training, effectively sampling different subnetworks:

$$x_{\text{dropout}} = x \cdot m, \quad m_i \sim \text{Bernoulli}(1 - p)$$

where (x) is the input activation and (m) is a binary mask. Dropout is applied in the attention and feed-forward layers of the transformer. During inference, dropout is disabled, and activations are scaled by $(1/(1 - p))$ to maintain expected values.

Why Dropout Helps

By forcing the model to learn robust representations that do not rely on specific neurons, dropout reduces overfitting and encourages redundancy in the network. This is particularly important for LLMs, where the high capacity could otherwise lead to memorization of training data.

Weight Decay

Penalizing Large Weights

Weight decay adds an L2 regularization term to the loss, penalizing large parameter values:

$$\mathcal{L}_{\text{total}} = \mathcal{L} + \lambda ||\theta||_2^2$$

where (λ) (e.g., 0.01) controls the regularization strength, and $(||\theta||_2^2 = \sum \theta_i^2)$ is the squared L2 norm of the parameters. This encourages simpler models by keeping weights small, reducing the risk of overfitting.

Implementation in Adam

In practice, weight decay is often decoupled from the optimizer (e.g., in AdamW, a variant of Adam), where the L2 penalty is applied directly to the parameter updates rather than the gradient moments, improving stability.

Label Smoothing

Softening Target Distributions

Label smoothing modifies the target distribution to prevent the model from becoming overconfident in its predictions. Instead of using a one-hot vector (where the true token has probability 1 and others 0), the target is smoothed:

$$y_{\text{smooth}}(w) = (1 - \epsilon) \cdot \mathbb{K}_{w=w_i} + \frac{\epsilon}{|V|}$$

where $(\epsilon \approx 0.1), and(\mathbb{K}_{w=w_i}) is 1 if (w = w_i)$ and 0 otherwise. This assigns a small probability $(\epsilon/|V|)$ to all other tokens, encouraging the model to generalize better by avoiding extreme probability assignments.

Impact on Training

Label smoothing reduces the sharpness of the loss landscape, making optimization more stable and improving performance on unseen data. It is particularly effective for large vocabularies, where overconfidence can lead to poor generalization.

Fine-Tuning and RLHF

Adapting Pretrained Models

After pretraining on a large, general corpus, LLMs are often fine-tuned to improve performance on specific tasks or to align with human preferences.

Supervised Fine-Tuning (SFT)

Task-Specific Optimization

Supervised fine-tuning (SFT) involves training the model on a curated dataset of input-output pairs, such as question-answer pairs or instruction-response pairs. The objective is to minimize the cross-entropy loss over the fine-tuning dataset:

$$\mathcal{L}_{\text{SFT}} = -\sum_i \log P(y_i|x_i)$$

where (x_i) is the input prompt and (y_i) is the target response. SFT adapts the model to specific tasks, improving its performance on targeted applications like chatbots or code generation.

Differences from Pretraining

Unlike pretraining, which uses vast, diverse datasets, SFT typically involves smaller, high-quality datasets tailored to the task. The learning rate is often lower (e.g., (10^{-5}))) to preserve the pretrained knowledge while making targeted adjustments.

Reinforcement Learning from Human Feedback (RLHF)

Aligning with Human Preferences

RLHF aligns the model with human values or preferences, addressing limitations of purely supervised training (e.g., generating unsafe or unhelpful responses). RLHF involves:

Reward Model Training: A reward model $(R(y, x))$ is trained on human feedback (e.g., pairwise comparisons of model outputs) to assign scores to responses based on quality, safety, or helpfulness.

Policy Optimization: The LLM, treated as a policy $(\pi_\theta(y|x))$, is optimized to maximize the expected reward using a policy gradient method like Proximal Policy Optimization (PPO):

$$\mathcal{L}_{\text{RLHF}} = -\mathbb{E}_{y \sim \pi_\theta}[\log \pi_\theta(y|x) \cdot R(y, x)]$$

To prevent the model from diverging too far from the pretrained distribution, a KL-divergence penalty is added:

$$\mathcal{L}_{\text{total}} = \mathcal{L}_{\text{RLHF}} + \lambda \text{KL}(\pi_\theta || \pi_{\text{ref}})$$

where (π_{ref}) is the pretrained policy, and (λ) https://www.codecogs.com/eqnedit.php?latex=(%20%5Clambda%20%5C)#0 controls the penalty strength.

Why RLHF Is Effective

RLHF allows the model to learn nuanced preferences that are difficult to capture with supervised data alone, such as tone, coherence, or ethical considerations. However, it is computationally intensive and requires high-quality human feedback.

Summarizing the Mathematical Framework

Training large language models is a complex, mathematically driven process that integrates:

- Modeling token sequences as conditional distributions using cross-entropy loss.
- Gradient-based methods like Adam to navigate high-dimensional loss landscapes.
- Techniques like dropout, weight decay, and label smoothing to ensure generalization.
- Mixed-precision training, gradient accumulation, and distributed training to handle computational demands.
- Empirical guidelines for balancing model size, data, and compute.
- SFT and RLHF to adapt models to specific tasks or human preferences.

These components work together to enable LLMs to achieve state-of-the-art performance in natural language tasks.

Modern LLM Architecture Characteristics As of the End of 2024 and During 2025

The period from late 2024 through 2025 marks one of the most rapid and transformative phases in the development of large language models. During these years, model design has shifted toward architectures that are vastly more efficient, more expressive, and capable of processing much longer spans of information than any generation before them. Three major themes define this era: the rise of Mixture-of-Experts (MoE) systems, the normalization of multimodal capabilities, and the dramatic expansion of context windows into the hundreds of thousands or even millions of tokens. Together, these advances have reshaped what LLMs can do and who can build with them.

A good entry point into this landscape is **Alibaba's Qwen 3 model**, released in April 2025. Qwen 3 is often highlighted as one of the strongest open-source models of the period. It uses a Mixture-of-Experts architecture, enabling it to scale to roughly 235 billion parameters while activating only the parts of the network needed for a given task. This approach allows the model to maintain high performance while consuming significantly less computation compared to dense models of similar size. Qwen 3 supports both text and vision inputs and provides a 128,000-token context window—large enough to handle long documents, research reports, or extensive dialogue. Its release under an open-source license made it widely accessible to research institutions and independent developers.

In contrast, **OpenAI's GPT-4.5**, introduced in February 2025, is a proprietary system. Although specific architectural details are undisclosed, it is generally understood to continue the transformer lineage while incorporating sparse activation or MoE-like strategies. GPT-4.5 offers multimodal capability across text, images, and audio, along with a 128,000-token context window. It was positioned as a refinement over GPT-4, particularly in reasoning, stability, and conversational fluidity. However, its closed-source nature reflects an ongoing tension in the field: the most capable models often offer limited visibility into their internal workings.

Anthropic's Claude 3.7 Sonnet, released only days before GPT-4.5, offers another perspective on advanced LLM design. Claude 3.7 uses a Mixture-of-Experts system and emphasizes safe, reflective, and controllable behavior. Its context window extends to 200,000 tokens, allowing it to analyze extremely long texts—legal documents, books, software repositories, or complex instructions—within a single prompt. Although the exact parameter count is not publicly disclosed, estimates place it above 200 billion. Like earlier Claude models, it reflects Anthropic's focus on "constitutional AI," prioritizing predictable and transparent reasoning processes within the constraints of a closed-source model.

The largest shifts in context capacity come from **Google's Gemini series**. Gemini 2.5 Pro, introduced in March 2025, supports a context window of approximately one million tokens, with future versions aiming for two million. This leap enables the model to process entire textbooks, corporate knowledge bases, and huge sequences of code without fragmentation. Along with strong multimodal support and reasoning tools such as self-fact-checking, Gemini 2.5 demonstrates how transformer architectures continue to scale both in depth and breadth. While parts of the Gemini ecosystem are proprietary, Google has also supported open-source releases like Gemma 3, which inherits some of the architectural lessons from its larger siblings.

Meanwhile, **DeepSeek R1,** released at the start of 2025, stands out as a model built specifically around large-scale MoE design. It contains an enormous 671 billion parameters in total, but only about 37 billion are active for each token, striking a balance between scale and efficiency. This makes DeepSeek R1 particularly strong in mathematical reasoning, algorithmic tasks, and code generation. As an open-weight model, it attracted attention for demonstrating that state-of-the-art reasoning performance is achievable without requiring an exclusively proprietary ecosystem.

Meta's Llama 3.3, building on the Llama 3 and Llama 3.1 generations, is another pillar of the open-source movement. Released in late 2024 and widely adopted into 2025, Llama 3.3 provides multimodal capabilities and a 128,000-token context window. Earlier variants such as Llama 3.1 reached up to 405 billion parameters, while Meta's announced successor, Llama 4, extends context windows into the multi-million range. The Llama series has become foundational for academic and commercial experimentation, largely because of its permissive licensing and extensive tooling ecosystem.

Alongside these major families, several smaller yet significant models contributed to the diversity of the landscape. Grok-3 from xAI focuses on real-time data access and spontaneous reasoning, though its closed-source nature limits deeper study. **Mistral's Small 3**, with roughly 24 billion parameters, shows how efficient transformer designs can deliver strong performance at lower computational cost. Models like these appeal to organizations seeking speed, cost-effectiveness, and deployability rather than maximum raw capability.

Across all of these systems, a set of consistent trends emerges. Mixture-of-Experts is becoming a dominant architecture because it lets models scale to extraordinary parameter counts without linear increases in activation cost. Multimodality has transitioned from a novelty to a basic expectation, allowing LLMs to interpret and generate across text, imagery, and audio in more integrated ways. Context windows have expanded so dramatically that many tasks once impossible—such as analyzing multiple books at once or maintaining a persistent, uninterrupted conversational history—are now routine. And the divide between open and proprietary models remains a defining feature of the field: open-source releases such as Llama, Qwen, and DeepSeek democratize experimentation, while proprietary models continue to push performance boundaries behind closed doors.

In sum, the LLM architectures of 2024–2025 represent a turning point. They deliver higher efficiency, deeper reasoning, richer modality support, and vastly expanded memory compared to earlier generations. These innovations have reshaped what individuals, companies, and researchers can build, setting the stage for even more capable and adaptive AI systems in the years to come.

In Summary

Chapter 2 provides an in-depth exploration of the core architectural principles and mathematical foundations behind modern Large Language Models (LLMs). It begins by explaining how the **transformer architecture**, introduced in 2017, revolutionized natural language processing by replacing sequential models like RNNs with self-attention mechanisms, enabling parallel processing and more efficient modeling of long-range dependencies.

Key components of the transformer—**input embeddings**, **positional encodings**, **multi-head attention**, **feed-forward networks**, **residual connections**, and **layer normalization**—are thoroughly explained both conceptually and mathematically. The chapter also outlines **decoder-specific mechanisms** like masked attention for autoregressive tasks and discusses the benefits and limitations of the architecture.

The second half delves into the **training process** of LLMs, including loss functions (like cross-entropy), optimization strategies (e.g., the Adam optimizer), and regularization techniques (e.g., dropout, label smoothing, and weight decay). It explains practical strategies for efficient large-scale training, including **mixed-precision training**, **gradient accumulation**, and **distributed computing**.

Finally, the chapter surveys the **latest LLM architectures as of 2024–2025**, highlighting models like GPT-4.5, Claude 3.7, and Qwen 3. These models emphasize **Mixture-of-Experts (MoE)** architectures, **multimodal inputs**, and **extended context windows**, reflecting trends toward more efficient, scalable, and versatile language models.

In the next chapter, we will **install all the necessary tools** and **begin building the embedding model (tokenizer)** that serves as a foundational component of our LLM. This step marks the start of our hands-on journey into model development.

CHAPTER 3

Building a Tokenizer for the Transformers Architecture Model

Tokenization lies at the heart of natural language processing (NLP), serving as the critical first step in transforming raw, unstructured text into structured units that computational models can process. These units, known as tokens, are the foundation upon which tasks such as sentiment analysis, machine translation, text generation, and large-scale language modeling are built.

A well-designed tokenizer can significantly influence the performance of NLP systems, balancing the trade-offs between vocabulary size, computational efficiency, and the ability to capture linguistic nuances. Among the many tokenization techniques available, byte-pair encoding (BPE) stands out as a versatile and widely adopted approach, particularly in modern language models like BERT, GPT, and T5. BPE's strength lies in its ability to create a compact yet expressive vocabulary, adeptly handling rare words, morphological variations, and even multilingual text.

In this chapter, we will dive into the fascinating world of tokenization by building a BPE tokenizer from the ground up. Through a step-by-step process, we will demystify the inner workings of BPE, exploring its origins in data compression and its adaptation to NLP. We will begin by understanding the theoretical foundations of tokenization and why BPE has become a cornerstone of modern NLP pipelines.

From there, we will implement the BPE algorithm, starting with basic text preprocessing and progressing to the iterative merging of frequent character pairs to form a vocabulary. Along the way, we will address practical challenges, such as handling special characters, managing vocabulary size, and ensuring the tokenizer is robust across diverse datasets.

D. Grigorov, *Building Large Language Models from Scratch*, https://doi.org/10.1007/979-8-8688-2297-1_3

By the end of this chapter, you will not only have a fully functional BPE tokenizer but also a deep appreciation for the design choices that make it effective. Whether you are a beginner eager to understand the mechanics of NLP or an experienced practitioner looking to customize tokenization for specific applications, this hands-on journey will equip you with the knowledge and skills to create and adapt tokenizers for your own projects. Let's embark on this exciting exploration of building a BPE tokenizer, step by step, and unlock the power of transforming text into tokens that machines can understand.

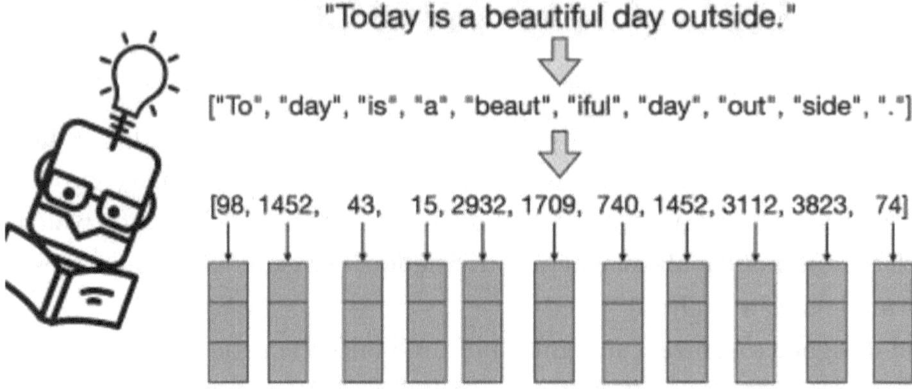

Figure 3-1. *How LLM Tokenizers Work, Source: LinkedIn, Nikitha R*

What Is Tokenization?

Tokenization is the process of splitting a piece of text—such as a sentence, paragraph, or document—into discrete units called tokens. These tokens can represent words, subwords, characters, or even punctuation marks, depending on the tokenization strategy. For example, consider the sentence "I love to code!" A simple word-based tokenizer might break it into the tokens ["I", "love", "to", "code", "!"].

The primary goal of tokenization is to convert unstructured text into a format that machines can process. Language models, whether based on traditional statistical methods or modern neural networks, operate on numerical data. Tokenizers bridge the gap between human-readable text and numerical representations by mapping tokens to unique identifiers (IDs) in a vocabulary. These IDs are then used as input for models to perform tasks like predicting the next word, classifying sentiment, or translating languages.

Tokenization is not a one-size-fits-all process. The granularity of tokens (e.g., words, subwords, or characters) and the rules for splitting text depend on the application, language, and model architecture. A poorly designed tokenizer can lead to issues like an overly large vocabulary, difficulty handling rare words, or loss of semantic meaning, all of which can degrade model performance.

Why Is Tokenization Important?

Tokenization is a critical step in NLP for several reasons:

1. **Standardization**: Raw text is inherently messy, with variations in punctuation, capitalization, and formatting. Tokenizers standardize text by defining consistent rules for splitting and processing it, ensuring uniformity across datasets.

2. **Vocabulary Creation**: Tokenizers define the vocabulary that a language model uses. A well-designed vocabulary balances size and expressiveness, capturing the most relevant linguistic units without overwhelming computational resources.

3. **Handling Linguistic Diversity**: Different languages have unique characteristics, such as complex morphology (e.g., German compound words) or no clear word boundaries (e.g., Chinese). Tokenizers must adapt to these variations to ensure effective processing.

4. **Efficiency**: By breaking text into smaller units, tokenizers reduce the complexity of input data, enabling models to process text more efficiently. Subword tokenization, for instance, can represent rare words as combinations of smaller units, reducing the need for an excessively large vocabulary.

5. **Semantic Preservation**: A good tokenizer preserves the semantic meaning of text. For example, splitting "unhappiness" into subword tokens like ["un", "happi", "ness"] allows models to learn relationships between related words (e.g., "happy" and "unhappiness").

Types of Tokenizers

There are several types of tokenizers, each with its own strengths and weaknesses. The choice of tokenizer depends on the task, language, and model requirements. Below are the most common types:

Word-Based Tokenizers

Word-based tokenizers split text into individual words, typically using whitespace and punctuation as delimiters. For example, the sentence "The cat is sleeping." might be tokenized as ["The", "cat", "is", "sleeping", "."].

Advantages:

- Intuitive and aligned with human understanding of language.
- Simple to implement for languages with clear word boundaries (e.g., English).

Disadvantages:

- Large vocabularies, as every unique word (including misspellings and morphological variants) requires a separate token.
- Poor handling of rare or out-of-vocabulary (OOV) words.
- Ineffective for languages without clear word boundaries (e.g., Chinese, Japanese).

Character-Based Tokenizers

Character-based tokenizers treat each character (including spaces and punctuation) as a token. For example, "The cat" becomes ["T", "h", "e", " ", "c", "a", "t"].

Advantages:

- Minimal vocabulary size, as it only includes characters.
- Handles rare words and misspellings naturally.
- Language-agnostic, suitable for multilingual applications.

Disadvantages:

- Produces long sequences of tokens, increasing computational cost.
- Loses higher-level semantic information (e.g., word boundaries).
- Requires models to learn complex patterns from low-level units.

Subword-Based Tokenizers

Subword-based tokenizers strike a balance between word-based and character-based approaches by splitting text into units smaller than words but larger than characters. Common subword tokenization algorithms include byte-pair encoding (BPE), WordPiece, and SentencePiece.

Advantages:

- Compact vocabularies that handle rare words effectively.
- Captures morphological patterns (e.g., "un" + "happiness").
- Language-agnostic, especially when using algorithms like SentencePiece.
- Reduces sequence length compared to character-based tokenization.

Disadvantages:

- More complex to implement than word-based tokenization.
- Requires training on a corpus to build the vocabulary.
- May produce unintuitive splits for some words.

Rule-Based Tokenizers

Rule-based tokenizers use predefined rules or regular expressions to split text. For example, a rule-based tokenizer might split text at specific punctuation marks or handle contractions like "don't" as ["do", "n't"].

Advantages:

- Highly customizable for specific domains or languages.
- Deterministic and interpretable.

Disadvantages:

- Requires manual rule creation, which is time-consuming and error-prone.
- Less adaptable to new or diverse datasets.

Byte-Pair Encoding (BPE): A Deep Dive

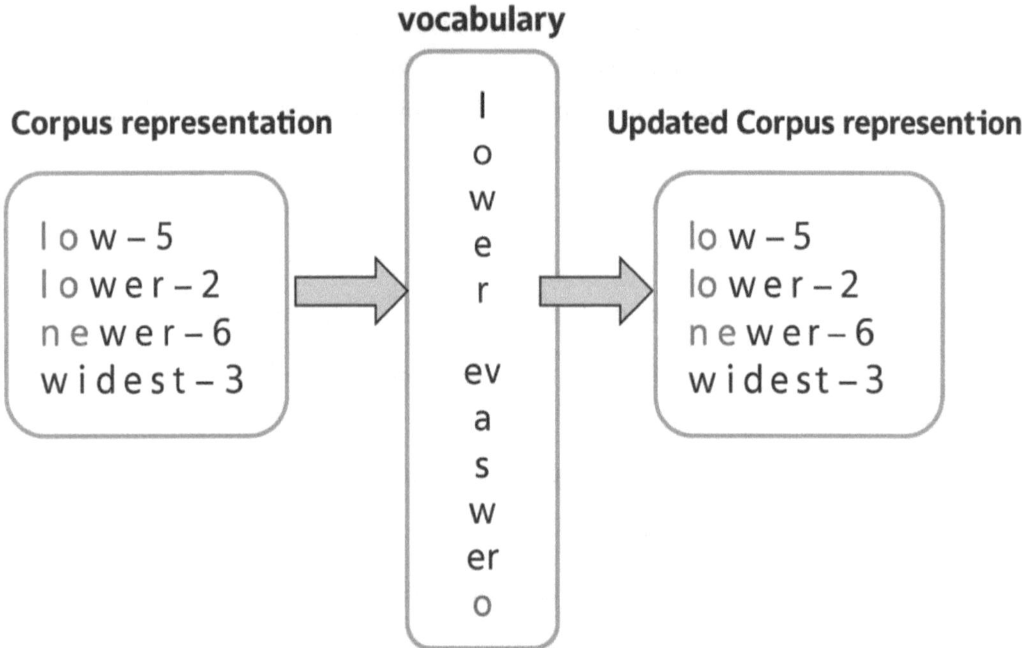

Figure 3-2. *Conception Behind BPE Tokenizer, Source: Toward Data Science*

Origins of BPE

Byte-pair encoding was first introduced by Philip Gage in 1994 as a data compression algorithm. In its original form, BPE aimed to reduce the size of text files by iteratively replacing the most frequent pair of adjacent bytes with a single, unused byte. This process compressed data while preserving the ability to reconstruct the original text. The algorithm's simplicity and effectiveness made it a compelling choice for compression tasks.

In 2015, Rico Sennrich, Barry Haddow, and Alexandra Birch adapted BPE for NLP in their paper "Neural Machine Translation of Rare Words with Subword Units." They recognized that BPE's iterative merging approach could address key challenges in NLP, such as handling rare words and creating compact vocabularies. By applying BPE to text tokenization, they enabled models to represent words as combinations of subword units, improving performance in tasks like machine translation. Since then, BPE has become a standard tokenization method in NLP, widely adopted in state-of-the-art language models.

BPE Algorithm Outline and Mathematical Formulation

BPE is a subword tokenization algorithm that iteratively merges the most frequent pair of adjacent tokens to build a vocabulary. Below is a detailed outline of the BPE algorithm, as adapted for NLP.

Identify Frequent Pairs

In each iteration, scan the text corpus to identify the most commonly occurring pair of adjacent tokens (initially characters or bytes). For example, in a text like "the cat in the hat," the pair "th" might appear frequently.

Replace and Record

- **Replace**: Replace all occurrences of the most frequent pair with a new placeholder ID not already in use. For instance, if the initial tokens are characters (mapped to IDs 0–255), the first new token might be assigned ID 256.
- **Record**: Store the mapping of the new ID to the merged pair in a lookup table. This table, often called the merge table or vocabulary, tracks all merge rules.
- **Vocabulary Size**: The size of the lookup table is a hyperparameter, also known as the vocabulary size. For example, GPT-2 uses a vocabulary size of 50,257 tokens.

Repeat Until No Gains

Continue iterating steps, merging the most frequent pairs in each iteration. Stop when no further compression is possible (e.g., no pair occurs more than once) or when the desired vocabulary size is reached.

Decompression (Decoding)

To restore the original text, reverse the process by substituting each token ID with its corresponding pair, using the lookup table in reverse order. This ensures lossless reconstruction of the text.

BPE Algorithm Example

To illustrate the BPE algorithm, let’s apply it to a small training dataset: “the cat in the hat.” We will demonstrate both the encoding (training) and decoding processes.

BPE operates on a corpus of text, represented as a sequence of words or characters, to construct a vocabulary of tokens. Let’s denote the corpus as $C = \{w_1, w_2, \ldots, w_n\}$, where each w_i is a word (or sequence of characters) in the corpus, and n is the number of words. The goal is to transform C into a tokenized representation by building a vocabulary V and a set of merge rules R.

Concrete Example of the Encoding Part

Suppose we want to build a BPE tokenizer vocabulary from the text “the cat in the hat.” We assume each word is preprocessed with an end-of-word marker (e.g., “”) to preserve word boundaries.

Each word w_i is initially split into a sequence of characters, augmented with an end-of-word marker (e.g., “</w>”). Formally, for a word $w_i = c_1 c_2 \cdots c_{l_i}$, where c_j are characters and l_i is the length of w_i, the initial token sequence is $s_i = [c_1, c_2, \ldots, c_{l_i}, ¡/\mathrm{w}¿]$.

The initial vocabulary V_0 is the set of all unique characters in C plus the end-of-word marker:

$$V_0 = \{c \mid c \text{ appears in } C\} \cup \{< /w >\}$$

For the corpus $C = \{\text{the}, \text{cat}, \text{in}, \text{the}, \text{hat}\}$, the initial token sequences are:

$$s_1 = [t, h, e, < /w >], \quad s_2 = [c, a, t, < /w >], \quad s_3 = [i, n, < /w >], \quad s_4 = [t, h, e, < /w >], \quad s_5 = [h, a, t, < /w >]$$

The initial vocabulary is:

$$V_0 = \{t, h, e, c, a, i, n, < /w >\}$$

In each iteration, BPE identifies the most frequent pair of adjacent tokens. Define a token sequence for the corpus at iteration k as $S_k = \{s_{k,1}, s_{k,2}, \ldots, s_{k,n}\}$, where $s_{k,i}$ is the token sequence for word w_i. For each pair of adjacent tokens (t_a, t_b) in S_k, compute the frequency:

$$f_k(t_a, t_b) = \sum_{i=1}^{n} \sum_{j=1}^{|s_{k,i}|-1} \mathbb{1}\{s_{k,i}[j] = t_a, s_{k,i}[j+1] = t_b\}$$

where $\mathbb{1}$ is the indicator function, and $|s_{k,i}|$ is the length of the token sequence $s_{k,i}$. The pair with the highest frequency:

$$(t_a^*, t_b^*) = \arg\max_{(t_a, t_b)} f_k(t_a, t_b)$$

For the example corpus at iteration 1 ($k = 0$), pairs include (t, h), (h, e), and $(e, < /w >)$, each with frequency 2. Suppose we select (t, h).

Iteration 1

- **Identify Frequent Pairs**: Split the text into characters with word boundaries: ["t h e", "c a t", "i n", "t h e", "h a t"]. Count adjacent pairs across the corpus:
    - "t h" appears twice (in "the" and "the").
    - "h e" appears twice.
    - "e" appears twice.
    - Other pairs (e.g., "c a", "a t", "i n") appear once.

- **Replace and Record**: Choose the pair "t h" (arbitrarily among the tied pairs) and replace it with a new token ID, e.g., 256. The new text becomes
    - "<256>e c a t i n <256>e h a t"
    - Update the vocabulary: {0: ..., 256: "th"}.

Iteration 2

- **Identify Frequent Pairs**: In the new text "<256>e c a t i n <256>e h a t", count pairs:
    - "<256> e" appears twice.
    - "e" appears twice.
    - Other pairs (e.g., "c a", "a t") appear once.
- **Replace and Record**: Merge "<256> e" into a new token ID, e.g., 257. The new text becomes
    - "<257> c a t i n <257> h a t"
    - Update the vocabulary: {0: ..., 256: "th", 257: "<256>e"}.

Iteration 3

- **Identify Frequent Pairs**: In the text "<257> c a t i n <257> h a t", count pairs:
    - "<257>" appears twice.
    - Other pairs (e.g., "c a", "i n") appear once.
- **Replace and Record**: Merge "<257>" into a new token ID, e.g., 258. The new text becomes
    - "<258> c a t i n <258> h a t".
    - Update the vocabulary: {0: ..., 256: "th", 257: "<256>e", 258: "<257>"}.

This process continues for a set number of merges or until the desired vocabulary size is reached.

Replace all occurrences of (t_a^*, t_b^*) with a new token t_{new}, typically assigned a unique ID (e.g., starting from 256 to avoid ASCII character IDs 0–255). The new token is:

$$t_{\text{new}} = t_a^* t_b^*$$

Update each token sequence $s_{k,i}$ to $s_{k+1,i}$ by replacing (t_a^*, t_b^*) with t_{new}. The merge rule is recorded as:

$$r_k = (t_a^*, t_b^* \rightarrow t_{\text{new}})$$

The vocabulary is updated:

$$V_{k+1} = V_k \cup \{t_{\text{new}}\}$$

For the pair (t, h), assign $t_{\text{new}} = 256$, representing "th". The updated sequences are $s_{1,1} = [256, e, < /w >], \quad s_{1,4} = [256, e, < /w >], \quad s_{1,2}, s_{1,3}, s_{1,5} unchanged$

The vocabulary becomes:

$$V_1 = V_0 \cup \{256\} \quad \text{where} \quad 256 \mapsto \text{th}$$

Repeat the process for m merges, where is a hyperparameter determining the vocabulary size $|V| = |V_0| + m$. The final vocabulary is:

$$V_m = V_0 \cup \{t_{new,1}, t_{new,2}, \ldots, t_{new,m}\}$$

The merge rules are:

$$R = \{r_1, r_2, \ldots, r_m\}$$

For our example, after three merges (e.g., $(t, h) \rightarrow 256$, $(256, e) \rightarrow 257$, $(257, ¡/\text{w}¿) \rightarrow 258$), the text becomes:

$$s_{3,1} = [258], \quad s_{3,2} = [c, a, t, < /w >], \quad s_{3,3} = [i, n, < /w >], \quad s_{3,4} = [258], \quad s_{3,5} = [h, a, t, < /w >]$$

The vocabulary is:

$$V_3 = \{t, h, e, c, a, i, n, < /w >, 256, 257, 258\}$$

where $256 \mapsto \text{th} \, 257 \mapsto 256 \text{ e}, 258 \mapsto 257 < /w >$.

Encoding New Text

To encode a new word $w = c_1 c_2 \cdots c_l$, initialize the token sequence:

$$s_0 = [c_1, c_2, \ldots, c_l, < /w >]$$

Apply each merge rule $r_k = (t_a, t_b \rightarrow t_{\text{new}})$ in order, replacing occurrences of (t_a, t_b) with t_{new}. The final sequence s_m is the encoded representation. For example, to encode "the":

$$s_0 = [t, h, e, < /w >]$$

Apply merges:

$$[t, h, e, < /w >] \xrightarrow{r1} [256, e, < /w >] \xrightarrow{r2} [257, < /w >] \xrightarrow{r3} [258]$$

The encoded form is $[258]$.

Concrete Example of the Decoding Part

Decoding reverses the encoding process. Given a sequence of tokens $s = [t_1, t_2, \ldots, t_k]$, apply the merge rules in reverse order. For each rule $r_i = (t_a, t_b \rightarrow t_{\text{new}})$, replace t_{new} with $t_a t_b$ `https://www.codecogs.com/eqnedit.php?latex=t_a%20t_b#0`. Finally, remove the "</w>" marker and join characters. For the token $[258]$:

$$[258] \xrightarrow{r_3^{-1}} [257, < /w >] \xrightarrow{r_2^{-1}} [256, e, < /w >] \xrightarrow{r_1^{-1}} [t, h, e, < /w >]$$

Remove "</w>" and join: "the".

To decode the compressed text "<258> c a t i n <258> h a t" back to the original text, reverse the merge process using the vocabulary in reverse order:

- **Substitute <258>**: According to the vocabulary, <258> maps to "<257>". The text becomes
  - "<257> c a t i n <257> h a t"
- **Substitute <257>**: <257> maps to "<256>e". The text becomes
  - "<256>e c a t i n <256>e h a t"

- **Substitute <256>**: <256> maps to "th". The text becomes
    - "t h e c a t i n t h e h a t"
- **Remove and Join**: Remove the end-of-word markers and join characters within each word:
    - "the cat in the hat"

This process restores the original text losslessly.

Optimization Objective

BPE implicitly optimizes for compression by minimizing the length of the tokenized corpus. Define the length of the tokenized corpus at iteration k

$$L_k = \sum_{i=1}^{n} |s_{k,i}|$$

Each merge reduces L_k by replacing multiple occurrences of a pair (t_a, t_b) with a single token. If $f_k(t_a, t_b)$ is the frequency of the pair, the reduction in length is:

$$\Delta L_k = f_k(t_a, t_b)$$

The total reduction over m merges approximates:

$$L_0 - L_m \approx \sum_{k=1}^{m} f_k(t_a^*, t_b^*)$$

This reflects BPE's compression roots, aiming to minimize L_m while maintaining a vocabulary size $|V_m| \leq |V_0| + m$.

Vocabulary Size and Hyperparameters

The vocabulary size $|V_m| \leq |V_0| + m$ is a critical hyperparameter. For example, GPT-2 uses $|V| = 50,257$.

The choice of m balances:

- **Expressiveness:** Larger m captures more complex tokens, improving semantic representation.
- **Efficiency:** Smaller m reduces memory and computational costs.

The frequency threshold for stopping merges can be formalized as: $f_k(t_a^*, t_b^*) \leq 1$, indicating no pair occurs more than once, though in practice, m is fixed to control $|V_m|$.

Step-by-Step Process of Building a BPE Tokenizer

To generalize the BPE process, let's outline the steps for building a tokenizer on a larger corpus, such as ["low", "lower", "lowest", "widest"], as discussed in previous contexts.

Initialize the Vocabulary

Split each word into characters, adding an end-of-word marker (e.g., ""). For the corpus:

- "low" → ["l", "o", "w", ""]
- "lower" → ["l", "o", "w", "e", "r", ""]
- "lowest" → ["l", "o", "w", "e", "s", "t", ""]
- "widest" → ["w", "i", "d", "e", "s", "t", ""]
- Initial vocabulary: ["l", "o", "w", "e", "r", "s", "t", "i", "d", ""]

Count Pair Frequencies

Count adjacent pairs:

- "l o" and "o w" appear three times each.
- "w e", "e s", "s t" appear two times each.
- Others (e.g., "w", "e r") appear once.

Merge the Most Frequent Pair

Merge “l o” → “lo”:

- “low” → [“lo”, “w”, “”]
- “lower” → [“lo”, “w”, “e”, “r”, “”]
- “lowest” → [“lo”, “w”, “e”, “s”, “t”, “”]
- “widest” → [“w”, “i”, “d”, “e”, “s”, “t”, “”]
- Vocabulary: [“l”, “o”, “w”, “e”, “r”, “s”, “t”, “i”, “d”, “”, “lo”]
- Merge rule: (“l”, “o”) → “lo”

Repeat Merging

Merge “lo w” → “low”:

- “low” → [“low”, “”]
- “lower” → [“low”, “e”, “r”, “”]
- “lowest” → [“low”, “e”, “s”, “t”, “”]
- “widest” → [“w”, “i”, “d”, “e”, “s”, “t”, “”]
- Vocabulary adds “low”
- Merge rule: (“lo”, “w”) → “low”

Continue merging (e.g., “e s” → “es”, “es t” → “est”) until the desired vocabulary size is reached.

Tokenize New Text

To tokenize “lowest”:

- Start: [“l”, “o”, “w”, “e”, “s”, “t”, “”]
- Apply merges: [“lo”, “w”, “e”, “s”, “t”, “”] → [“low”, “e”, “s”, “t”, “”] → [“low”, “es”, “t”, “”] → [“low”, “est”, “”]
- Final tokens: [“low”, “est”]

Map Tokens to IDs

Assign IDs (e.g., "low" → 1, "est" → 2) for model input.

Applications of BPE

Byte-pair encoding finds extensive use in natural language processing due to its efficiency and adaptability across a wide range of tasks. In machine translation, BPE excels at handling rare words and morphological variations, allowing models to produce more accurate translations by breaking complex words into manageable subword units. For language modeling, BPE enables models like GPT to process text efficiently by creating compact vocabularies that reduce memory and computational requirements while maintaining expressiveness.

In text classification tasks, such as sentiment analysis or spam detection, BPE provides robust token representations that capture essential linguistic patterns, enhancing model performance. Its language-agnostic nature makes it particularly valuable for multilingual NLP, where it seamlessly adapts to diverse languages, including those with complex scripts or no clear word boundaries. Additionally, BPE can be tailored to domain-specific applications by training on specialized corpora, such as medical or legal texts, enabling the creation of vocabularies that reflect the unique terminology of those fields.

Advantages of BPE

Byte-pair encoding offers several key benefits that contribute to its widespread adoption in NLP. It creates a compact vocabulary that balances size and expressiveness, reducing computational costs while maintaining the ability to represent a wide range of linguistic patterns. BPE is particularly effective at handling rare or unknown words by breaking them into known subword units, which improves model generalization and robustness. Its ability to capture morphological patterns, such as prefixes like "un-" or suffixes like "-ing," allows models to understand relationships between related words, enhancing semantic understanding.

As a language-agnostic method, BPE performs well across diverse languages, including those without clear word boundaries, making it suitable for multilingual applications. Furthermore, the iterative merging process is computationally efficient, enabling BPE to scale effectively to large corpora, which is critical for training modern language models.

Limitations of BPE

Despite its strengths, byte-pair encoding has notable limitations that practitioners must consider. The algorithm relies heavily on the training corpus to learn effective merge rules, and an unrepresentative or biased corpus can lead to suboptimal tokenization. BPE may produce token splits that are not linguistically intuitive, such as splitting "playing" into ["pla", "ying"], which can complicate interpretation or downstream processing. Once trained, the vocabulary is fixed, making it less adaptable to new domains or vocabularies without retraining, which can be resource-intensive.

The performance of BPE is sensitive to hyperparameters, such as vocabulary size and the number of merges, requiring careful tuning to achieve optimal results. Additionally, BPE can sometimes over-segment words, resulting in longer token sequences that increase computational costs, particularly for languages with complex morphology.

Comparison with Other Tokenization Methods

Byte-pair encoding can be compared to other tokenization approaches to understand its unique strengths and trade-offs. Compared to word-based tokenization, BPE handles rare words more effectively by breaking them into subword units and uses a smaller vocabulary, reducing memory requirements, though it is more complex to implement. In contrast to character-based tokenization, BPE produces shorter token sequences, which lowers computational costs, but it requires a training phase to build the vocabulary, unlike the simpler character-based approach.

When compared to other subword tokenization methods like WordPiece and SentencePiece, BPE is simpler in its merging strategy, relying on frequency-based pair selection, whereas WordPiece uses a likelihood-based criterion to optimize merges, potentially improving token quality. SentencePiece, designed for language-agnostic tokenization, excels in scenarios where presegmented words are unavailable, such as in languages like Chinese, but BPE remains a robust and straightforward choice for many applications.

Practical Considerations

When implementing byte-pair encoding, several practical factors should be taken into account to ensure effective tokenization. Selecting a representative corpus is critical, as the vocabulary must capture the linguistic patterns relevant to the target application. The vocabulary size should be carefully tuned to balance model performance and computational efficiency, with sizes like 30,000 to 50,000 tokens (as used in GPT-2's 50,257-token vocabulary) often serving as a practical range for large models.

Preprocessing steps, such as lowercasing text or removing special characters, are essential to ensure consistency and reduce noise in the tokenization process. For multilingual applications, training BPE on a diverse corpus that includes multiple languages is necessary to support varied scripts and linguistic structures. Finally, evaluating the tokenizer's performance is crucial, which involves assessing metrics like model accuracy, token sequence length, and the ability to handle rare or out-of-vocabulary words effectively.

Byte-pair encoding (BPE) is a transformative tokenization algorithm that bridges the gap between word-based and character-based approaches. By iteratively merging frequent token pairs, BPE creates a compact and expressive vocabulary capable of handling diverse linguistic patterns. Its applications in machine translation, language modeling, and beyond highlight its importance in modern NLP.

While challenges like training dependency and unintuitive splits exist, BPE's advantages make it a go-to choice for state-of-the-art models. Through the detailed algorithm outline and example provided, we hope this article equips you with the knowledge and tools to implement BPE and harness its power in your NLP projects.

The effectiveness of BPE depends on the corpus. Let $P(C)$ denote the probability distribution over words in C. The expected frequency of a pair (t_a, t_b) is:

$$4\mathbb{E}[f(t_a, t_b)] = \sum_{w \in C} P(w) \cdot count(t_a, t_b; w)$$

A representative corpus ensures $P(C)$ approximates the target language distribution. Preprocessing (e.g., lowercasing) standardizes C, reducing $|V_0|$.

BPE Implementation Walkthrough

As I mentioned, in the realm of natural language processing, tokenization serves as a foundational step, transforming raw text into a format suitable for machine learning models. The byte-pair encoding (BPE) tokenizer offers a robust and efficient approach to this task. **Inspired by the GPT OSS model**, this tokenizer incorporates a sophisticated mechanism for handling text by merging frequently occurring character pairs, enabling it to adapt to diverse vocabularies while maintaining efficiency.

Key features include the use of the "Ġ" space-prefix convention to denote word boundaries, support for special tokens like <pad>, <unk>, <s>, </s>, and <mask>, and the ability to train from scratch or load pretrained vocabularies and merges. This subheading explores the mechanics of the BPE tokenizer, detailing its training process, encoding and decoding capabilities, and its ability to save and load configurations, providing a comprehensive understanding of its role in text processing.

Let's implement it together.

```
from __future__ import annotations
import json, re
from collections import Counter, deque
from typing import List, Tuple, Dict
```

This block of code above sets up the necessary imports and configurations for the BPETokenizer class.

Here's a breakdown of what happens:

1. **`from __future__ import annotations:`**

    - This enables postponed evaluation of type annotations (introduced in Python 3.7+). It allows the use of forward references in type hints, meaning you can reference types (like List, Dict, or even the class itself) in annotations before they are fully defined. This is particularly useful for type hints in the class methods, ensuring compatibility and cleaner code.

2. **`import json, re:`**

   - **json**: Imports the json module, which is used later for saving and loading the tokenizer's vocabulary and merge rules to/from JSON files.
   - **re**: Imports the remodule for regular expression operations, used in text preprocessing (e.g., handling spaces, special tokens, and cleaning up decoded text).

3. **`from collections import Counter, deque:`**

   - **Counter**: Imported from the collections module, Counter is used in the train method to count the frequency of characters and pairs of token IDs during vocabulary construction and merge learning.
   - **deque**: A double-ended queue from the collections module, used in the _replace_pair method to efficiently process and replace pairs of token IDs during training.

4. **`from typing import List, Tuple, Dict:`**

   - Imports type hinting classes from the typing module to specify the expected types for function arguments and return values (e.g., List[str], Tuple[int, int], Dict[str, int]). This enhances code readability and enables static type checking, ensuring methods like train, encode, and save operate on the correct data types.

This setup prepares the environment for the **BPETokenizer class** by providing tools for file I/O, text processing, efficient data structures, and type safety, laying the groundwork for the tokenizer's functionality.

Let's define the class

The BPETokenizer class below implements a BPE tokenizer tailored for NLP tasks, following GPT conventions. The docstring outlines its key features: it uses the "Ġ" (Unicode U+0120) character to mark word boundaries by replacing spaces, supports training from a corpus or loading from files, and includes special tokens, padding, unknown tokens, sequence boundaries, and masking in language modeling.

The encode method can optionally add beginning-of-sequence (BOS) and end-of-sequence (EOS) tokens if add_bos_eos=True. The tokenizer saves its vocabulary to {prefix}_vocab.json and merge rules to {prefix}_merges.json, and can load them from these files, ensuring flexibility and compatibility.

```
class BPETokenizer:
    """
    GPT-2 style BPE tokenizer with:
      - "Ġ" space-prefix convention
      - train-from-scratch or load-from-files
      - special tokens: <pad>, <unk>, <s>, </s>, <mask>
      - BOS/EOS optional in encode(add_bos_eos=True)
    Saves/loads to: {prefix}_vocab.json, {prefix}_merges.json
    """

    def __init__(self):
        # string <-> id
        self.token_to_id: Dict[str, int] = {}
        self.id_to_token: Dict[int, str] = {}

        # merges:
        # - if trained-from-scratch: self.bpe_merges={(id_a,id_b):new_id}
        # - if ranked (GPT-2-style): self.bpe_ranks={(str_a,str_b):rank}
        self.bpe_merges: Dict[Tuple[int, int], int] = {}
        self.bpe_ranks: Dict[Tuple[str, str], int] = {}

        # specials (kept same names as your original tokenizer)
        self.special_tokens = ["<pad>", "<unk>", "<s>", "</s>", "<mask>"]

    # handy ids
    @property
    def pad_id(self):  return self.token_to_id["<pad>"]
    @property
    def unk_id(self):  return self.token_to_id["<unk>"]
    @property
    def bos_id(self):  return self.token_to_id["<s>"]
    @property
    def eos_id(self):  return self.token_to_id["</s>"]
```

The __init__ method initializes the tokenizer's core structures:

1. **Vocabulary Maps**

    - token_to_id and id_to_token start empty; they are filled during training or loading. They convert between string tokens and their integer IDs.

2. **Merge Rules**

    - **bpe_merges (ID-based)**: Used **only during training from scratch**. It maps (id_a, id_b) → new_id, making it easy to track and create new token IDs as merges are learned.
    - **bpe_ranks (string-based)**: Used **for inference and for loading GPT-2–style merges**. It stores (str_a, str_b) → merge rank, ensuring stable and deterministic BPE lookup following the GPT-2 convention. After training, merges are typically converted into this format.

3. **Special Tokens**

    A predefined list (<pad>, <unk>, <s>, </s>, <mask>) is added to the vocabulary.

    @property accessors (e.g., pad_id, unk_id) provide clean, direct access to their IDs via token_to_id. These raise KeyError if called before the vocabulary is populated.

    Training

    The train method learns a vocabulary and BPE merges from a text corpus up to a target vocab_size, building tokens and bpe_merges (later convertible to bpe_ranks) from scratch.

```
def train(self, corpus: List[str], vocab_size: int = 1000):
    """
    Train BPE from scratch on a small corpus using GPT-2-style spacing.
      - Replaces spaces with 'Ġ' (via _preprocess_gpt2)
      - Seeds vocab with printable chars seen in corpus (+ 'Ġ' and '\n')
      - Adds special tokens FIRST (stable IDs)
```

```
      - Reserves headroom for merges (avoid stuffing all base chars)
    """

# 1) Preprocess (space -> 'Ġ', keep '\n')
    text = "\n".join(line.rstrip("\n") for line in corpus)
    processed = self._preprocess_gpt2(text)

    # 2) Collect printable base chars actually present in corpus
    #    (keeps 'Ġ' and '\n' even if they're not "printable")
    base_chars = [
        ch for ch in processed
        if (ch == "Ġ" or ch == "\n" or ch.isprintable())
    ]
    freq = Counter(base_chars)
    unique_chars = list(freq.keys())

    # 3) Build initial vocab with specials FIRST
    tokens = list(self.special_tokens)

    # Reserve some headroom for merges (≈10% of vocab or at least 32)
    headroom = max(32, int(0.10 * vocab_size))
    max_base = max(1, vocab_size - len(tokens) - headroom)

    # Sort base chars by frequency (desc), then lexicographically
    base_sorted = sorted(unique_chars, key=lambda c: (-freq[c], c))
    base_trimmed = base_sorted[:max_base]

    tokens += base_trimmed
    self._rebuild_vocab(tokens)

    # 4) Map processed text to initial token ids (char-level now)
    token_ids = [self.token_to_id.get(ch, self.unk_id) for ch in
    processed]

    # 5) Learn merges up to vocab_size
    self.bpe_merges.clear()
    next_id = len(self.token_to_id)
```

```
    while next_id < vocab_size:
        pair = self._find_freq_pair(token_ids)
        if pair is None:
            break

        token_ids = self._replace_pair(token_ids, pair, next_id)
        self.bpe_merges[pair] = next_id

        # Register merged string token
        a, b = pair
        merged_str = self.id_to_token[a] + self.id_to_token[b]
        self.token_to_id[merged_str] = next_id
        self.id_to_token[next_id] = merged_str
        next_id += 1

    # 6) Build GPT-2-style rank view of merges (sorted by creation order)
    self.bpe_ranks = {}
    for rank, (pair, new_id) in enumerate(sorted(self.bpe_merges.items(),
    key=lambda kv: kv[1])):
        a, b = pair
        self.bpe_ranks[(self.id_to_token[a], self.id_to_token[b])] = rank
```

Train Method Code Breakdown—Step by Step

Step 1: Preprocess the Corpus

```
# 1) Preprocess (space -> 'Ġ', keep '\n')
  text = "\n".join(line.rstrip("\n") for line in corpus)
  processed = self._preprocess_gpt2(text)
```

- **Joining the Corpus**: All input strings are stripped of trailing newlines and joined with \n to form one clean text block. This normalizes newline handling and prepares the corpus for uniform processing.
- **GPT-2 Preprocessing (_preprocess_gpt2)**: The text is transformed so that spaces become the special marker **"Ġ"** (except at line starts), while newlines are kept intact. This GPT-2 convention encodes word boundaries explicitly—for example, "hello world" → "helloĠworld"—producing a string ready for BPE tokenization.

Step 2: Collect Base Characters

```
# 2) Collect printable base chars actually present in corpus
#    (keeps 'Ġ' and '\n' even if they're not "printable")
base_chars = [
    ch for ch in processed
    if (ch == "Ġ" or ch == "\n" or ch.isprintable())
]
freq = Counter(base_chars)
unique_chars = list(freq.keys())
```

- **Extract Characters:** The processed text is scanned into base_chars, keeping only characters that are printable, "Ġ", or "\n". This forms the true set of characters present in the corpus, including GPT-2's space marker and newlines.
- **Count Frequencies**: A Counter tallies how often each character appears (e.g., "a" → 100). These counts guide which merges are useful during training.
- **Collect Unique Characters**: unique_chars is simply the set of all characters found, providing the initial vocabulary before any BPE merging begins.

Step 3: Build Initial Vocabulary

```
# 3) Build initial vocab with specials FIRST
tokens = list(self.special_tokens)

# Reserve some headroom for merges (≈10% of vocab or at least 32)
headroom = max(32, int(0.10 * vocab_size))
max_base = max(1, vocab_size - len(tokens) - headroom)

# Sort base chars by frequency (desc), then lexicographically
base_sorted = sorted(unique_chars, key=lambda c: (-freq[c], c))
base_trimmed = base_sorted[:max_base]

tokens += base_trimmed
self._rebuild_vocab(tokens)
```

- **Initialize with Special Tokens:** Start the tokens list with <pad>, <unk>, <s>, </s>, and <mask> so they receive stable, low-numbered IDs.
- **Reserve Headroom:** Reserve room in the vocabulary for future merges—at least 32 entries or 10% of vocab_size. Compute max_base = vocab_size - len(tokens) – headroom, so the vocab isn't consumed entirely by base characters.
- **Sort and Trim Characters:** Sort unique_chars by descending frequency (using -freq[c]) and then lexicographically to break ties. Keep the top max_base as base_trimmed, prioritizing the most frequent characters.
- **Build Vocabulary:** Combine the special tokens and trimmed base characters into the final tokens list, then call _rebuild_vocab to populate token_to_id and id_to_token.

Step 4: Map Text to Initial Token IDs

```
# 4) Map processed text to initial token ids (char-level now)
token_ids = [self.token_to_id.get(ch, self.unk_id) for ch in
processed]
```

The processed text is converted to a list of token IDs (token_ids) by mapping each character to its ID in self.token_to_id. If a character isn't in the vocabulary (unlikely since base_chars includes all seen characters), it uses self.unk_id (the ID for unknown characters). At this stage, the tokenization is character-level, as merges haven't been applied yet.

Step 5: Learn Merges

```
# 5) Learn merges up to vocab_size
 self.bpe_merges.clear()
 next_id = len(self.token_to_id)

 while next_id < vocab_size:
     pair = self._find_freq_pair(token_ids)
     if pair is None:
         break
```

```
        token_ids = self._replace_pair(token_ids, pair, next_id)
        self.bpe_merges[pair] = next_id

        # Register merged string token
        a, b = pair
        merged_str = self.id_to_token[a] + self.id_to_token[b]
        self.token_to_id[merged_str] = next_id
        self.id_to_token[next_id] = merged_str
        next_id += 1
```

- ***Clear Merges**: self.bpe_merges is emptied so training starts with no existing merge rules.*
- ***Initialize Next ID**: next_id is set to the current vocab size, defining the ID for the next new merged token.*
- ***Merge Loop**: Repeats merges until vocab_size is reached or no pairs remain:*
    - ***Find Frequent Pair**: _find_freq_pair returns the most common adjacent ID pair; if none, training stops.*
    - ***Replace Pair**: _replace_pair replaces every occurrence of the pair in token_ids with next_id.*
    - ***Store Merge**: The pair → next_id mapping is added to self.bpe_merges.*
    - ***Update Vocabulary**: The merged token string (token_a + token_b) is added to the vocab with ID next_id.*
    - ***Increment ID**: next_id increases so the next merge receives a new ID.*

Step 6: Build GPT-2-Style Merge Ranks

```
# 6) Build GPT-2-style rank view of merges (sorted by creation order)
self.bpe_ranks = {}
for rank, (pair, new_id) in enumerate(sorted(self.bpe_merges.items(),
key=lambda kv: kv[1])):
    a, b = pair
    self.bpe_ranks[(self.id_to_token[a], self.id_to_token[b])] = rank
```

Create Ranked Merges: The self.bpe_ranks dictionary is populated to store GPT-2-style merge ranks. The self.bpe_merges items are sorted by their new ID (kv[1]), reflecting the order in which merges were created.

Map to Strings: For each merge (pair, new_id), the pair's token IDs (a, b) are converted to their string representations (self.id_to_token[a], self.id_to_token[b]), and the pair is assigned a rank (its index in the sorted list).

Store Ranks: The self.bpe_ranks dictionary maps string pairs to their rank (e.g., ("t", "h"): 0), enabling GPT-2-style encoding where merges are applied in rank order (lower rank first).

The Encode Method

The encode method of the BPETokenizer class is responsible for converting input text into a sequence of numerical token IDs using byte-pair encoding (BPE) with GPT-2-style conventions. I'll break down the method's docstring and code step by step, explaining its purpose and functionality in a clear and concise manner.

```
def encode(self, text: str, add_bos_eos: bool = False, allowed_special:
set[str] | None = None) -> List[int]:
        """
        Encode with GPT-2-style "Ġ" spacing. If allowed_special is
        provided,
        those exact substrings pass through as single tokens (if present
        in vocab).
        """
        ids: List[int] = []
        if add_bos_eos:
            ids.append(self.bos_id)

        for tid in self._encode_core(text, allowed_special=allowed_
        special):
            ids.append(tid)

        if add_bos_eos:
            ids.append(self.eos_id)
        return ids
```

Encode Method: Converts an input string into a list of integer token IDs following GPT-2 conventions (notably "Ġ" for word boundaries):

- **text**—the string to encode.
- **add_bos_eos**—if True, add <s> at the start and </s> at the end.
- **allowed_special**—optional set of special tokens to treat as single units.

Initialize IDs—start with an empty ids: List[int].

Add BOS (optional)—if add_bos_eos is True, append self.bos_id.

Run _encode_core—handles GPT-2 preprocessing, token splitting, BPE merges, and respecting allowed special tokens; yields token IDs.

Append Tokens—each ID from _encode_core is appended to ids.

Add EOS (optional)—if add_bos_eos is True, append self.eos_id.

Return IDs—the final list of token IDs representing the encoded text.

The Decode Method: One of the Most Important Ones

The decode method of the BPETokenizer class is one of the most critical components of the tokenizer! This method transforms a sequence of numerical token IDs back into human-readable text, reversing the encoding process while ensuring the output is clean and natural.

Its importance lies in its role as the bridge between the numerical representations used by machine learning models and the text that humans or downstream applications can understand. I'll break down the method's code step by step, highlighting its functionality and significance.

```
def decode(self, ids: list[int]) -> str:
    """
    Decode token IDs to readable text:
      - 'Ġ' -> space
      - keep '\n', skip specials
      - Unicode normalize, strip control chars
      - tidy underscores used as markup (_word_ -> word)
      - fix spacing around punctuation and newlines
    """
    import re, unicodedata
```

```
# 1) Reconstruct raw string from tokens
pieces = []
for tid in ids:
    tok = self.id_to_token.get(int(tid), "")
    if not tok or tok in getattr(self, "special_tokens", []):
        continue
    if tok == "\n":
        pieces.append("\n")
    elif tok.startswith("Ġ"):      # GPT-2 space marker
        pieces.append(" " + tok[1:])
    else:
        pieces.append(tok)
text = "".join(pieces)

# 2) Normalize unicode and drop control chars (keep \n/\t)
text = unicodedata.normalize("NFKC", text)
text = "".join(
    ch for ch in text
    if ch in ("\n", "\t") or unicodedata.category(ch)[0] != "C"
)

# 3) Tidy underscores used as pseudo-markup
#    _word_ -> word, and remove stray standalone underscores
text = re.sub(r"\b_([^\W_][^_]*)_\b", r"\1", text)  # wrap pair
text = re.sub(r"(?<=\s)_(?=\s)|^_$", "", text)       # lone underscores

# 4) Whitespace + punctuation cleanup
text = re.sub(r"[ \t]+", " ", text)                  # collapse spaces/tabs
text = re.sub(r"\s*\n\s*", "\n", text).strip()       # clean newlines
text = re.sub(r"\s+([,.;:!?])", r"\1", text)         # no space before punct
text = re.sub(r"([.?!])([^\s])", r"\1 \2", text)     # ensure space after .?!

return text
```

The decode method is a cornerstone of the tokenizer, converting a list of integer token IDs back into a readable string. It takes one parameter:

- **ids: list[int]**: A list of integer token IDs, typically produced by the encode method or a model's output.

This method is critical because it ensures the tokenized output of a model is converted back into coherent, readable text, making it essential for applications like text generation, translation, or any task requiring human-interpretable output.

Step 1: Reconstruct Raw String from Tokens

```
# 1) Reconstruct raw string from tokens
pieces = []
for tid in ids:
    tok = self.id_to_token.get(int(tid), "")
    if not tok or tok in getattr(self, "special_tokens", []):
        continue
    if tok == "\n":
        pieces.append("\n")
    elif tok.startswith("Ġ"):        # GPT-2 space marker
        pieces.append(" " + tok[1:])
    else:
        pieces.append(tok)
text = "".join(pieces)
```

The method imports **re** (regex) and **unicodedata** for text cleanup and normalization. It initializes a pieces list to collect reconstructed text fragments. For each token ID (tid) in ids:

- **Lookup Token**: Convert the ID to its token string via self.id_to_token.get(tid, ""), using an empty string as a safe fallback.
- **Skip Special Tokens**: Ignore <pad>, <unk>, <s>, </s>, <mask>, so they don't appear in the output.
- **Preserve Newlines**: Keep \n exactly as-is.
- **Handle GPT-2 Spaces**: If the token starts with "Ġ", replace it with a leading space plus the remainder of the token.
- **Append Normal Tokens**: Add all other tokens unchanged.

Finally, the pieces list is joined into text, restoring natural word boundaries and producing the decoded string.

Step 2: Normalize Unicode and Remove Control Characters

```
# 2) Normalize unicode and drop control chars (keep \n/\t)
text = unicodedata.normalize("NFKC", text)
text = "".join(
    ch for ch in text
    if ch in ("\n", "\t") or unicodedata.category(ch)[0] != "C"
)
```

Unicode Normalization: unicodedata.normalize("NFKC", text) standardizes characters into a consistent form, merging compatibility variants and combining marks into single composed characters.

Filter Control Chars: They remove all Unicode control characters (categories starting with "C") except \n and \t, keeping the output clean and readable.

Why: Ensures the decoded text is stable, standardized, and free of artifacts, improving reliability for downstream use.

Step 3: Clean Up Underscore Markup

```
# 3) Tidy underscores used as pseudo-markup
#    _word_ -> word, and remove stray standalone underscores
text = re.sub(r"\b_([^\W_][^_]*)_\b", r"\1", text)  # wrap pair
text = re.sub(r"(?<=\s)_(?=\s)|^_$", "", text)       # lone
                                                       underscores
```

Remove Underscored Markup: re.sub(r"\b_([^\W_][^_]*)_\b", r"\1", text) strips underscores used as simple markup (e.g., _word_ → word) by capturing the inner text and removing the surrounding underscores.

Remove Stray Underscores: re.sub(r"(?<=\s)_(?=\s)|^_$", "", text) deletes isolated underscores that appear between spaces or as the entire string, cleaning up leftover artifacts.

Why: Ensures the final text is free of tokenizer-style markup and reads naturally, improving clarity and usability.

Step 4: Clean Up Whitespace and Punctuation

```
# 4) Whitespace + punctuation cleanup
text = re.sub(r"[ \t]+", " ", text)                   # collapse spaces/tabs
text = re.sub(r"\s*\n\s*", "\n", text).strip()        # clean newlines
text = re.sub(r"\s+([,.;:!?])", r"\1", text)          # no space before punct
text = re.sub(r"([.?!])([^\s])", r"\1 \2", text)      # ensure space after .?!
```

- **Collapse Whitespace**: `re.sub(r"[ \t]+", " ", text)` reduces runs of spaces or tabs to a single space for consistent spacing.
- **Clean Newlines**: `re.sub(r"\s*\n\s*", "\n", text).strip()` removes extra whitespace around newlines and trims the text, producing clean line breaks.
- **Punctuation Spacing**
  - `re.sub(r"\s+([,.;:!?])", r"\1", text)` removes unwanted spaces before punctuation.
  - `re.sub(r"([.?!])([^\s])", r"\1 \2", text)` inserts a space after sentence-ending punctuation when missing.
- **Significance**: These formatting fixes refine the text to standard, readable conventions, making the output polished and natural.

Why This Method Is One of the Most Important

- **Readability**: Decoding removes artifacts (e.g., "Ġ", underscore markup) and reconstructs text that reads naturally for humans.
- **Robustness**: Unicode normalization, control-char filtering, and punctuation cleanup ensure stable, clean output across diverse inputs.

- **GPT Compatibility**: Properly handles GPT-2 conventions like "Ġ" and special tokens, preserving fidelity with common modern tokenizers.
- **User-Facing Output**: As the bridge from token IDs back to text, decoding directly shapes the final user-visible result, making it fundamental to real-world usability.

Save and Load Methods

```
# ----------------------
    # Save / Load
    # ----------------------
    def save(self, path_prefix: str = "bpe_tokenizer"):
        """
        Saves:
          - {prefix}_vocab.json   : dict[str,int]  (token_to_id)
          - {prefix}_merges.json  : list[[str,str]] in rank order
                                    (GPT-2 style)
        """
        # vocab
        with open(f"{path_prefix}_vocab.json", "w", encoding="utf-8") as f:
            json.dump(self.token_to_id, f, ensure_ascii=False)

        # merges (rank order)
        merges_list: List[Tuple[str, str]] = []
        if self.bpe_ranks:
            # already ranked (str,str)->rank
            pairs_sorted = sorted(self.bpe_ranks.items(), key=lambda
            kv: kv[1])
            merges_list = [list(p) for p, _ in pairs_sorted]
        elif self.bpe_merges:
            # convert (id,id)->new_id to a stable order by new_id
            pairs_sorted = sorted(self.bpe_merges.items(), key=lambda
            kv: kv[1])
            merges_list = [[self.id_to_token[a], self.id_to_token[b]]
            for (a, b), _ in pairs_sorted]
```

```
        with open(f"{path_prefix}_merges.json", "w",
        encoding="utf-8") as f:
            json.dump(merges_list, f, ensure_ascii=False)

    def load(self, path_prefix: str = "bpe_tokenizer"):
        """
        Loads tokenizer. Accepts:
          - vocab json as dict[str,int]  or dict[int,str]
          - merges json as list[[str,str]] (rank order)
        """
        # vocab
        with open(f"{path_prefix}_vocab.json", "r", encoding="utf-8") as f:
            voc = json.load(f)
        if all(isinstance(k, str) for k in voc.keys()):
            # token_to_id
            self.token_to_id = {k: int(v) for k, v in voc.items()}
            self.id_to_token = {int(v): k for k, v in voc.items()}
        else:
            # id->token
            self.id_to_token = {int(k): v for k, v in voc.items()}
            self.token_to_id = {v: int(k) for k, v in voc.items()}

        # merges
        with open(f"{path_prefix}_merges.json", "r",
        encoding="utf-8") as f:
            merges_list = json.load(f)

        # rebuild ranks from list order
        self.bpe_ranks = {}
        for rank, pair in enumerate(merges_list):
            a, b = pair
            self.bpe_ranks[(a, b)] = rank

        # clear id-based merges (we'll use ranks for encoding)
        self.bpe_merges = {}
```

Save Method: Serializes the tokenizer using a prefix (default "bpe_tokenizer").

- **Save Vocabulary**: Writes token_to_id to {prefix}_vocab.json with ensure_ascii=False so Unicode characters are preserved.
- **Prepare Merges**
  - If bpe_ranks exists (GPT-2-style), sort by rank and extract the string pairs.
  - Otherwise, use bpe_merges (ID-based), sort by new ID, and convert ID pairs to token strings via id_to_token.
- **Save Merges**: Writes the ordered list of merge pairs to {prefix}_merges.json, producing a stable GPT-2-compatible merge file.

Result: The tokenizer is saved in a portable, reconstructible format.
Load Method: Restores the tokenizer from the JSON files.

- **Load Vocabulary**: Read {prefix}_vocab.json and detect the format:
  - If keys are **strings** → data is token→ID; populate token_to_id and build id_to_token.
  - If keys are **integers** → data is ID→token; reverse accordingly, ensuring IDs are ints for consistency.
- **Load Merges**: Read merge pairs from {prefix}_merges.json and rebuild bpe_ranks by enumerating them (index = rank).
- **Reset ID-Based Merges**: Clear bpe_merges so the tokenizer relies solely on GPT-2-style rank-based merges during encoding.

Result: The loaded tokenizer behaves identically to a trained GPT-2-style tokenizer, with robust handling of multiple possible save formats.

Helping Functions

These methods below handle **critical low-level operations for vocabulary management**, text preprocessing, encoding, and training, enabling the tokenizer's core functionality with GPT-2-style conventions. I'll describe each method's purpose and behavior concisely, focusing on its role in the tokenizer's pipeline.

```
# ------------------------
    # Internal helpers
    # ------------------------
    def _rebuild_vocab(self, tokens: List[str]):
        self.token_to_id = {tok: i for i, tok in enumerate(tokens)}
        self.id_to_token = {i: tok for tok, i in self.token_to_id.items()}

    def _preprocess_gpt2(self, text: str) -> str:
        """
        Convert spaces to 'Ġ' (except at very beginning), preserve
        newlines.
        """
        out = []
        at_line_start = True
        for ch in text:
            if ch == "\n":
                out.append("\n")
                at_line_start = True
            elif ch == " ":
                # represent an inter-word space with a literal 'Ġ'
                if not at_line_start:
                    out.append("Ġ")
            else:
                out.append(ch)
                at_line_start = False
        return "".join(out)

    def _encode_core(self, text: str, allowed_special: set[str] | None):
        """
        Core encoder:
          - pass through allowed special tokens
          - split into tokens with 'Ġ' prefix for word starts
            (GPT-2 style)
          - apply merges (ranked if available; else local bpe_merges)
        """
        token_ids: List[int] = []
```

```
        # handle allowed specials by regex split
        if allowed_special:
            pat = "(" + "|".join(re.escape(tok) for tok in sorted(allowed_
            special, key=len, reverse=True)) + ")"
            last = 0
            for m in re.finditer(pat, text):
                prefix = text[last:m.start()]
                token_ids.extend(self._encode_core(prefix, allowed_
                special=None))
                sp = m.group(0)
                if sp in self.token_to_id:
                    token_ids.append(self.token_to_id[sp])
                else:
                    raise ValueError(f"Special token {sp!r} not in vocab")
                last = m.end()
            text = text[last:]  # remainder

            # sanity: disallow other <|...|> if not allowed
            disallowed = re.findall(r"<\|[^|]*\|>", text)
            if any(tok not in allowed_special for tok in disallowed):
                raise ValueError(f"Disallowed special tokens in text:
                {disallowed}")

        # split into GPT-2 style surface tokens using lines & words
        tokens = []
        for li, line in enumerate(text.split("\n")):
            if li > 0:
                tokens.append("\n")
            words = line.split()
            for j, word in enumerate(words):
                if j == 0:
                    tokens.append(word)
                else:
                    tokens.append("Ġ" + word)
```

```
        # apply BPE to each surface token
        for tok in tokens:
            if tok in self.token_to_id:
                token_ids.append(self.token_to_id[tok])
            else:
                token_ids.extend(self._bpe_tokenize(tok))
        return token_ids

    def _bpe_tokenize(self, token: str) -> List[int]:
        """
        Apply BPE merges to a single surface token.
        Prefers ranked GPT-2-style merges if present; otherwise uses local
        (id,id)->new_id mapping.
        """
        # characters to ids
        ids = []
        for ch in token:
            tid = self.token_to_id.get(ch)
            if tid is None:
                # unseen char -> <unk>
                tid = self.unk_id
            ids.append(tid)

        if self.bpe_ranks:
            # ranked merges over strings
            syms = [self.id_to_token[i] for i in ids]
            while True:
                pairs = set(zip(syms, syms[1:]))
                if not pairs:
                    break
                best = None
                best_rank = float("inf")
                for p in pairs:
                    r = self.bpe_ranks.get(p, float("inf"))
                    if r < best_rank:
                        best_rank = r
                        best = p
```

```
            if best is None or best_rank == float("inf"):
                break
            first, second = best
            # merge all occurrences
            new_syms = []
            i = 0
            while i < len(syms):
                if i < len(syms) - 1 and syms[i] == first and syms[i+1]
                == second:
                    new_syms.append(first + second)
                    i += 2
                else:
                    new_syms.append(syms[i]); i += 1
            syms = new_syms
            if len(syms) == 1:
                break
        return [self.token_to_id.get(s, self.unk_id) for s in syms]

    # fallback: local (id,id)->new_id merges
    can_merge = True
    while can_merge and len(ids) > 1:
        can_merge = False
        new_ids = []
        i = 0
        while i < len(ids) - 1:
            pair = (ids[i], ids[i+1])
            if pair in self.bpe_merges:
                new_ids.append(self.bpe_merges[pair])
                i += 2
                can_merge = True
            else:
                new_ids.append(ids[i]); i += 1
        if i < len(ids):
            new_ids.append(ids[i])
        ids = new_ids
    return ids
```

```
# -----------------------
# Pair ops for training
# -----------------------
@staticmethod
def _find_freq_pair(token_ids: List[int]):
    pairs = Counter(zip(token_ids, token_ids[1:]))
    if not pairs:
        return None
    return max(pairs.items(), key=lambda kv: kv[1])[0]

@staticmethod
def _replace_pair(token_ids: List[int], pair_id: Tuple[int, int],
new_id: int) -> List[int]:
    dq = deque(token_ids)
    out = []
    while dq:
        a = dq.popleft()
        if dq and (a, dq[0]) == pair_id:
            out.append(new_id)
            dq.popleft()
        else:
            out.append(a)
    return out
```

`_rebuild_vocab(self, tokens: List[str])`

Purpose: Rebuilds the tokenizer's vocab maps from a token list.

How It Works: Uses enumerate to assign each token an ID based on position:

- token_to_id = {token: i}
- id_to_token = {i: token}

When Used: During training after creating the initial token list and when adding new merged tokens.

Efficiency: O(n), fast even for large vocabularies.

Edge Cases: Duplicate tokens overwrite earlier ones; empty lists produce empty vocab maps.

Importance: Correct mappings are essential for stable encoding and decoding.

`_preprocess_gpt2(self, text: str)`

Purpose: Applies GPT-2 spacing rules by replacing spaces with "Ġ" except at line starts.

How It Works

- Build output char-by-char.
- Track at_line_start to avoid "Ġ" at line beginnings.
- Preserve newlines; treat all other chars normally.

Performance: O(m) for m characters.

Example: "Hello world\n New line" → "HelloĠworld\nĠNewĠline".

Importance: Provides explicit word-boundary markers without a separate space token.

`_encode_core(self, text, allowed_special) -> List[int]`

Purpose: Main encoder that converts preprocessed text into token IDs, with optional special-token handling.

Special-Token Mode

- Build a regex matching longest specials first.
- Recursively encode normal sections.
- Insert the special token's ID directly.

Regular Mode

- Split into lines, preserve \n.
- Split words; prefix non-first words with "Ġ".
- For each surface token: direct lookup or _bpe_tokenize.

Output: A flat list of IDs.

Importance: Combines GPT-2 whitespace rules with flexible special-token behavior.

`_bpe_tokenize(self, token: str) -> List[int]`

Purpose: Applies BPE to a token using ranked merges (GPT-2) or training merges.

Rank-Based Mode

- Convert chars → IDs → strings.
- **Loop**: Find the lowest-ranked pair, merge, repeat.
- Convert final strings back to IDs.

Training Mode

- Merge directly on ID pairs using bpe_merges.

Edge Cases: Single-char tokens, unknown chars, tokens with no applicable merges.
Importance: Enables compact subword vocabularies that still cover rare words.
`_find_freq_pair(self, token_ids)`
Purpose: Finds the most frequent adjacent pair for the next training merge.
How It Works

- Use Counter(zip(ids, ids[1:])).
- Return the most common pair, or None if no pairs.

Importance: Drives BPE's data-driven merge decisions.
`_replace_pair(self, token_ids, pair_id, new_id)`
Purpose: Replaces all occurrences of a given pair with a new ID during training.
How It Works

- Scan left-to-right using a deque.
- If the next two IDs match the pair → append new_id and skip one.
- Otherwise → append the current ID.

Edge Cases: Overlapping pairs, no matches, or fully-merged sequences.
Importance: Compresses the sequence after each merge step, speeding up training.

Testing Our Tokenizer for Accuracy

The script trains a BPETokenizer on a small toy corpus by starting at the character level and repeatedly merging the most frequent adjacent pairs until the vocabulary reaches the target size, mirroring GPT-2's subword-building process.

After training, it prints portions of the vocabulary and early merges so you can see how characters like "H" and "e" form "He," then "Hel," and so on. It then demonstrates encoding and decoding: encoding maps text to token IDs (and shows the tokens themselves), while decoding reconstructs human-readable text, correctly handling newlines and the "Ġ" space marker.

Examples include standard text, sequences with BOS/EOS tokens, and cases where <mask> is treated as a protected special token. A BPE trace focuses on a single word to illustrate each merge step applied during tokenization. Finally, the script saves the trained tokenizer to JSON files, reloads it into a fresh instance, and verifies that the reloaded tokenizer reproduces the same encodings.

```
from typing import List, Tuple, Optional

# If your class lives elsewhere, import it:
# from your_module import BPETokenizer

# --- Paste your BPETokenizer class above this line or import it ---

# -------------------------
# Pretty-printing utilities
# -------------------------
def tokens_from_ids(tok: "BPETokenizer", ids: List[int]) -> List[str]:
    return [tok.id_to_token.get(int(i), "<NA>") for i in ids]

def show_example(
    tok: "BPETokenizer",
    text: str,
    title: str,
    add_bos_eos: bool = False,
    allowed_special: Optional[set[str]] = None,
):
    print("\n" + "=" * 72)
    print(title)
    print("=" * 72)
    print(f"INPUT TEXT:\n{text!r}")
    ids = tok.encode(text, add_bos_eos=add_bos_eos, allowed_
    special=allowed_special)
    toks = tokens_from_ids(tok, ids)
    print("\nENCODED IDS:\n", ids)
    print("ENCODED TOKENS:\n", toks)
    out = tok.decode(ids)
    print("\nDECODED TEXT:\n", repr(out))
```

```
def print_merges(tok: "BPETokenizer", k: int = 20):
    """
    Show the first k learned merges in (token_a, token_b) -> rank order.
    """
    if not tok.bpe_ranks:
        print("\n[!] No ranked merges found. Train or load a tokenizer
        first.")
        return
    print("\nTop learned merges (ranked, lowest rank = earlier merge):")
    pairs_by_rank = sorted(tok.bpe_ranks.items(), key=lambda kv: kv[1])
    for (a, b), r in pairs_by_rank[:k]:
        print(f"  rank {r:3d}: {a!r} + {b!r} -> {a+b!r}")

def bpe_trace_word(tok: "BPETokenizer", word: str):
    """
    Illustrate how GPT-2 style BPE merges apply to a *single*
    surface token.
    Uses tok.bpe_ranks (the ranked (str,str)->rank mapping).
    """
    print("\nBPE TRACE for:", repr(word))
    # Turn word into initial character symbols using tokenizer's base vocab
    syms = []
    for ch in word:
        if ch in tok.token_to_id:
            syms.append(ch)
        else:
            # out-of-vocab char becomes <unk> for this trace
            syms.append("<unk>")
    print("Start syms:", syms)

    if not tok.bpe_ranks:
        print("[!] No ranked merges; trace not available (train or load
        first).")
        return
```

```
step = 0
while True:
    pairs = list(zip(syms, syms[1:]))
    if not pairs:
        break
    # choose the best-ranked pair present
    best_pair: Optional[Tuple[str, str]] = None
    best_rank = float("inf")
    for p in set(pairs):
        r = tok.bpe_ranks.get(p, float("inf"))
        if r < best_rank:
            best_pair, best_rank = p, r
    if best_pair is None or best_rank == float("inf"):
        break

    print(f"Step {step}: best pair {best_pair} @ rank {best_rank}")
    # Merge every occurrence of that best pair
    merged = []
    i = 0
    while i < len(syms):
        if i < len(syms) - 1 and (syms[i], syms[i+1]) == best_pair:
            merged.append(syms[i] + syms[i+1])
            i += 2
        else:
            merged.append(syms[i])
            i += 1
    syms = merged
    print("  ->", syms)
    step += 1
    if len(syms) == 1:
        break

print("Final BPE pieces:", syms)
print("Token IDs:", [tok.token_to_id.get(s, tok.unk_id) for s in syms])
```

```
def main():
    # -----------------------------
    # 1) Train on a tiny toy corpus
    # -----------------------------
    corpus = [
        "Hello world.",
        "Hello there, world!",
        "New line\nanother line",
        "mask me <mask> maybe",
        "Under_scores _word_ lone _  end_",
    ]

    vocab_size = 160  # small but big enough to learn a few merges
    tok = BPETokenizer()
    tok.train(corpus, vocab_size=vocab_size)

    print("\n" + "#" * 72)
    print("TRAINED BPETokenizer")
    print("#" * 72)
    print(f"Vocab size: {len(tok.token_to_id)} (requested {vocab_size})")
    print("Special tokens and IDs:")
    for s in tok.special_tokens:
        print(f"  {s:>6} : {tok.token_to_id[s]}")
    print("Examples of base tokens present (first 30 by ID):")
    for i in range(min(30, len(tok.id_to_token))):
        print(f"  id {i:3d} -> {repr(tok.id_to_token[i])}")

    # ---------------------------------------
    # 2) Inspect the first few learned merges
    # ---------------------------------------
    print_merges(tok, k=20)

    # -----------------------------------------------------
    # 3) Encode/Decode examples (with and without BOS/EOS)
    # -----------------------------------------------------
    show_example(tok, "Hello world.", "Example A: Simple sentence")
    show_example(tok, "Hello world.\nNew line", "Example B: With newline")
```

```
show_example(
    tok,
    "Please, <mask> only this token.",
    "Example C: Allowed special passes through",
    allowed_special={"<mask>"},
)
show_example(
    tok,
    "Hello there, world!",
    "Example D: With BOS/EOS",
    add_bos_eos=True,
)

# -----------------------------------------------------
# 4) BPE trace on a *single* word (didactic view)
# -----------------------------------------------------
bpe_trace_word(tok, "Hello")
bpe_trace_word(tok, "world!")

# -----------------------------------------------------
# 5) Save & Load and show same encoding after reload
# -----------------------------------------------------
prefix = "demo_tok"
tok.save(prefix)
print(f"\nSaved files:\n  {prefix}_vocab.json\n  {prefix}_merges.json")

sample = "Hello there, world!"
before_ids = tok.encode(sample)

tok2 = BPETokenizer()
tok2.load(prefix)
after_ids = tok2.encode(sample)

print("\nRe-encode after load:")
print("  Before IDs:", before_ids)
print("  After  IDs:", after_ids)
print("  Match?     ", before_ids == after_ids)

print("\nDone.")
```

```
if __name__ == "__main__":
    main()
```

Tokenizer Output

```
##########################################################################
TRAINED BPETokenizer
##########################################################################
Vocab size: 111 (requested 160)
Special tokens and IDs:
   <pad> : 0
   <unk> : 1
     <s> : 2
    </s> : 3
  <mask> : 4
Examples of base tokens present (first 30 by ID):
  id   0 -> '<pad>'
  id   1 -> '<unk>'
  id   2 -> '<s>'
  id   3 -> '</s>'
  id   4 -> '<mask>'
  id   5 -> 'e'
  id   6 -> 'Ġ'
  id   7 -> 'l'
  id   8 -> 'o'
  id   9 -> 'r'
  id  10 -> 'n'
  id  11 -> '\n'
  id  12 -> '_'
  id  13 -> 'd'
  id  14 -> 'a'
  id  15 -> 'm'
  id  16 -> 's'
  id  17 -> 'w'
  id  18 -> 'H'
```

```
  id  19 -> 'h'
  id  20 -> 'i'
  id  21 -> 'k'
  id  22 -> 't'
  id  23 -> '!'
  id  24 -> ','
  id  25 -> '.'
  id  26 -> '<'
  id  27 -> '>'
  id  28 -> 'N'
  id  29 -> 'U'

Top learned merges (ranked, lowest rank = earlier merge):
  rank   0: 'o' + 'r' -> 'or'
  rank   1: 'l' + 'o' -> 'lo'
  rank   2: 'w' + 'or' -> 'wor'
  rank   3: 'e' + 'r' -> 'er'
  rank   4: 'n' + 'e' -> 'ne'
  rank   5: 'm' + 'a' -> 'ma'
  rank   6: 'H' + 'e' -> 'He'
  rank   7: 'He' + 'l' -> 'Hel'
  rank   8: 'Hel' + 'lo' -> 'Hello'
  rank   9: 'Hello' + 'Ġ' -> 'HelloĠ'
  rank  10: 'wor' + 'l' -> 'worl'
  rank  11: 'worl' + 'd' -> 'world'
  rank  12: 't' + 'h' -> 'th'
  rank  13: 'th' + 'er' -> 'ther'
  rank  14: 'Ġ' + 'l' -> 'Ġl'
  rank  15: 'Ġl' + 'i' -> 'Ġli'
  rank  16: 'Ġli' + 'ne' -> 'Ġline'
  rank  17: 'Ġline' + '\n' -> 'Ġline\n'
  rank  18: 'ma' + 's' -> 'mas'
  rank  19: 'mas' + 'k' -> 'mask'
```

```
======================================================================
Example A: Simple sentence
======================================================================
INPUT TEXT:
'Hello world.'

ENCODED IDS:
 [41, 6, 44, 25]
ENCODED TOKENS:
 ['Hello', 'Ġ', 'world', '.']

DECODED TEXT:
 'Hello world.'

======================================================================
Example B: With newline
======================================================================
INPUT TEXT:
'Hello world.\nNew line'

ENCODED IDS:
 [41, 6, 44, 25, 11, 28, 5, 17, 49]
ENCODED TOKENS:
 ['Hello', 'Ġ', 'world', '.', '\n', 'N', 'e', 'w', 'Ġline']

DECODED TEXT:
 'Hello world.\nNew line'

======================================================================
Example C: Allowed special passes through
======================================================================
INPUT TEXT:
'Please, <mask> only this token.'

ENCODED IDS:
 [1, 7, 5, 14, 16, 5, 24, 4, 8, 10, 7, 32, 6, 45, 20, 16, 6, 22, 8, 21,
5, 10, 25]
```

```
ENCODED TOKENS:
 ['<unk>', 'l', 'e', 'a', 's', 'e', ',', '<mask>', 'o', 'n', 'l', 'y',
'Ġ', 'th', 'i', 's', 'Ġ', 't', 'o', 'k', 'e', 'n', '.']

DECODED TEXT:
 'lease,only this token.'

==========================================================================
Example D: With BOS/EOS
==========================================================================
INPUT TEXT:
'Hello there, world!'

ENCODED IDS:
 [2, 41, 6, 46, 5, 24, 6, 44, 23, 3]
ENCODED TOKENS:
 ['<s>', 'Hello', 'Ġ', 'ther', 'e', ',', 'Ġ', 'world', '!', '</s>']

DECODED TEXT:
 'Hello there, world!'

BPE TRACE for: 'Hello'
Start syms: ['H', 'e', 'l', 'l', 'o']
Step 0: best pair ('l', 'o') @ rank 1
  -> ['H', 'e', 'l', 'lo']
Step 1: best pair ('H', 'e') @ rank 6
  -> ['He', 'l', 'lo']
Step 2: best pair ('He', 'l') @ rank 7
  -> ['Hel', 'lo']
Step 3: best pair ('Hel', 'lo') @ rank 8
  -> ['Hello']
Final BPE pieces: ['Hello']
Token IDs: [41]

BPE TRACE for: 'world!'
Start syms: ['w', 'o', 'r', 'l', 'd', '!']
Step 0: best pair ('o', 'r') @ rank 0
  -> ['w', 'or', 'l', 'd', '!']
```

```
Step 1: best pair ('w', 'or') @ rank 2
  -> ['wor', 'l', 'd', '!']
Step 2: best pair ('wor', 'l') @ rank 10
  -> ['worl', 'd', '!']
Step 3: best pair ('worl', 'd') @ rank 11
  -> ['world', '!']
Final BPE pieces: ['world', '!']
Token IDs: [44, 23]

Saved files:
  demo_tok_vocab.json
  demo_tok_merges.json

Re-encode after load:
  Before IDs: [41, 6, 46, 5, 24, 6, 44, 23]
  After  IDs: [41, 6, 46, 5, 24, 6, 44, 23]
  Match?      True

Done.
```

Analysis of the Output

Training the Tokenizer

We start by training the tokenizer on a **very small toy corpus**—just a handful of lines such as "Hello world." and "mask me <mask> maybe". For demonstration purposes, we ask for a vocabulary of size 160:

```
##########################################################################
TRAINED BPETokenizer
##########################################################################
Vocab size: 111 (requested 160)
Special tokens and IDs:
   <pad> : 0
   <unk> : 1
     <s> : 2
    </s> : 3
  <mask> : 4
```

Notice that even though we requested 160 tokens, the tokenizer only managed to create 111. This is because the corpus is so tiny that there simply weren't enough frequent pairs to merge into new tokens. BPE stops once it runs out of meaningful merges, so we end up with fewer tokens than the maximum we allowed.

The **special tokens** always occupy the first five positions in the vocabulary. These give us consistent IDs for padding (`<pad>`), unknown characters (`<unk>`), beginning-of-sentence (`<s>`), end-of-sentence (`</s>`), and masking (`<mask>`). By reserving these at the front, we ensure that our model can always recognize and handle them.

After the specials, the vocabulary consists of characters actually found in the corpus, plus merged tokens discovered during training. For example, here are the first few base entries:

```
id  5 -> 'e'
id  6 -> 'Ġ'
id  7 -> 'l'
id  8 -> 'o'
id  9 -> 'r'
id 10 -> 'n'
id 11 -> '\n'
```

The symbol Ġ is particularly important: this is GPT-2's convention for marking the beginning of a new word (it represents a space). Instead of treating space as an independent token, GPT-2 attaches it to the next word. This means "ĠHello" is a distinct token from "Hello". In our case, the tokenizer keeps Ġ as a standalone entry, but merges often fold it into full tokens such as "HelloĠ".

Learned Merges

Training BPE is essentially about **merging the most frequent neighboring pairs**. The following is the beginning of our learned merge list:

```
rank  0: 'o' + 'r' -> 'or'
rank  1: 'l' + 'o' -> 'lo'
rank  2: 'w' + 'or' -> 'wor'
rank  3: 'e' + 'r' -> 'er'
rank  6: 'H' + 'e' -> 'He'
```

```
rank   7: 'He' + 'l' -> 'Hel'
rank   8: 'Hel' + 'lo' -> 'Hello'
rank  11: 'worl' + 'd' -> 'world'
```

This illustrates how larger words gradually emerge. First, "o" and "r" combine into "or". Then, "l" and "o" become "lo", and eventually, "Hel" + "lo" yields the full word "Hello". Similarly, "w" + "or" → "wor", which then grows into "world".

These merges reveal the distributional structure of the corpus: because "Hello world" appears several times, the tokenizer naturally learns to treat it as stable units.

An interesting detail is that the tokenizer even learned to merge across boundaries, such as `HelloĠ` (the word plus the space after it) and `Ġline\n` (a word ending in a newline). These are valid, but if you wanted to avoid merges crossing line breaks, you could forbid any merge where either side is `\n`.

Encoding and Decoding Examples

Let's see how this vocabulary is used on actual text.

Example A: Simple Sentence

```
INPUT TEXT:     'Hello world.'
ENCODED TOKENS: ['Hello', 'Ġ', 'world', '.']
DECODED TEXT:   'Hello world.'
```

Here, the tokenizer recognizes both "Hello" and "world" as complete tokens. The space in between is represented by Ġ, and the period remains its own token. Decoding brings us back to the original string without any changes.

Example B: Sentence with Newline

```
INPUT TEXT:     'Hello world.\nNew line'
ENCODED TOKENS: ['Hello', 'Ġ', 'world', '.', '\n', 'N', 'e', 'w', 'Ġline']
DECODED TEXT:   'Hello world.\nNew line'
```

This time, the newline character `\n` is tokenized separately, while "line" is often seen after a space, so the tokenizer merges it into the single token `Ġline`. The string is again perfectly reconstructed.

Example C: Special Token <mask>

```
INPUT TEXT:  'Please, <mask> only this token.'
ENCODED TOKENS: ['<unk>', 'l', 'e', 'a', 's', 'e', ',', '<mask>', 'o', 'n',
'l', 'y', 'Ġ', 'th', 'i', 's', 'Ġ', 't', 'o', 'k', 'e', 'n', '.']
DECODED TEXT:   'lease,only this token.'
```

This example shows two subtle points.

First, because our toy corpus never contained the capital letter **"P"**, the tokenizer does not know it. The "P" is replaced with the `<unk>` token, which is later dropped in decoding, leaving us with "lease" instead of "Please". With a larger corpus, the vocabulary would naturally include all capital letters.

Second, `<mask>` is allowed to pass through during encoding, but all special tokens are removed on decoding. As a result, the comma is followed directly by "only", giving "lease,only". In practice, this is desirable: models treat `<mask>` as a control symbol, but the human-readable text doesn't display it.

Example D: With BOS and EOS

```
INPUT TEXT:  'Hello there, world!'
ENCODED TOKENS: ['<s>', 'Hello', 'Ġ', 'ther', 'e', ',', 'Ġ', 'world',
'!', '</s>']
DECODED TEXT:   'Hello there, world!'
```

When we set `add_bos_eos=True`, the tokenizer automatically adds `<s>` at the beginning and `</s>` at the end. These tokens are useful for models that need to mark sentence boundaries explicitly. They are ignored during decoding, so the output text is unaffected.

Step-by-Step BPE Trace

To really understand how BPE works, let's trace the tokenization of individual words.

For "Hello":

```
Start: ['H', 'e', 'l', 'l', 'o']
Step 0: merge ('l', 'o') → ['H', 'e', 'l', 'lo']
Step 1: merge ('H', 'e') → ['He', 'l', 'lo']
```

```
Step 2: merge ('He', 'l') → ['Hel', 'lo']
Step 3: merge ('Hel', 'lo') → ['Hello']
Final pieces: ['Hello']
```

For "world!":

```
Start: ['w', 'o', 'r', 'l', 'd', '!']
Step 0: merge ('o', 'r') → ['w', 'or', 'l', 'd', '!']
Step 1: merge ('w', 'or') → ['wor', 'l', 'd', '!']
Step 2: merge ('wor', 'l') → ['worl', 'd', '!']
Step 3: merge ('worl', 'd') → ['world', '!']
Final pieces: ['world', '!']
```

These traces make the mechanics visible. Each time, the tokenizer looks at all possible neighboring pairs, chooses the one with the **highest priority (lowest rank)**, merges it, and repeats until no more merges apply. In this way, frequent words like "Hello" and "world" eventually collapse into single tokens.

Saving and Loading

Finally, the tokenizer is saved to disk. Two files are produced:

- `demo_tok_vocab.json`: The vocabulary mapping tokens to IDs
- `demo_tok_merges.json`: The list of merges, in order

Reloading these files into a fresh tokenizer reproduces the exact same behavior:

```
Before IDs: [41, 6, 46, 5, 24, 6, 44, 23]
After  IDs: [41, 6, 46, 5, 24, 6, 44, 23]
Match? True
```

This guarantees that once trained, a tokenizer can be frozen, shipped, and reused anywhere, ensuring consistent preprocessing for both training and inference.

This demonstration shows the life cycle of a BPE tokenizer: from training on raw text, through merge learning, to encoding/decoding and reproducibility. Even on a tiny corpus, the behavior is easy to see: words like "Hello" and "world" are promoted to single tokens, while rarer patterns stay at the character level.

The important conclusions are:

- Special tokens are reserved up front.
- Frequent character sequences become whole tokens through merges.
- Unknown characters map to <unk> and vanish on decoding.
- BOS/EOS markers can be added automatically when needed.
- Saving and loading preserves exact behavior.

I hope this hands-on walkthrough gives you both an intuition for how BPE builds subwords and a concrete tool you can use in practice.

In this chapter, we built a **byte-pair encoding (BPE) tokenizer**, the first essential step in any GPT pipeline. We examined how raw text is converted into tokens, implemented encoding and decoding methods, and addressed key design choices such as vocabulary size, special tokens, and word-boundary handling. With this, we now have a reliable bridge between natural language and numerical input IDs.

The next chapter moves beyond preprocessing to the **core model architecture**. We will explore the fundamental building blocks of a GPT-style transformer—embeddings, positional encodings, multi-head attention, feedforward layers, and normalization—and then assemble them into a complete system capable of learning and generating language.

CHAPTER 4

RMS Normalization and Model Configuration

In the rapidly evolving world of large-scale neural networks, progress often depends not on grand architectural overhauls but on the careful engineering of small but powerful building blocks. Every component—from the way we normalize activations to how we structure the model's configuration—shapes the stability, efficiency, and adaptability of the system we are trying to create.

This chapter begins our journey into constructing a custom transformer model, piece by piece, by examining the foundational code that makes such systems both trainable and extensible.

We start with configuration: a blueprint that determines the depth of the model, the number of experts available for routing, the scale of embeddings, and even the nature of the positional encoding scheme. You'll see a simple `dataclass` providing an elegant way to encapsulate these hyperparameters, ensuring that design choices remain explicit, transparent, and easy to adapt. Behind each parameter—whether it is the vocabulary size or the sliding window length—lies a critical decision about computational trade-offs, memory constraints, and the types of tasks the model is expected to perform.

Next, we explore normalization, a subtle but essential ingredient in stabilizing deep architectures. Rather than relying on the more common LayerNorm, we turn to **Root Mean Square Normalization (RMSNorm)**, a lightweight alternative that avoids unnecessary bias terms while keeping variance under control. The forward pass, compact yet precise, reveals an important principle: sophisticated behavior in neural networks often emerges from deceptively simple formulas.

Together, configuration and normalization form the backbone of our model. They serve as a reminder that the path toward large-scale intelligence begins not with complexity, but with clarity: defining exactly what the system should be and ensuring that its signals remain balanced as information flows through its layers. In the pages

D. Grigorov, *Building Large Language Models from Scratch*, https://doi.org/10.1007/979-8-8688-2297-1_4

ahead, we will see how these fundamental pieces connect to attention, routing, and eventually, to the emergence of structured reasoning in modern transformer-based architectures.

Model Parameters Configuration and Mathematical Foundations

```
import json
import math
import os
from dataclasses import dataclass

import torch
import torch.distributed as dist
# --- CONFIGURATION ---

@dataclass
class ModelConfig:
    num_hidden_layers: int = 4
    num_experts: int = 16
    experts_per_token: int = 2
    vocab_size: int = 8000
    hidden_size: int = 256
    intermediate_size: int = 256
    head_dim: int = 64
    num_attention_heads: int = 4
    num_key_value_heads: int = 2
    sliding_window: int = 0
    initial_context_length: int = 4096
    rope_theta: float = 10000.0
    rope_scaling_factor: float = 1.0
    rope_ntk_alpha: float = 1.0
    rope_ntk_beta: float = 32.0
    swiglu_limit: float = 7.0
```

When we think about a neural network, it is easy to imagine a vast and opaque machine: layers upon layers of linear transformations, billions of parameters, and endless streams of tokens being transformed into meaning. But underneath this complexity lies something deceptively simple: a list of numbers, each representing a **design decision**. These numbers form the blueprint of the model. They determine how tall the structure will be, how wide its hallways are, how much memory it consumes, and what kind of knowledge it can reasonably acquire.

In our implementation, this blueprint is captured in a Python `dataclass` named `ModelConfig`. Each field of this class is a window into a choice that reflects decades of collective research in deep learning. To appreciate the power of these design choices, we will explore each parameter in detail.

Mathematical Notation

Let B represent the batch size (number of sequences processed in parallel), L the sequence length (number of tokens per sequence), d the hidden size (dimensionality of embeddings and hidden states), h the number of attention heads (parallel attention mechanisms), d_h the per-head dimension (size of each attention head's subspace), h_{kv} the number of key/value heads (for grouped query attention (GQA)), m the intermediate feed-forward network (FFN) size (expansion factor), E the number of experts (in mixture of experts layers), and k the experts-per-token (sparsity in MoE routing). Assume fp16 or bf16 activations, where each element uses $bytes_{per\_elt}=2$ bytes of memory.

Global Tensor Shapes and Consistency

The following equations define the core tensor shapes used throughout the model, ensuring dimensional compatibility across layers and operations.

Activation and Projection Shapes:

$$x \in \mathbb{R}^{B \times L \times d}$$

This tensor represents the input activations with batch size B, sequence length L, and hidden dimension d. This shape is maintained across layers, enabling efficient batch processing and consistent dimensionality for downstream operations like attention and FFNs.

num_hidden_layers—The Depth of Abstraction

At the very heart of a transformer lies its depth: the number of times information is repeatedly passed through a block of operations consisting of normalization, attention, and feed-forward computation. This parameter, `num_hidden_layers`, decides just how deep the model's reasoning stack will be. The number of transformer blocks, or depth, determines how many sequential layers process the input, enabling hierarchical feature extraction.

Why does this matter? Because depth is what allows the model to build **hierarchies of abstraction**. In the first layer, the model might only detect very surface-level features: whether a token looks like a number, or whether two words are adjacent. By the middle layers, it begins to understand syntax, dependencies, and phrase-level relationships. By the final layers, it can encode abstract semantics—ideas, intentions, or logical dependencies spanning long distances.

Shallow models, with only a few hidden layers, are computationally cheap but often brittle. They may recognize local patterns but fail to integrate them into coherent structures. Deep models, on the other hand, are powerful but fragile; every extra layer introduces the risk of vanishing gradients, exploding values, or overfitting. The art of choosing `num_hidden_layers` is therefore about balance: *How much depth can we afford, and how much do we truly need?*

Composition Through Blocks:

$$x^{(L_{\text{blk}})} = f^{(L_{\text{blk}})} \circ f^{(L_{\text{blk}}-1)} \circ \cdots \circ f^{(1)}(x^{(0)}), \quad f^{(l)} : \mathbb{R}^d \to \mathbb{R}^d$$

Each block $f^{(l)}$ applies self-attention and an FFN, typically with residual connections and normalization. More blocks (L_{blk}) allow deeper transformations but increase computational cost and risk issues like vanishing gradients, requiring careful initialization.

vocab_size—The Universe of Symbols

At first glance, `vocab_size` seems mundane. It simply states how many entries exist in the model's embedding table. But in truth, it defines the very **language** in which the model thinks.

A vocabulary of 8,000 tokens means the model can directly represent 8,000 distinct symbols—typically subword units produced by a tokenizer. This choice has ripple effects:

- A small vocabulary compresses words into many smaller parts, lengthening sequences but reducing memory footprint.
- A large vocabulary shortens sequences but inflates the size of embeddings and softmax layers.

It is, in essence, a question of granularity. Do we want the model to think in coarse blocks, like full words, or in fine fragments, like syllables? Choosing 8,000 here reflects a compromise: small enough for efficiency, large enough to handle linguistic variety.

Embedding and Output Projection:

$$W_{\text{emb}} \in \mathbb{R}^{V \times d}, \quad W_{\text{out}} \in \mathbb{R}^{d \times V}$$

The embedding matrix maps tokens to d -dimensional vectors, and the output projection maps hidden states to logits over V tokens.

Embedding Parameters:

$$\text{Params}(\text{emb}) = Vd$$

Parameters scale with vocabulary size and hidden dimension, often significant for large V (e.g., 50,000+ tokens).

Logits Cost Per Token:

$$z = h, W_{\text{out}} \in \mathbb{R}^{V}, \quad \text{cost} \sim dV$$

Computing logits involves a matrix-vector product, with cost proportional to dV . Large vocabulary increases this cost, impacting inference latency.

`hidden_size`—The Resolution of Thought

The `hidden_size` parameter defines the dimensionality of the model's core representation. Every token embedding, every intermediate activation, every projection lives in this space. If we think of the model as a camera, then `hidden_size` is its resolution.

At 256 dimensions, the model's internal world is neither too coarse nor too fine. With higher dimensions, representations become richer, allowing subtle distinctions between contexts to be encoded. With fewer, the model risks blurring distinctions, treating very different tokens as if they were the same.

But higher resolution also means heavier computation: every layer must manipulate larger matrices, every training step requires more memory. This parameter exemplifies the eternal trade-off between expressivity and efficiency.

Core Projection Matrices:

$$W_Q, W_K, W_V, W_O \in \mathbb{R}^{d \times d}, \quad W_1 \in \mathbb{R}^{d \times m}, \quad W_2 \in \mathbb{R}^{m \times d}$$

These matrices handle attention (query, key, value, output) and FFN projections. Their size drives parameter and compute costs.

Per-Block Parameters (Dense, Ignoring Biases/Norms):

$$\text{Params/block} \approx 4d^2 + 2dm$$

Each block includes four $d \times d$ attention matrices and two FFN matrices ($d \to m$, $m \to d$), dominating parameter counts.

intermediate_size—The Breathing Room of the Network

Each transformer block contains a **feed-forward network** that temporarily expands the dimensionality of the hidden representation before compressing it back down. The size of this expanded space is `intermediate_size`.

Why expand at all? Because sometimes the only way to express complex interactions is to give the network room to "breathe." Imagine trying to solve an equation in your head without scratch paper—you might manage for simple cases, but for anything subtle, you need to expand into a larger workspace before compressing the final answer back down.

Here, `intermediate_size` is equal to `hidden_size`. This is a lightweight choice, but larger models often expand by a factor of two or four, allowing their feed-forward layers to act as high-dimensional workshops where features can mix more freely.

FFN Mapping:

$$\text{FFN}(x) = W_2, \phi(W_1 x + b_1) + b_2, \quad W_1 \in \mathbb{R}^{d \times m}, \quad W_2 \in \mathbb{R}^{m \times d}$$

The FFN projects to a larger space ($m > d$), applies a nonlinearity ϕ (e.g., GELU), and projects back. This expansion enhances expressiveness.

FFN Cost Per Token:

$$\text{MACs/token (FFN)} \approx 2dm$$

MACs come from up and down projections, scaling with d and m. Larger m increases capacity but also compute.

head_dim—The Grain of Attention

When tokens attend to one another, their hidden representations are divided into smaller subspaces called **heads**. Each head looks at the input through a slightly different lens. The size of each lens, or grain, is `head_dim`.

This parameter dictates how detailed each head's view is. A larger `head_dim` allows a head to capture more information about each relationship, while a smaller one encourages heads to specialize narrowly. Together with the number of heads, it defines how the total representational space is carved up for attention.

num_attention_heads—The Many Eyes of the Model

Attention is not a single gaze but a collection of simultaneous perspectives. Each attention head looks at the sequence differently, focusing on different dependencies: one may track subject-verb agreement, another might align pronouns, while yet another watches for long-range discourse.

The parameter `num_attention_heads` specifies how many such eyes the model has. More heads mean more perspectives and a richer ensemble of relational cues. But each head is small, and their usefulness depends on how well the overall space (`hidden_size`) is divided among them.

These parameters define the multihead attention structure.

Head Consistency:

$$d = h, d_h$$

The hidden size splits across h heads, each with dimension d_h, ensuring efficient parallel attention.

Per-Head Attention:

$$\text{head}_i = \text{Softmax}\left(\frac{Q_i K_i^\top}{\sqrt{d_h}}\right), V_i, \quad Q_i, K_i, V_i \in \mathbb{R}^{B \times L \times d_h}$$

Each head computes scaled dot-product attention, with normalization by $\sqrt{d_h}$ to stabilize gradients.

Projection MACs Per Token:

$$\text{MACs/token (QKV+O)} \approx 4d^2$$

Four $d \times d$ projections (Q, K, V, output) dominate attention compute.

Attention Core Complexity (Global):

$$\text{Scores+AV (global)} \approx 2, h, L^2, d_h = 2L^2 d$$

Computing attention scores and applying them to values scales quadratically with L, a bottleneck for long sequences.

Local Attention with Window w:

$$\text{Scores+AV (local)} \approx 2, h, L, (2w+1), d_h = 2L(2w+1)d$$

A sliding window reduces the complexity to linear in L, attending only to $2w + 1$ tokens per position.

num_key_value_heads—Sharing the Burden

Traditional transformers give each query, key, and value its own head. But newer designs realize that keys and values don't always need as much diversity as queries. After all, many tokens share common roles in context.

This parameter, `num_key_value_heads`, allows the model to use fewer key-value projections than queries. The result is memory savings and faster computation without significant loss of performance. It is a small adjustment, but it reflects a broader principle of engineering: *efficiency often comes from knowing where to economize.*

GQA reduces memory by sharing key/value heads.

Projection Shapes:

$$Q \in \mathbb{R}^{B \times h \times L \times d_h}, \quad K, V \in \mathbb{R}^{B \times h_{kv} \times L \times d_h}, \quad r = \frac{h}{h_{kv}}$$

Fewer key/value heads ($h_{kv} < h$) reduce memory and compute for K and V.

KV-Cache Memory:

$$\mathrm{BytesKV} = 2 \cdot B \cdot L \cdot hkv \cdot d_h \cdot \mathrm{bytes}_{\mathrm{p}}\mathrm{er}_{\mathrm{e}}\mathrm{lt}$$

GQA lowers h_{kv}, reducing KV-cache memory compared to standard multihead attention.

`sliding_window`—Constraining Attention

Full attention is powerful but costly. For long sequences, it requires quadratic time and memory. The `sliding_window` parameter introduces a practical alternative: restrict each token to attend only to a fixed window of recent tokens.

This mirrors how humans often operate. When reading a book, we don't constantly rescan every page from the beginning; we focus on the last few paragraphs to ground our current understanding. Setting this parameter to a positive value enforces locality, which can be crucial for scaling to extremely long contexts.

Complexity Reduction:

$$\mathcal{O}(L^2 d) \implies \mathcal{O}(L, w, d)$$

Local attention reduces quadratic complexity to linear, critical for long sequences.

Attention Window:

$$[i - w, i + w]$$

Each token attends to a fixed window of $2w + 1$ tokens, balancing context and efficiency.

initial_context_length—How Far the Model Can See

The initial_context_length is the maximum sequence length the model is designed to handle at initialization. Here, it is set to 4,096 tokens, which is enough to process several pages of text at once.

This number has profound implications. Too short, and the model is blind to long-range dependencies. Too long, and memory consumption skyrockets. Moreover, the positional encoding scheme must adapt to represent such distances without losing precision. Choosing this number is like deciding the maximum horizon of the model's vision.

Sequence Length Bound:

$$L \leq L_{\max}$$

Training is capped at $L_{\max}$, ensuring manageable memory and compute.

KV-Cache Memory Per Layer at Inference:

$$\mathrm{BytesKV} = 2 \cdot B \cdot L \cdot hkv \cdot d_h \cdot \mathrm{bytes_per_elt}$$

During inference, the KV-cache stores keys and values to avoid recomputation in autoregressive generation. The memory cost accounts for two tensors (K and V), batch B, context length L, key/value heads h_{kv}, per-head dimension d_h, and two bytes per element. This scales linearly with L, making long-context models memory-intensive.

KV-Cache and RoPE Planning:

$$[0, L_{\max}]$$

The KV-cache and RoPE frequencies are designed for sequences up to $L_{\max}$, impacting memory planning.

The Geometry of Position—rope_theta, rope_scaling_factor, rope_ntk_alpha, rope_ntk_beta

When we read a sentence, word order matters. "The dog chased the cat" does not mean the same thing as "The cat chased the dog." Transformers, however, don't have an inherent sense of order; they see tokens as a bag of embeddings. Positional encoding is what tells the model *where* each token belongs in the sequence.

Rotary position embeddings (RoPE) are a particularly elegant solution. Instead of adding a separate positional vector, RoPE *rotates* each token's representation in a way that encodes its position. This makes relative positions naturally meaningful: the "angle" between two tokens encodes how far apart they are.

Now, let's look at the four parameters that control this geometry.

`rope_theta`—The Base Frequency

Think of `rope_theta` as the *lens* through which the model perceives position. A small value means the model focuses tightly on local details—it can tell the difference between adjacent tokens very precisely, but it loses awareness of long-range structure. A large value, by contrast, stretches positions apart, so the model keeps track of distant relationships better, though nearby tokens may look blurrier.

In essence:

- Low `rope_theta` = microscope (sharp local view, short horizon).
- High `rope_theta` = telescope (broad horizon, fuzzier close-up view).

`rope_scaling_factor`—Stretching the Map

If `rope_theta` sets the lens, `rope_scaling_factor` is like zooming in or out on the entire positional map. By increasing this factor, we "stretch" the embedding space so that the model can handle sequences longer than what it was originally trained on.

Without scaling, a model trained on 4,000 tokens may struggle at 16,000. With scaling, we can extend its range gracefully. The price is that local precision is diluted—zooming out helps us see the big picture, but we lose some fine-grained detail.

`rope_ntk_alpha`—Balancing Long-Context Stability

When we extend transformers to very long sequences, one of the dangers is that positional encodings start to break down: rotations become too extreme, and attention weights destabilize. `rope_ntk_alpha` acts like a stabilizer, carefully adjusting how much the rotations grow with context length.

Think of it like tuning a guitar string: too tight and it snaps, too loose and it loses tone. `rope_ntk_alpha` is the tension that keeps positional encoding "in tune" as the model is stretched to longer contexts.

rope_ntk_beta—Controlling the Rate of Growth

If rope_ntk_alpha is the tension of the string, then rope_ntk_beta is the **curve of adjustment**. It decides how quickly the stabilizing effect is applied as context length increases. A small beta makes the correction kick in early (good for moderate sequence extensions), while a larger beta delays adjustment until very long contexts (better if we really push beyond the training window).

Together, alpha and beta form a fine-tuning system for long-context reasoning. They embody a lesson from both theory and practice: encoding position is not a trivial matter, but a geometric balancing act between short-range clarity and long-range stability.

Together, these parameters illustrate the hidden sophistication of positional encoding. Far from being an afterthought, they are the quiet engineers of long-context understanding.

swiglu_limit—Taming the Activations

Finally, we encounter swiglu_limit, a safeguard built into the activation function. The SwiGLU nonlinearity is powerful, but without constraints, activations can grow explosively, destabilizing training. By limiting their magnitude, we enforce discipline on the network's internal signals.

This may seem like a minor parameter, but in practice, it can be the difference between a model that converges smoothly and one that spirals into numerical chaos. It reflects a simple truth: sometimes, intelligence requires boundaries.

swiglu_limit (Activation Control)

What Is SwiGLU?

Swish-Gated Linear Unit (SwiGLU) is an activation function used in modern neural network architectures, particularly in transformer-based large language models (e.g., PaLM or LLaMA variants). It is a variant of the Gated Linear Unit (GLU), replacing the sigmoid gating mechanism with the Swish activation function to improve gradient flow and model expressiveness.

Background on Related Concepts

GLU uses one part of the input as a signal and another part as a gate that decides how much of that signal gets through. Swish is a smooth activation function that avoids some of the problems seen in harsher functions like ReLU, helping gradients flow more reliably.

SwiGLU blends these ideas by replacing GLU's usual gate with Swish, often leading to more stable training and better model performance.

How SwiGLU Works

In neural network feed-forward layers, SwiGLU processes an input vector x by projecting it into a higher-dimensional space using two weight matrices, W and V, with optional biases. The output is the element-wise product of one projection and the Swish-applied version of the other.

The mathematical formulation is:

$$\text{SwiGLU}(x) = (xW) \odot \text{Swish}(xV)$$

where

- $\odot$ represents element-wise multiplication (Hadamard product).
- W and V are learnable weight matrices, often of the same shape.
- Swish is defined as $\text{Swish}(z) = z \cdot \sigma(z) = z \cdot \frac{1}{1+e^{-z}}$.

In practice, the input is often up-projected to twice the desired output dimension, split into two parts (say a and b), and computed as $a \odot \text{Swish}(b)$ for efficiency, avoiding separate matrices.

Advantages and Usage

- **Why Use SwiGLU?** It balances expressivity and computational efficiency. Compared to ReLU or GLU, SwiGLU captures more nuanced non-linearities, improving generalization in large-scale models, as shown in empirical studies for language modeling.

- **Common Applications**: SwiGLU is used in the MLP (multilayer perceptron) blocks of transformer models, often in the feed-forward network (FFN) layer: linear projection → SwiGLU → linear projection back to the original dimension.
- In frameworks like PyTorch, it might be implemented as `import torch.nn.functional as F; return x[:, :dim] * F.silu(x[:, dim:])` after a single linear layer doubling the dimension.

SwiGLU excels in high-dimensional, deep learning scenarios where gating enhances information flow control, though it's not a universal replacement for all activation functions.

SwiGLU Enhances FFN Expressiveness in GPT Architectures with Gated Activations

SwiGLU:

$$\mathrm{SwiGLU}(x) = (\mathrm{SiLU}(xW_g)) \odot (xW_v), W_o, \quad \mathrm{SiLU}(u) = u, \sigma(u)$$

SiLU combines sigmoid gating with linear scaling, and element-wise multiplication $\odot$ adds sparsity.

Hard Clipping:

$$y = \mathrm{clip}(\mathrm{SwiGLU}(x), -L_{\mathrm{clip}}, L_{\mathrm{clip}})$$

Clipping bounds activations, preventing numerical instability.

Soft Saturation:

$$y = L_{\mathrm{clip}}, \tanh\left(\frac{\mathrm{SwiGLU}(x)}{L_{\mathrm{clip}}}\right)$$

Tanh provides smooth saturation, preserving gradients while controlling output magnitude.

RMS Normalization in GPT Architectures

RMS normalization (RMSNorm) is a normalization technique used in neural networks, particularly in transformer-based architectures like those underlying GPT models. It serves as a simplified alternative to layer normalization (LayerNorm), focusing on stabilizing the training process by normalizing activations while reducing computational overhead.

What Is RMS Normalization?

RMSNorm normalizes the inputs to a layer by dividing them by the root mean square (RMS) of the inputs, without subtracting the mean (a step included in LayerNorm). This makes it a rescaling operation that ensures the outputs are invariant to scaling in the inputs or weights.

Mathematically, for a vector of summed inputs $a = [a_1, a_2, \ldots, a_n]$ to a neuron (where n is the dimensionality), RMSNorm computes:

$$\bar{a}_i = \frac{a_i}{\text{RMS}(a)} \cdot g_i$$

where

- $\text{RMS}(a) = \sqrt{\frac{1}{n}\sum_{i=1}^{n} a_i^2}$ (the root mean square, essentially a scaled L2 norm).
- g_i is a learnable gain parameter (often initialized to 1) that allows the model to rescale the normalized values if needed.

In contrast, LayerNorm includes mean centering:

$$\bar{a}_i = \frac{a_i - \mu}{\sigma} \cdot g_i$$

where μ 'is the mean and σ is the standard deviation. RMSNorm skips the mean (μ) computation, hypothesizing that recentering is less critical for performance than rescaling.

A variant called partial RMSNorm (pRMSNorm) estimates the RMS using only a subset (e.g., the first 6.25%) of the elements, further reducing computation while preserving rescaling invariance.

Why Is RMSNorm Used in GPT Architectures?

GPT architectures (and transformer models in general) rely on normalization to address issues like vanishing or exploding gradients, internal covariate shift, and unstable training dynamics in deep networks. Normalization helps accelerate convergence and improves generalization by keeping activations within a stable range.

RMSNorm was proposed as an improvement over LayerNorm because:

- **Computational Efficiency**: By eliminating the mean subtraction, RMSNorm reduces overhead. Experiments show speed-ups of 7–64% in training time across models like RNNs and Transformers, without sacrificing performance. This is especially valuable for large-scale models like GPT, where training involves billions of parameters and massive datasets.
- **Simpler and Effective**: The recentering step in LayerNorm is often dispensable, as rescaling alone stabilizes magnitudes. RMSNorm maintains comparable or better results in tasks like machine translation (e.g., similar BLEU scores) and reading comprehension (faster convergence and lower error rates).
- **Implicit Learning Rate Adaptation**: Gradients in RMSNorm naturally adapt based on input magnitudes, aiding optimization.
- **Adoption in Modern LLMs**: While early GPT models (e.g., GPT-3) used LayerNorm, RMSNorm has become popular in subsequent GPT-like architectures, such as LLaMA, GPT-J, and other open-source large language models (LLMs). It's credited with fueling efficiency in next-generation LLMs by simplifying normalization without performance loss.

In essence, RMSNorm is used to make training faster and more stable, particularly in deep transformer stacks where normalization layers are applied frequently.

How Is RMSNorm Used in GPT Architectures?

In GPT architectures (decoder-only transformers), normalization is applied within each transformer block to the outputs of sub-layers like multihead self-attention and feed-forward networks (FFNs). RMSNorm acts as a drop-in replacement for LayerNorm in these positions.

Step-by-Step Implementation in a Transformer Block:

1. **Input to Normalization**: After a sub-layer (e.g., attention or FFN), the residual connection adds the sub-layer output to the input: $a = x + \text{sub-layer}(x)$.
2. **Compute RMS**: Calculate the RMS across the last dimension (typically the embedding dimension) for each token in the sequence.
3. **Normalize**: Divide each element by the RMS and multiply by the gain g.
4. **Output**: The normalized vector $\bar{a}$ `https://www.codecogs.com/eqnedit.php?latex=%20%5Cbar%7Ba%7D%20#0` proceeds to the next sub-layer or block.

This is typically done in a prenormalization setup (normalize before the sub-layer) or post-normalization (after), depending on the model variant. For example:

- In models like LLaMA (a GPT-style architecture), RMSNorm is applied before the attention and FFN layers.

In practice, it's implemented in libraries like PyTorch (via `torch.nn.RMSNorm`) and used in transformer layers for efficiency.

Experimental results from the original paper show RMSNorm achieving similar performance to LayerNorm on Transformer-based machine translation (BLEU scores of ~26.8–27.7) with 7–9% faster training, and even better convergence in reading comprehension tasks.

RMS Normalization for Our Custom Large Language Model

```
# --- RMS NORMALIZATION ---
class RMSNorm(torch.nn.Module):
    def __init__(self, num_features: int, eps: float = 1e-5, device: torch.
    device | None = None):
        super().__init__()
        self.eps = eps
        self.num_features = num_features
        self.scale = torch.nn.Parameter(torch.ones(num_features,
        device=device))

    def forward(self, x: torch.Tensor) -> torch.Tensor:
        # x: [..., H]
        if x.shape[-1] != self.num_features:
            raise ValueError(f"Expected input with last dim {self.num_
            features}, got {x.shape[-1]}")
        # Standard RMSNorm: x / sqrt(mean(x^2) + eps)
        var = x.pow(2).mean(dim=-1, keepdim=True)
        x = x * torch.rsqrt(var + self.eps)
        return x * self.scale
```

Comprehensive Explanation of RMSNorm Code

The provided Python code implements RMS Normalization (RMSNorm) as a PyTorch module, a technique widely used in transformer-based architectures like GPT to normalize layer inputs, ensuring stable and efficient training. This detailed explanation focuses exclusively on the code, breaking down each component, operation, and design choice comprehensively.

```
import torch

class RMSNorm(torch.nn.Module):
    def __init__(self, num_features: int, eps: float = 1e-5, device: torch.
    device | None = None):
```

```
        super().__init__()
        self.eps = eps
        self.num_features = num_features
        self.scale = torch.nn.Parameter(torch.ones(num_features,
        device=device))

    def forward(self, x: torch.Tensor) -> torch.Tensor:
        # x: [..., H]
        if x.shape[-1] != self.num_features:
            raise ValueError(f"Expected input with last dim {self.num_
            features}, got {x.shape[-1]}")
        # Standard RMSNorm: x / sqrt(mean(x^2) + eps)
        var = x.pow(2).mean(dim=-1, keepdim=True)
        x = x * torch.rsqrt(var + self.eps)
        return x * self.scale
```

Detailed Code Explanation

The `RMSNorm` class is a PyTorch module (`torch.nn.Module`) designed to normalize input tensors by dividing them by their root mean square (RMS) and applying a learnable scaling factor. This approach stabilizes training in deep neural networks, particularly in transformers, by ensuring activations remain within a consistent magnitude range. Unlike Layer Normalization (LayerNorm), RMSNorm skips mean-centering, reducing computational overhead while maintaining performance. Below, each part of the code is dissected with a focus on its purpose, implementation, and mathematical underpinnings.

1. Import and Class Definition

```
import torch

class RMSNorm(torch.nn.Module):
```

- **Import**: The line `import torch` brings in the PyTorch library, which provides tensor operations, automatic differentiation (autograd), and neural network utilities essential for implementing RMSNorm.

- **Class Definition**: `RMSNorm` is defined as a subclass of `torch.nn.Module`, PyTorch's base class for neural network modules. This inheritance enables integration with PyTorch's ecosystem, including parameter management, device handling (CPU/GPU), and compatibility with training pipelines. By subclassing `torch.nn.Module`, `RMSNorm` can be seamlessly embedded in transformer models, such as those used in GPT architectures, where normalization is applied within each transformer block.

2. Initialization (`__init__`)

```
def __init__(self, num_features: int, eps: float = 1e-5, device: torch.
device | None = None):
    super().__init__()
    self.eps = eps
    self.num_features = num_features
    self.scale = torch.nn.Parameter(torch.ones(num_features,
    device=device))
```

- **Method Purpose**: The `__init__` method configures the `RMSNorm` module by defining its parameters and initializing its state, preparing it to process input tensors of a specific shape.
- **Parameters**:
  - `num_features: int`: Specifies the size of the input tensor's last dimension, denoted H. In transformer models like GPT, H represents the embedding or hidden dimension (e.g., 768 for BERT, 4096 for larger GPT models). This ensures the module normalizes over the correct feature dimension.
  - `eps: float = 1e-5`: A small constant $\epsilon = 10^{-5}$ added to the RMS denominator to prevent division by zero, ensuring numerical stability during computation. This value is chosen to be small enough to avoid significant perturbation but large enough to handle cases where the variance is near zero.

- `device: torch.device | None = None`: Specifies the device (e.g., `torch.device('cuda')` for GPU or `'cpu'`) for the module's parameters. If `None`, PyTorch uses the default device, allowing flexibility in deployment across hardware.

- **Functionality:**
  - `super().__init__()`: Calls the constructor of the parent `torch.nn.Module` class, initializing the module's core functionality, such as parameter tracking and device management.
  - `self.eps = eps`: Stores the ϵ value as an instance variable, making it accessible in the forward pass for numerical stability.
  - `self.num_features = num_features`: Stores H to validate the input tensor's shape during the forward pass, ensuring the input's last dimension matches the expected feature size.
  - `self.scale = torch.nn.Parameter(torch.ones(num_features, device=device))`: Creates a learnable parameter tensor of shape $[H]$ H, initialized to ones (i.e., $g_i = 1$ for each feature). The `torch.nn.Parameter` wrapper marks `self.scale` as a trainable parameter, enabling gradient computation and optimization during training. The `device` argument ensures the tensor is allocated on the specified device, aligning with the model's hardware configuration.

- **Purpose of `self.scale`**: The learnable scale parameter g_i allows the model to adjust the magnitude of the normalized output, preserving representational capacity. Initializing to ones ensures that, at the start of training, the normalization is a simple division by the RMS, which the model can refine through learning.

This initialization sets up the module to handle inputs with a last dimension of H and prepares the necessary parameters for RMSNorm computation.

3. Forward Pass (forward)

```
def forward(self, x: torch.Tensor) -> torch.Tensor:
    # x: [..., H]
    if x.shape[-1] != self.num_features:
        raise ValueError(f"Expected input with last dim {self.num_
        features}, got {x.shape[-1]}")
    # Standard RMSNorm: x / sqrt(mean(x^2) + eps)
    var = x.pow(2).mean(dim=-1, keepdim=True)
    x = x * torch.rsqrt(var + self.eps)
    return x * self.scale
```

- **Method Purpose**: The forward method defines how the input tensor x https://www.codecogs.com/eqnedit.php?latex=%20x%20#0 is processed to produce a normalized output. The input x has shape $[\ldots, H]$, where $\ldots$ represents arbitrary leading dimensions (e.g., $[B, S, H]$ for batch size B, sequence length S, and feature dimension H in a transformer). The method applies RMSNorm to normalize the input over the last dimension and scales it with a learnable parameter.

- **Step-by-Step Breakdown:**

  Input Shape Validation:

  ```
  if x.shape[-1] != self.num_features:
      raise ValueError(f"Expected input with last dim {self.num_
      features}, got {x.shape[-1]}")
  ```

  1.

     - Checks if the last dimension of the input tensor x (accessed via `x.shape[-1]`) equals `self.num_features` (H). For example, if `num_features = 768`, the input must have a last dimension of 768 (e.g., $[B, S, 768]$).

     - If the dimensions mismatch, a `ValueError` is raised with a descriptive message, preventing incorrect normalization and aiding debugging. This ensures the module processes inputs consistent with its initialization.

- This check is critical in transformers, where the feature dimension must align across layers (e.g., attention and feed-forward outputs).

Compute Variance:

```
var = x.pow(2).mean(dim=-1, keepdim=True)
```

2.

- Calculates the mean of the squared input values along the last dimension to compute the variance:

$$\text{var} = \frac{1}{H}\sum_{i=1}^{H} x_i^2$$

- `x.pow(2)` computes the element-wise square of the input tensor (x_i^2), producing a tensor of the same shape $[\ldots, H]$.
- `mean(dim=-1, keepdim=True)` averages these squared values over the last dimension (H), resulting in a tensor of shape $[\ldots, 1]$ (e.g., $[B, S, 1]$ for a batched input). The `keepdim=True` argument ensures the last dimension is retained as a singleton dimension, enabling broadcasting in the next step.
- This variance represents the squared magnitude of the input vector along the feature dimension, a key component of the RMS calculation.

Normalize with RMS:

```
x = x * torch.rsqrt(var + self.eps)
```

3. Normalizes the input by dividing it by the root mean square, defined as:

$$\text{RMS}(x) = \sqrt{\text{var} + \epsilon}$$

- `var + self.eps` adds the small constant $\epsilon = 10^{-5}$ to the variance to prevent division by zero in cases where is very small or zero.

- `torch.rsqrt(var + self.eps)` computes the reciprocal square root:

$$\frac{1}{\sqrt{\text{var} + \epsilon}}$$

  This is a numerically efficient operation in PyTorch, avoiding the need to compute the square root and then perform division.

- The input x is multiplied by this reciprocal, equivalent to:

$$x_i = \frac{x_i}{\sqrt{\text{var} + \epsilon}}$$

- This operation normalizes each element in the feature dimension, scaling the input vector to have a unit RMS (adjusted by ϵ). The broadcasting of $[\ldots, 1]$ (from `var`) with $[\ldots, H]$ (from `x`) ensures each feature is divided by the appropriate RMS value.
- The result is a tensor of shape $[\ldots, H]$, with normalized magnitudes across the feature dimension.

Apply Learnable Scale:

```
return x * self.scale
```

- Multiplies the normalized tensor by `self.scale`, a learnable parameter of shape $[H]$ (g_i for each feature).
- This operation applies element-wise scaling:

$$\bar{x}_i = x_i \cdot \frac{1}{\sqrt{\text{var} + \epsilon}} \cdot g_i$$

- The `self.scale` tensor is broadcast across the leading dimensions of x (e.g., batch and sequence dimensions), applying the same scaling factor to each feature across all tokens and batches.

- This step allows the model to learn the optimal scaling for the normalized outputs, preserving the network's expressiveness. Without `self.scale`, the output would be strictly normalized to a fixed magnitude, potentially limiting the model's ability to represent complex patterns.
- The output tensor retains the input shape $[\ldots, H]$ (e.g., $[B, S, H]$).

4. **Return Output:**
   - The final output $\bar{x}$ is returned, ready for further processing in the transformer pipeline (e.g., as input to the next sub-layer, such as attention or feed-forward networks).
   - The output has the same shape as the input ($[\ldots, H]$), ensuring compatibility with subsequent layers.

Mathematical Summary

The `forward` method implements the RMSNorm formula:

$$\bar{x}i = \frac{x_i}{\sqrt{\frac{1}{H} \sum j = 1^H x_j^2 + \epsilon}} \cdot g_i$$

- The variance $\text{var} = \frac{1}{H} \sum_{j=1}^{H} x_j^2$ is computed over the feature dimension.
- The RMS is approximated as $\sqrt{\text{var} + \epsilon}$, with ϵ ensuring stability.
- Each element is divided by the RMS and scaled by g_i (from `self.scale`), producing a normalized output with learned scaling.

Implementation Details and Design Choices

- **Efficiency**: The use of `torch.rsqrt` instead of computing the square root and dividing is performance optimization, reducing computational steps. This is critical in large-scale models like GPT, where normalization is applied multiple times per transformer block.

- **Broadcasting**: PyTorch's broadcasting handles the shape mismatch between x ($[\ldots, H]$) and `var` ($[\ldots, 1]$) or `self.scale` ($[H]$), ensuring efficient element-wise operations without explicit looping.
- **Numerical Stability**: The inclusion of $\epsilon = 10^{-5}$ prevents division by zero, a common issue when input variances are small, especially early in training or with certain data distributions.
- **Learnable Parameters**: The `self.scale` parameter (g_i) is trainable, allowing the model to adapt the normalization's effect during optimization, unlike fixed normalization schemes that might constrain the output range.
- **Shape Flexibility**: The code supports arbitrary leading dimensions ($\ldots$), making it versatile for inputs like $[B, H]$ (for single-token inputs) or $[B, S, H]$ (for batched sequences in transformers).
- **Error Handling**: The shape validation ensures robustness, catching errors early if the input tensor's feature dimension does not match the module's configuration.

The `RMSNorm` module is a concise yet powerful implementation of RMS Normalization in PyTorch. It normalizes input tensors by dividing by the RMS ($\sqrt{\frac{1}{H}\sum_{i=1}^{H} x_i^2 + \epsilon}$) and applies a learnable scale g_i, producing outputs that are stable and optimized for transformer training. The code's design leverages PyTorch's efficient tensor operations, robust error handling, and flexible shape handling, making it ideal for large-scale models like GPT where normalization is critical for performance and stability.

Summary—RMS Normalization and Model Configuration

In this chapter, we established the foundations of our custom transformer by focusing on two essential components: **model configuration** and **RMS normalization**.

- **Model Configuration:**

  We introduced a structured way to define a transformer's blueprint using a Python dataclass. Each hyperparameter—such as the number of hidden layers, experts per token, vocabulary size, hidden size, attention heads, and context length—encodes a design decision that directly affects capacity, efficiency, and the model's ability to reason across different contexts. We explored how these parameters balance trade-offs between depth, memory, specialization (via mixture of experts), and scalability to long sequences.

- **RMS Normalization:**

  Instead of the more common LayerNorm, we implemented **RMSNorm** to stabilize activations with reduced computational overhead. By scaling activations based on their root mean square (rather than subtracting the mean), RMSNorm ensures smoother training dynamics in deep transformer stacks. We examined its mathematical foundation, code implementation, and role in modern GPT-style architectures, emphasizing its efficiency and stability benefits.

Together, configuration and normalization form the **structural and numerical backbone** of our model. They ensure that the system is not only well-defined but also stable enough to support more complex mechanisms like attention, routing, and large-context reasoning.

While normalization and configuration give the model stability and structure, transformers also require a **sense of order**—a way to understand where tokens are positioned in a sequence. This is where **rotary positional embeddings (RoPE)** come in.

In the next chapter, we will:

- Explore **how RoPE encodes position** through elegant geometric rotations.
- Study the **limitations of standard RoPE** when extending context length far beyond training.

- Introduce **NTK scaling**, a method inspired by Neural Tangent Kernel theory, that stabilizes positional encodings in very long contexts.
- Discuss **YaRN scaling**, a complementary approach that stretches positional representations to gracefully extrapolate beyond the model's original horizon.
- Combine RoPE + NTK + YaRN into a **unified long-context strategy**, enabling transformers to maintain both short-range precision and long-range stability.

This sets the stage for building models capable of **handling sequences much longer than their training context**, a critical requirement for modern applications such as document understanding, multiturn reasoning, and real-time streaming inputs.

CHAPTER 5

Rotary Positional Embeddings: Integrating NTK and YaRN Scaling

Next, we continue building in our custom large language model rotary positional embeddings. Rotary positional embeddings (RoPE) have become a cornerstone of modern large language models (LLMs) due to their ability to address critical challenges in processing sequential data within transformer architectures.

Introduced in the 2021 paper "RoFormer: Enhanced Transformer with Rotary Position Embedding" by Jianlin Su et al.[1] RoPE provides a robust and efficient method for encoding positional information, which is vital for LLMs to understand and generate coherent, contextually accurate text.

Rotary Positional Embeddings: An In-Depth and Comprehensive Exploration

Rotary positional embeddings (RoPE) represent a transformative approach in transformer-based architectures, particularly for natural language processing (NLP) and large language models (LLMs). Introduced to address shortcomings in traditional positional encoding methods, RoPE elegantly integrates absolute and relative positional information, enabling superior handling of sequence order, enhanced generalization to longer contexts, and improved model performance.

[1] Su, J., Lu, Y., Pan, S., Murtadha, A., Wen, B., & Liu, Y. (2021). "RoFormer: Enhanced Transformer with Rotary Position Embedding."

D. Grigorov, *Building Large Language Models from Scratch*, https://doi.org/10.1007/979-8-8688-2297-1_5

RoPE encodes absolute positions via rotation matrices, naturally embedding relative position dependencies in self-attention. Unlike additive embeddings, RoPE applies rotations to token vectors, preserving magnitudes and separating positional from semantic information.

The intuition: Treat pairs of embedding dimensions as 2D coordinates and rotate them by angles proportional to position. This encodes absolute positions while making attention scores depend only on relative distances, unifying absolute and relative approaches. RoPE's properties—length extrapolation, decaying attention with distance, and linear attention compatibility—make it a cornerstone of modern LLMs.

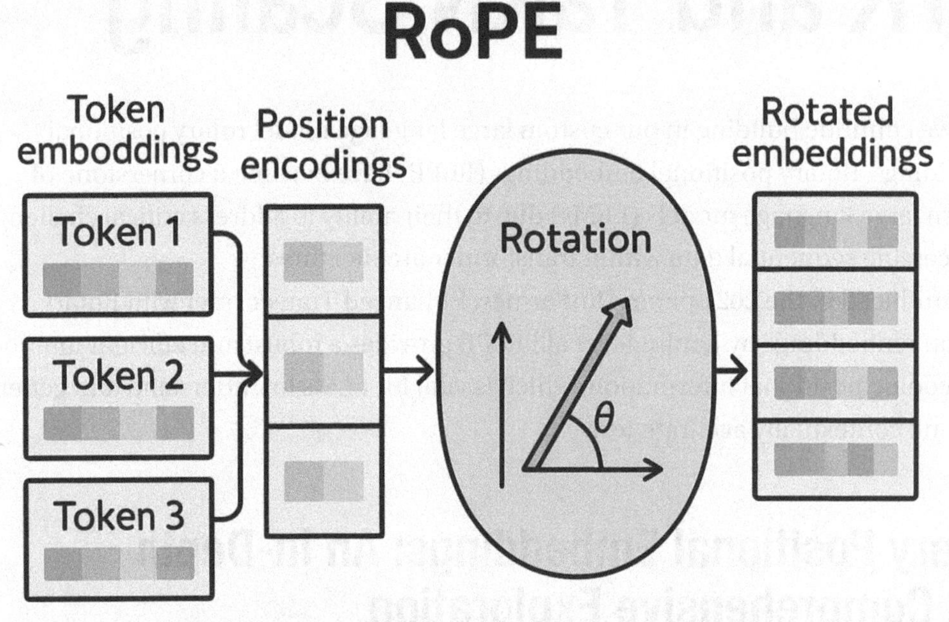

***Figure 5-1.** Visualizing RoPE*

The Fundamental Need for Positional Embeddings in Transformers

The transformer model, introduced in the 2017 paper "Attention Is All You Need"[2] by Vaswani et al., revolutionized NLP through its self-attention mechanism, which processes sequences in parallel. However, self-attention is permutation-invariant, treating input tokens as an unordered set and ignoring their sequential order. This is a critical limitation, as word order significantly impacts meaning (e.g., "the dog bites the man" vs. "the man bites the dog"). Without positional information, transformers cannot distinguish such sequences, leading to poor performance in tasks requiring syntactic or semantic understanding.

Positional embeddings address this by injecting sequence order information into token embeddings. Effective embeddings must balance absolute positions (a token's exact location in the sequence) and relative positions (distances between tokens), while being efficient and scalable. RoPE achieves this balance, overcoming limitations of earlier methods.

Historical Evolution of Positional Encodings

The development of positional encodings has been a critical aspect of advancing natural language processing (NLP), particularly with the rise of transformer-based architectures. As models evolved from sequential processing to parallelized attention mechanisms, the need to effectively encode the positional information of tokens in a sequence became paramount. Rotary positional embeddings (RoPE) represent a significant milestone in this evolution, addressing limitations of earlier methods by combining absolute and relative positional information in a rotation-based framework.

Pretransformer Approaches: Sequential Processing

Before transformers, NLP relied heavily on recurrent neural networks (RNNs) and their variants, such as long short-term memory (LSTM) networks and gated recurrent units (GRUs). These models inherently capture sequence order by processing tokens

[2] Vaswani, A., Shazeer, N., Parmar, N., Uszoreit, J., Jones, L., Gomez, A. N., Kaiser, Ł., & Polosukhin, I. (2017). "Attention Is All You Need."

sequentially, with each step updating a hidden state based on the current input and previous context.

- **Mechanism**: RNNs process tokens one at a time, maintaining a hidden state (h_t) updated as: $h_t = f(h_{t-1}, x_t)$, where (x_t) is the input at time (t) and (f) is a non-linear function (e.g., tanh or LSTM cell). LSTMs mitigate vanishing gradient issues with memory cells and gates, preserving long-term dependencies.
- **Strengths**: Naturally encode token order without explicit positional embeddings, suitable for tasks like language modeling and sequence tagging.
- **Weaknesses**: Sequential processing prevents parallelization, leading to high computational costs for long sequences (($O(n)$) per step, where (n) is sequence length). Additionally, capturing very long dependencies is challenging due to gradient issues, even in LSTMs.
- **Impact**: While effective for short sequences, the computational inefficiency and limited scalability of RNNs spurred the development of transformers, which required explicit positional encodings.

Why Include: This sets the stage by contrasting sequential models with transformers, highlighting why positional encodings became necessary when parallel processing was introduced.

Early Transformer Positional Encodings

The transformer, introduced in the seminal 2017 paper "Attention Is All You Need" by Vaswani et al., revolutionized NLP by replacing sequential processing with self-attention, enabling parallel computation across tokens. However, self-attention is permutation-invariant, ignoring token order unless positional information is explicitly added. Early transformers addressed this with two primary approaches: sinusoidal and learned absolute positional embeddings.

Sinusoidal Positional Embeddings

The original transformer used fixed, non-learnable sinusoidal embeddings to encode absolute positions, defined as: $PE(m, 2i) = \sin\left(\frac{m}{10000^{2i/d}}\right)$ $PE(m, 2i+1) = \cos\left(\frac{m}{10000^{2i/d}}\right)$, where (m) is the position, (i) ranges from 0 to $(\lfloor d/2 \rfloor - 1)$, and (d) is the embedding dimension. These are added to token embeddings: $x'_m = x_m + PE(m)$.

- **Mechanism**: The sinusoidal functions create unique, periodic patterns for each position, with frequencies varying across dimensions $lower(i)$ for coarser scales, higher (i) for finer details. This allows the model to infer relative positions through linear combinations of these vectors.
- **Strengths**:
  - No trainable parameters, reducing memory usage
  - Deterministic, ensuring consistent encodings across models
  - Enables some relative position learning via attention mechanisms
- **Weaknesses**:
  - Poor extrapolation to sequences longer than those seen during training, as the periodic patterns do not generalize well.
  - Additive nature can interfere with the semantic content of token embeddings.
  - Limited explicit modeling of relative positions, relying on the model to learn these relationships.
- **Impact**: Widely adopted in early transformers (e.g., BERT, GPT-1) due to simplicity, but limitations in handling long sequences prompted further research.

Learned Absolute Positional Embeddings

Subsequent models, such as GPT-2 and GPT-3, introduced learned absolute positional embeddings, where each position (m) up to a maximum sequence length is assigned a trainable vector ($PE(m) \in \mathbb{R}^d$), optimized during training alongside token embeddings.

- **Mechanism**: The positional embedding is added to the token embedding: $x'_m = x_m + PE(m)$, where ($PE(m)$) is a parameter learned via backpropagation.
- **Strengths**:
  - Adaptable to specific tasks, potentially improving performance on datasets with unique positional patterns.
  - Simple to implement within existing transformer architectures.
- **Weaknesses**:
  - Requires storing one vector per position, increasing parameter count (e.g., $O(n_{\max} \cdot d)$ for maximum sequence length $(n_{\max})$).
  - Poor generalization to sequences longer than the training context, as unseen positions lack trained embeddings.
  - Similar to sinusoidal embeddings, additive mixing can dilute semantic information.
- **Impact**: Popular in large-scale models like GPT-3, but the parameter overhead and extrapolation issues limited their scalability for very long sequences.

Why Include: These methods establish the baseline for positional encodings in transformers, highlighting the trade-offs that RoPE addresses, such as parameter efficiency and length generalization.

Relative Positional Encodings

As transformer models were scaled to handle longer sequences and more complex tasks, researchers recognized the importance of explicitly modeling relative positions (i.e., distances between tokens) rather than relying solely on absolute positions. This led to the development of relative positional encodings, which focus on pairwise token relationships.

Transformer-XL Relative Embeddings

Introduced in Dai et al. (2019), Transformer-XL[3] extended transformers for long-term dependencies by incorporating relative positional encodings into the attention mechanism. Instead of adding positional embeddings to token embeddings, it modifies the attention score computation: $\text{Score}(q_m, k_n) = q_m^T k_n + q_m^T R_{m-n}$ where (R_{m-n}) is a learned relative positional embedding based on the distance $(m - n)$.

- **Mechanism**: The relative embedding (R_{m-n}) is a trainable vector for each possible relative distance, shared across positions. This allows the model to focus on pairwise relationships.
- **Strengths**:
  - Better captures long-range dependencies by emphasizing relative distances.
  - Supports longer contexts than absolute embeddings.
- **Weaknesses**:
  - Requires recomputing attention scores during inference, as relative distances shift with each new token, reducing caching efficiency.
  - Increased parameter count for storing relative embeddings.
- **Impact**: Effective for tasks requiring long-term dependencies (e.g., document-level modeling), but computational overhead spurred further innovations.

T5 Relative Bias

The T5 model (Raffel et al., 2020) introduced a relative positional bias approach, adding a learned scalar bias to the attention scores based on relative distances: $\text{Score}(q_m, k_n) = q_m^T k_n + b_{m-n}$, where b_{m-n} is a learned bias for the relative position $(m - n)$, typically bucketed to limit the number of parameters (e.g., biases for distances $-k, \ldots, k$).

[3] Dai, Z., Yang, Z., Yang, Y., Carbonell, J., Le, Q. V., & Salakhutdinov, R. (2019). "Transformer-XL: Attentive Language Models Beyond a Fixed-Length Context."

- **Mechanism**: Biases are shared across all positions with the same relative distance, added directly to the query-key dot product before softmax.
- **Strengths**:
  - Excellent length extrapolation, as biases are distance-based and independent of absolute position.
  - Fewer parameters than learned absolute embeddings, using a fixed set of biases.
- **Weaknesses**:
  - Inefficient during autoregressive inference, as shifting relative distances prevent caching of attention scores.
  - Lacks explicit absolute positional information, relying on the model to infer it indirectly.
- **Impact**: Widely used in models like T5 and BART, but inference inefficiencies motivated alternatives like RoPE.

Why Include: Relative encodings highlight the shift toward modeling pairwise relationships, setting the stage for RoPE's hybrid approach that combines absolute and relative information efficiently.

Emergence of Rotary Positional Embeddings (RoPE)

The limitations of absolute and relative positional encodings—poor length extrapolation, parameter overhead, semantic interference, and inference inefficiencies—prompted the development of RoPE in the 2021 paper "RoFormer: Enhanced Transformer with Rotary Position Embedding" by Su et al. RoPE encodes absolute positions using rotation matrices while ensuring attention scores depend on relative distances, unifying the strengths of prior methods.

- **Mechanism**: For a token embedding ($x_m \in \mathbb{R}^d$) at position (m), RoPE applies a rotation matrix: $R_m = \left(\cos(m\theta_0) \quad -\sin(m\theta_0) \quad 0 \quad \cdots \quad \sin(m\theta_0) \quad \cos(m\theta_0) \quad 0 \quad \cdots \quad 0 \quad 0 \quad \cos(m\theta_1) \quad \cdots \quad \vdots \quad \vdots \quad \vdots \quad \ddots \right)$, where ($\theta_i = 10000^{-2i/d}$).

- The attention score becomes: $q_m^T k_n = (W_q x_m)^T R_{m-n}(W_k x_n)$. This depends only on the relative distance $(m - n)$, combining absolute encoding (via rotations at position (m)) with relative dependency in attention.
- **Strengths**:
  - Parameter-free, using fixed rotations.
  - Excellent length extrapolation, supporting contexts up to 128k tokens (e.g., in LLaMA-3).
  - Norm-preserving, avoiding semantic distortion.
  - Efficient caching during inference, as rotated keys and queries can be reused.
  - Natural decay of attention scores with distance, aligning with linguistic priors.
- **Weaknesses**:
  - Requires careful frequency selection to balance coarse and fine positional scales.
  - Less flexible than learned embeddings for task-specific adaptation.
- **Impact**: RoPE has become a standard in modern LLMs (e.g., LLaMA, Mistral, Gemma) due to its efficiency, scalability, and performance, marking a significant leap in positional encoding design.

Summary of Traditional Positional Embedding Approaches

Before RoPE, two primary categories dominated: absolute and relative positional embeddings. Understanding these provides context for RoPE's innovations.

Absolute Positional Embeddings

Absolute embeddings assign a unique vector to each sequence position, added to the token embedding.

- **Sinusoidal Positional Embeddings**: Used in the original transformer, these are fixed, non-learnable embeddings defined by sine and cosine functions. For position m and embedding dimension d, the embedding is:

$$PE(m, 2i) = \sin\left(\frac{m}{10000^{2i/d}}\right)$$

$$PE(m, 2i+1) = \cos\left(\frac{m}{10000^{2i/d}}\right)$$

  where $i \in 0, 1, \ldots, \lfloor d/2 \rfloor - 1$. This creates unique, periodic patterns, enabling the model to infer relative positions via linear combinations. However, extrapolation to longer sequences is poor, as patterns don't generalize beyond training lengths.

Learned Absolute Embeddings: These treat positional vectors as trainable parameters, as in GPT-3. While adaptable, they increase parameter count and struggle with length extrapolation.

- **Advantages**: Simple, effective for absolute position encoding in fixed-length sequences.
- **Disadvantages**: Poor generalization to longer sequences, semantic interference from additive mixing, no explicit relative position encoding.

Relative Positional Embeddings

Relative embeddings focus on pairwise token distances, often integrated into the attention mechanism.

T5-Style Relative Bias: Adds a learned bias to the attention score based on relative offsets (e.g., $+1, -1$), as in T5. Biases are shared across positions, supporting length extrapolation.

- **Advantages**: Strong relative position encoding, good extrapolation.
- **Disadvantages**: Inefficient key/value caching in autoregressive inference (relative positions shift), no direct absolute position encoding.

Absolute vs. Relative Positional Embeddings

Aspect	Absolute (Sinusoidal/Learned)	Relative (e.g., T5 Bias)
Position Encoding	Unique vector per position	Bias per relative distance
Integration	Added to embeddings	Added to attention scores
Length Extrapolation	Poor	Good
Efficiency	High (cacheable)	Lower (recomputation)
Absolute/Relative Focus	Strong absolute, implicit relative	Strong relative, no absolute

Mathematical Formulation of RoPE

The core challenge in transformer self-attention is its permutation invariance, which ignores token order. To incorporate positional information, RoPE aims to define position-dependent transformations for query and key vectors such that their inner product in the attention mechanism depends only on the token embeddings and their relative positional distance. Formally, for a token embedding $x_m \in \mathbb{R}^d$ at position m, and query/key projections $W_q x_m$ and $W_k x_n$, RoPE seeks functions f_q and f_k such that:

$$\langle f_q(x_m, m), f_k(x_n, n) \rangle = g(x_m, x_n, m - n)$$

Here, g is a function that depends solely on the token embeddings x_m, x_n, and the relative distance $m - n$, ensuring that attention scores reflect relative positional relationships without explicit dependence on absolute positions m or n. This formulation addresses limitations of traditional methods like sinusoidal embeddings, which struggle with length extrapolation, and relative bias methods (e.g., T5), which incur inference inefficiencies.

2D Intuition and Derivation

RoPE encodes positional information through rotation rather than addition. Instead of adding position vectors to token embeddings, RoPE *rotates* them—a subtle shift with profound consequences.

The key insight is to treat pairs of embedding dimensions as coordinates in a 2D plane, applying rotations whose angles grow proportionally with position. This preserves embedding magnitudes while altering orientations, allowing attention scores to naturally reflect relative positional relationships.

Starting in Two Dimensions

For a token embedding $\mathbf{x}$ at position m, RoPE applies a rotation matrix with angle $m\theta$, where θ is a fixed frequency:

$$R_m = \left(\cos(m\theta) \quad -\sin(m\theta) \;\; \sin(m\theta) \quad \cos(m\theta)\right)$$

The position-encoded embedding becomes:

$$\mathbf{x}' = R_m \mathbf{x}$$

Geometrically, as position m increases, the vector traces a spiral path with linearly growing rotation angle.

For attention between a query at position m and key at position n:

$$\mathbf{q}'^{\top}\mathbf{k}' = (R_m\mathbf{q})^{\top}(R_n\mathbf{k}) = \mathbf{q}^{\top}R_m^{\top}R_n\mathbf{k}$$

Since rotation matrices satisfy $R_m^{\top} = R_{-m}$,

$$\mathbf{q}'^{\top}\mathbf{k}' = \mathbf{q}^{\top}R_{n-m}\mathbf{k}$$

The score depends only on relative distance $(n - m)$—closer tokens have more aligned vectors and higher dot products.

The Complex Number Perspective

An elegant alternative uses complex numbers. Position-encoded query and key at positions m and n:

$$q' = q \cdot e^{im\theta}, \quad k' = k \cdot e^{in\theta}$$

The attention score:

$$\text{score} = \text{Re}(q' \cdot \bar{k}') = \text{Re}(q\bar{k} \cdot e^{i(m-n)\theta})$$

The phase difference $e^{i(m-n)\theta}$ makes the relative-distance dependency immediately transparent.

Generalizing to High Dimensions

Real embeddings have dimensions like 512 or 1024. RoPE generalizes by applying independent 2D rotation blocks across $d/2$ dimension pairs, each with frequency:

$$\theta_i = \text{base}^{-2i/d}$$

The full rotation matrix is block-diagonal:

$$R_m = \begin{pmatrix} R_m^{(0)} & \ddots & R_m^{(d/2-1)} \end{pmatrix}$$

where each block is:

$$R_m^{(i)} = \begin{pmatrix} \cos(m\theta_i) & -\sin(m\theta_i) \ \sin(m\theta_i) & \cos(m\theta_i) \end{pmatrix}$$

Lower frequencies capture coarse positional relationships (“roughly in the first half”); higher frequencies capture fine-grained details (“exactly 3 positions after”).

Attention Decay and Frequency Selection

Attention scores include terms like $\cos((m-n)\theta)$: small distances keep the cosine near 1 (strong attention), while large distances cause oscillation and decay (weak attention). This matches linguistic intuition—nearby words typically have stronger dependencies.

The frequency schedule follows:

$$\theta_i = 10000^{-2i/d}$$

Lower dimensions rotate slowly (coarse encoding); higher dimensions rotate quickly (fine-grained details).

Efficient Implementation

Rather than computing the full block-diagonal matrix, rotations are applied per dimension pair:

$$\left(x'2i \ x'2i+1\right) = \left(\cos(m\theta_i) \quad -\sin(m\theta_i) \ \sin(m\theta_i) \quad \cos(m\theta_i)\right) \left(x_{2i} \ x_{2i+1}\right)$$

This reduces complexity to $O(d)$ per token. Precomputing $\cos(m\theta_i)$ and $\sin(m\theta_i)$ for all positions further optimizes inference.

Key Properties

1. **Relative Position Dependency**: Attention depends only on $(m-n)$, enabling position-equivariant pattern learning.
2. **Length Extrapolation**: Continuous rotations generalize to sequences longer than training data.
3. **Norm Preservation**: Orthogonal rotation matrices preserve embedding magnitudes and semantic information.
4. **Natural Attention Decay**: Vector misalignment at larger distances aligns with linguistic locality priors.

No Additional Parameters: Fixed rotations require no trainable parameters.

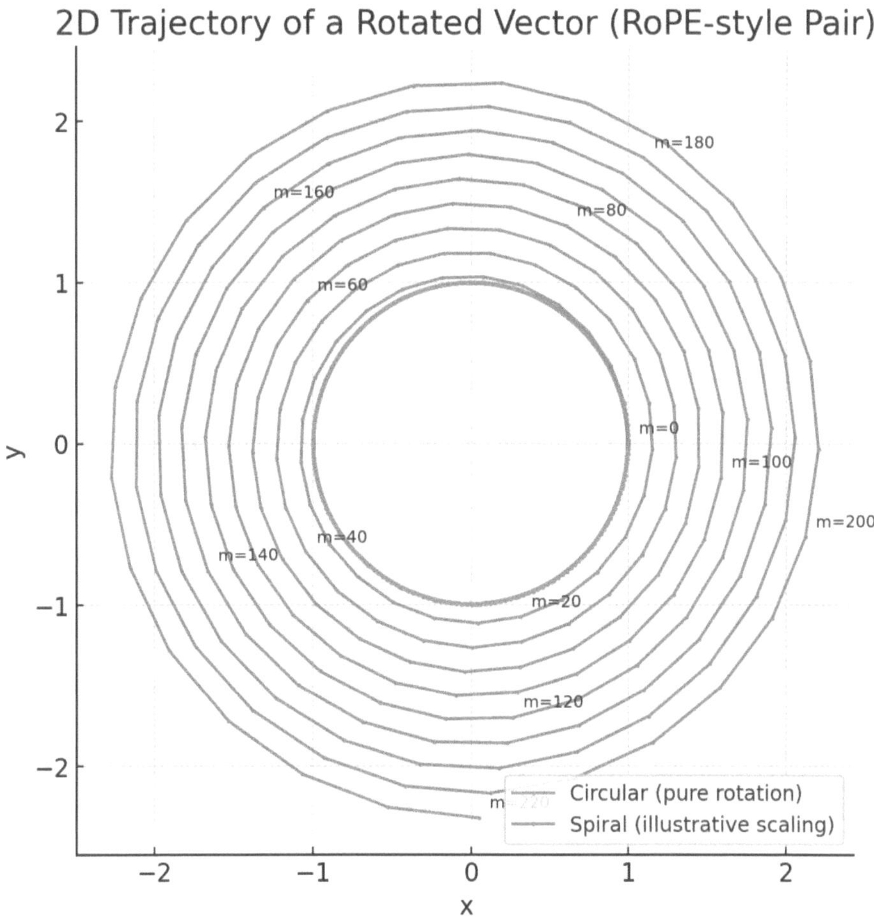

Figure 5-2. *2D Trajectory of a Rotated Vector*

Attention Matrix Heatmap

A heatmap of attention scores for a sequence is generated, where the score between positions m and n is dominated by $\cos((m - n)\theta_i)$ terms. This shows higher values near the diagonal (small $|m - n|$) and decay off-diagonal, reflecting RoPE's bias toward local dependencies.

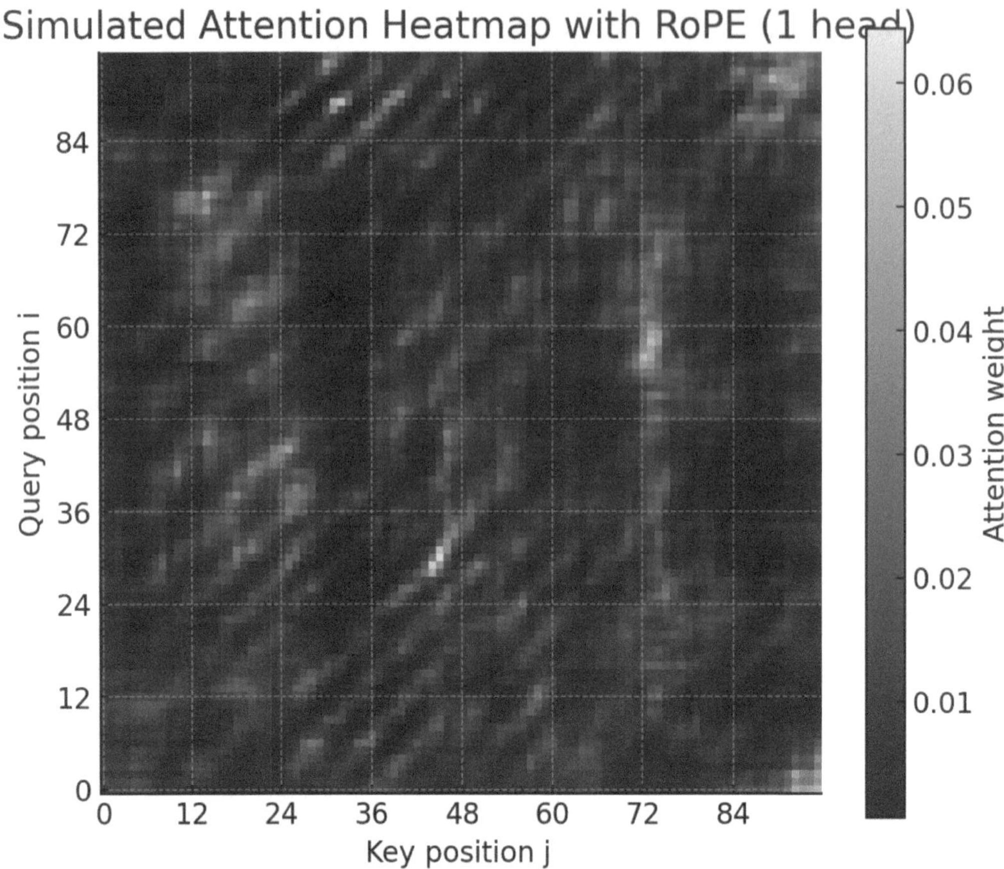

Figure 5-3. Simulated Attention Heatmap

High-Dimensional Projection: Used dimensionality reduction (e.g., PCA) to project rotated embeddings into 2D or 3D space, showing how tokens at different positions form distinct clusters or spirals. This can highlight how RoPE organizes embeddings in a geometrically meaningful way.

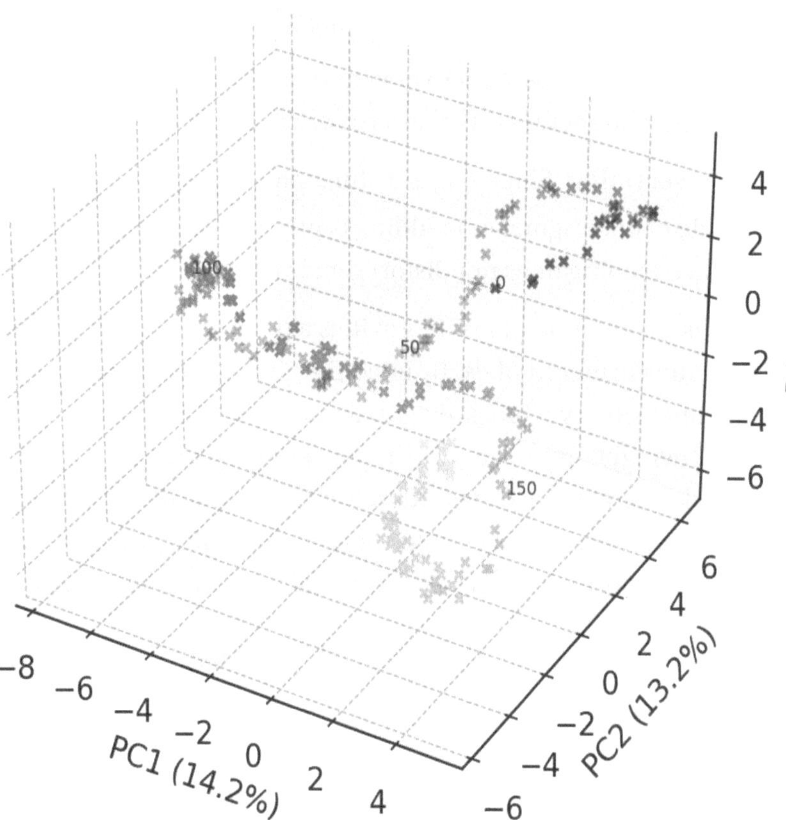

Figure 5-4. PCA of Q with RoPE

Connection to Linguistic Interpretability

The geometric nature of RoPE enhances its interpretability. The rotation-based encoding creates a structured embedding space where:

- **Proximity Reflects Similarity**: Tokens at nearby positions have similar rotation angles, leading to higher dot products and stronger attention scores, mirroring syntactic and semantic locality in language.

- **Multiscale Representation**: Different frequencies θ_i https://www.codecogs.com/eqnedit.php?latex=%20%5Ctheta_i%20#0 allow RoPE to capture both local (high θ_i https://www.codecogs.com/eqnedit.php?latex=%20%5Ctheta_i%20#0) and global (low θ_i) positional relationships, akin to how humans process short-term syntactic dependencies and long-term discourse structure.
- **Norm Preservation**: Since R_m is orthogonal ($R_m R_m^T = I$), the magnitude of the original embedding is preserved, ensuring that positional encoding does not distort semantic content.

The geometric interpretation of RoPE provides a powerful lens for understanding its effectiveness in transformer models. By encoding positions as rotations in 2D planes, extended to high dimensions via block-diagonal matrices, RoPE creates a structured embedding space where relative distances drive attention scores.

Visualizations like spiral trajectories, attention heatmaps, and high-dimensional projections make these concepts accessible, highlighting RoPE's alignment with linguistic priors and its role in enabling efficient, scalable, and interpretable positional encoding in modern LLMs.

Advantages of RoPE

RoPE offers significant improvements over traditional methods:

1. **Hybrid Encoding**: Encodes absolute positions while computing relative distances in attention, blending both paradigms.
2. **Length Extrapolation**: Generalizes to sequences far beyond training lengths (e.g., 128k tokens in LLaMA), with smooth performance degradation.
3. **Decaying Dependency**: Attention scores decrease with larger relative distances, aligning with linguistic priors (nearby tokens are more relevant).
4. **Efficiency**: No additional parameters, preserves vector norms, supports linear attention, and enables key/value caching.

5. **Stability**: Faster convergence (20–30% improvement) and lower validation loss in experiments.

6. **Interpretability**: Rotations create intuitive geometric patterns, with closer positions aligning better.

Advantage	RoPE vs. Absolute	RoPE vs. Relative
Length Generalization	Superior	Comparable
Relative Focus	Explicit	Explicit
Absolute Info	Encoded	Implicit/None
Inference Efficiency	High	Higher
Parameter Overhead	None	None

RoPE outperforms baselines on tasks like translation (slight BLEU gains) and GLUE benchmarks.

Applications in Modern Models

RoPE is integral to many state-of-the-art LLMs:

- **LLaMA Family (Meta)**: LLaMA-2 and LLaMA-3 use RoPE for 128k-token contexts.
- **Mistral AI**: Mistral-7B and Mixtral-8x7B combine RoPE with sliding window attention.
- **PaLM and Gemma (Google)**: Leverage RoPE for relative positioning and optimization.
- **CodeGen**: Uses RoPE for code generation.
- **Vision-Language Models**: RoPE extends to multimodal tasks, interleaving audio/visual tokens.

Variants and Extensions

- **YaRN**[4]: Enhances RoPE for extreme context lengths (128k+ tokens).
- **N-Dimensional RoPE**: Generalizes to 3D/4D for video or non-sequential data.
- **Interleaved RoPE**: Handles multimodal token sequences.
- **CRATE**: Interprets RoPE via coding theory for sparse manifolds.

Why Rotary Positional Embeddings Are Essential to Large Language Models

Solving Permutation Invariance

Self-attention is inherently permutation-invariant—it doesn't naturally account for token order. But in language, order is everything: "Alice loves Bob" means something quite different from "Bob loves Alice."

Without positional information, models can't distinguish these sequences. RoPE addresses this by encoding positions directly into embeddings through rotation matrices. For a token at position m, RoPE applies rotation matrix R_m to produce $\mathbf{x}' = R_m\mathbf{x}$. The attention computation becomes:

$$\text{Attention}(q, k) = (R_m q)^\top (R_n k) = q^\top R_{n-m} k$$

This encodes absolute positions while ensuring attention scores depend on relative distances—both properties essential for language understanding.

Superior Length Extrapolation

LLMs increasingly process sequences far exceeding training lengths (e.g., 128k tokens). Traditional sinusoidal embeddings struggle here:

$$PE_{(pos,2i)} = \sin(pos/10000^{2i/d}), \quad PE_{(pos,2i+1)} = \cos(pos/10000^{2i/d})$$

[4] YaRN: Yet another RoPE extensioN method

These fixed additive patterns don't extend gracefully to unseen lengths. RoPE's rotation-based approach, using frequencies $\theta_i = 10000^{-2i/d}$:

$$R_m^{(i)} = \begin{pmatrix} \cos(m\theta_i) & -\sin(m\theta_i) & \sin(m\theta_i) & \cos(m\theta_i) \end{pmatrix}$$

maintains consistent relative relationships regardless of sequence length. Models like LLaMA-2 and Mistral-7B demonstrate stable perplexity even at extended contexts.

Balancing Absolute and Relative Information

Absolute embeddings encode exact positions but struggle with relative dependencies. Relative embeddings (e.g., T5's relative bias) capture distances but lack absolute context and require costly recomputation during inference.

RoPE bridges this gap: rotations encode absolute positions explicitly, while the attention mechanism implicitly captures relative distances through R_{n-m}. This hybrid approach benefits tasks like machine translation (where word order and proximity both matter) and dialogue systems (where turn-taking relies on positional cues).

Efficiency and Scalability

RoPE introduces no trainable parameters—unlike learned embeddings requiring one vector per position. The fixed rotation matrices based on θ_i are computed once and cached. During autoregressive inference, rotations for new tokens at position m are computed incrementally without reprocessing prior positions. RoPE's compatibility with linear attention mechanisms (Performer, Linformer) further enhances scalability for billion-parameter models.

Natural Attention Decay

Linguistic intuition suggests nearby tokens matter more than distant ones. RoPE naturally produces this pattern: the dot product between rotated query and key vectors decreases as relative distance $|m - n|$ increases, prioritizing local context for coherent generation.

Training Benefits

Empirical studies show RoPE enables 20-30% faster convergence and lower validation loss compared to alternatives. On benchmarks like OpenWebText2, RoPE-based models achieve notably lower perplexity (e.g., 3.5 vs. 4.0)—significant savings when training large-scale models.

Industry Adoption

RoPE has become standard in leading LLMs:

- **LLaMA Family (Meta)**: LLaMA-2 and LLaMA-3 for efficient long-context processing
- **Mistral AI**: Mistral-7B and Mixtral-8x7B with sliding window attention
- **Gemma (Google)**: Combined with RMSNorm for robust generation
- **CodeGen**: Applied to code generation where sequence order is critical

RoPE Embeddings Implementation

Let's dive into the provided PyTorch code for applying rotary positional embeddings (RoPE) and explain its functionality step-by-step, focusing on its purpose, inputs, operations, and output.

The code snippets below are a concise implementation of RoPE's core operation, which rotates pairs of embedding dimensions to encode positional information in transformer models. I'll break down each component, explain the underlying mathematics, and highlight the role of safety features like dtype handling.

```
# --- RoPE with assert + dtype safety ---
def _apply_rotary_emb(x: torch.Tensor, cos: torch.Tensor, sin: torch.
Tensor) -> torch.Tensor:
    # x: [T, *, D]; D must be even
    x1, x2 = torch.chunk(x, 2, dim=-1)
    cos = cos.unsqueeze(-2).to(x.dtype)
    sin = sin.unsqueeze(-2).to(x.dtype)
    return torch.cat((x1 * cos - x2 * sin, x2 * cos + x1 * sin), dim=-1)
```

The function `_apply_rotary_emb` applies RoPE to a tensor of token embeddings using precomputed cosine and sine values to perform rotations. It is designed for efficiency and includes safety features like dtype casting to ensure numerical stability.

RoPE rotates each token's embedding vector based on its position m in the sequence, using a block-diagonal rotation matrix to encode positional information. This ensures that the self-attention mechanism in transformers captures relative positional dependencies, as the attention score $q_m^T k_n$ depends on the relative distance $m - n$. The function is typically applied to query and key vectors before computing attention scores.

Function Signature

- **Inputs:**
    - `x: torch.Tensor`: Input tensor of shape $[T, *, D]$, where:
        - T `https://www.codecogs.com/eqnedit.php?latex=%20T%20#0` is the sequence length (number of tokens).
        - $*$ represents optional dimensions (e.g., batch size, number of attention heads).
        - D `https://www.codecogs.com/eqnedit.php?latex=%20D%20#0` is the embedding dimension, which must be even to allow pairing of dimensions.
    - `cos: torch.Tensor`: Precomputed cosine values, typically of shape $[T, D/2]$, representing $\cos(m\theta_i)$ for positions $m = 0, \ldots, T-1$ and frequencies θ_i.
    - `sin: torch.Tensor`: Precomputed sine values, matching `cos` in shape, representing $\sin(m\theta_i)$.
- **Output**: A tensor of shape $[T, *, D]$, with RoPE applied to each embedding vector.

Step-by-Step Explanation

Let's break down each line of the code, connecting it to the mathematical formulation of RoPE and explaining its role in the positional encoding process.

1. Splitting the Embedding Tensor

```
x1, x2 = torch.chunk(x, 2, dim=-1)
```

- **Operation**: The input tensor x of shape $[T, *, D]$ is split along the last dimension (embedding dimension D) into two equal parts, x1 and x2, each with shape $[T, *, D/2]$.
- **Why**: RoPE pairs consecutive dimensions of the embedding vector to form 2D planes for rotation. For a token embedding at position m https://www.codecogs.com/eqnedit.php?latex=%20m%20#0, denoted $x_m = (x_0, x_1, \ldots, x_{D-1}) \in \mathbb{R}^D$, the dimensions are grouped as $(x_0, x_1), (x_2, x_3), \ldots, (x_{D-2}, x_{D-1})$. The split creates:
  - x1: Even-indexed dimensions $(x_0, x_2, \ldots, x_{D-2})$.
  - x2: Odd-indexed dimensions $(x_1, x_3, \ldots, x_{D-1})$.
- **Safety Note**: The code assumes D is even, as required by `torch.chunk(x, 2, dim=-1)`. The comment "D must be even" suggests an implicit requirement, and in a robust implementation, an assertion like `assert x.shape[-1] % 2 == 0` could be added to enforce this.

2. Preparing Cosine and Sine Tensors

```
cos = cos.unsqueeze(-2).to(x.dtype)
sin = sin.unsqueeze(-2).to(x.dtype)
```

- **Operation:**
  - `unsqueeze(-2)`: Adds a singleton dimension to the cos and sin tensors at the second-to-last axis. If cos and sin have shape $[T, D/2]$, this changes them to $[T, 1, D/2]$, enabling broadcasting over the optional dimensions $(*)$ in x.
  - `to(x.dtype)`: Converts cos and sin to the same data type as x (e.g., `torch.float32`, `torch.float16`), ensuring numerical compatibility.

- **Why:**
    - **Broadcasting**: The `unsqueeze(-2)` ensures that `cos` and `sin` can be applied element-wise to `x1` and `x2` across batch and head dimensions. For example, if `x` is $[T, B, H, D]$ (batch size B, heads H), `cos` broadcasts over B and H.
    - **Dtype Safety**: Matching the dtype prevents precision errors, especially in mixed-precision training (e.g., float16 on GPUs), where mismatches can cause numerical instability or incorrect results.

3. Applying Rotations

```
return torch.cat((x1 * cos - x2 * sin, x2 * cos + x1 * sin), dim=-1)
```

- **Operation:**
    - Compute the rotated components:
        - `x1 * cos - x2 * sin`: Calculates the new even-indexed dimensions.
        - `x2 * cos + x1 * sin`: Calculates the new odd-indexed dimensions.
    - `torch.cat(..., dim=-1)`: Concatenates the two resulting tensors along the last dimension, producing an output tensor of shape $[T, *, D]$, matching the input `x`.
- **Why**: This implements the 2D rotation for each pair of dimensions. For a pair (x_{2i}, x_{2i+1}) at position m, the rotation is:

$$(x'2i \; x'2i+1) = (\cos(m\theta_i) \quad -\sin(m\theta_i) \;\; \sin(m\theta_i) \quad \cos(m\theta_i)) \; (x_{2i} \; x_{2i+1})$$

Expanding:

$x'2i = x2i\cos(m\theta_i) - x_{2i+1}\sin(m\theta_i)$,

$x'2i+1 = x2i\sin(m\theta_i) + x_{2i+1}\cos(m\theta_i)$. In the code, `x1` corresponds to x_{2i}, `x2` to x_{2i+1}, `cos` to $\cos(m\theta_i)$, and `sin` to $\sin(m\theta_i)$. These operations are applied element-wise across all pairs $i = 0, 1, \ldots, D/2 - 1$.

Output

The function returns a tensor of shape $[T, *, D]$, where each embedding vector has been rotated according to its position m. This output is typically used as the query or key vectors in a transformer's self-attention mechanism, enabling position-aware attention computations.

Key Features and Safety Considerations

- **Efficiency**: The implementation avoids explicit matrix multiplication, using element-wise operations for $O(D)$ complexity per token. Precomputing `cos` and `sin` externally reduces computational overhead, especially in autoregressive inference where positions increment sequentially.
- **Dtype Safety**: The `.to(x.dtype)` ensures compatibility in mixed-precision settings (e.g., float16 on GPUs), preventing numerical errors due to type mismatches.
- **Broadcasting**: The `unsqueeze(-2)` enables flexible handling of batched inputs or multihead attention, making the function adaptable to various transformer architectures (e.g., shapes like $[T, B, H, D]$).
- **Assumption**: The comment "D must be even" indicates that D must be even for `torch.chunk` to work. While not implemented here, an assertion like `assert x.shape[-1] % 2 == 0` could enhance robustness in production code.

Example Workflow

Consider an example:

- `x` has shape $[T = 100, B = 32, H = 8, D = 64]$ (100 tokens, 32 batches, 8 attention heads, 64 dimensions).
- `cos` and `sin` have shape $[100, 32]$, containing $\cos(m\theta_i)$ and $\sin(m\theta_i)$ for $m = 0, \ldots, 99$ and $i = 0, \ldots, 31$.

- `torch.chunk` splits `x` into `x1` and `x2`, each $[100, 32, 8, 32]$.
- `cos.unsqueeze(-2)` and `sin.unsqueeze(-2)` yield $[100, 1, 32]$, broadcasting over batch and head dimensions.
- The output is $[100, 32, 8, 64]$, with each embedding rotated to encode positional information.

Geometric Interpretation

RoPE treats each pair of dimensions (x_{2i}, x_{2i+1}) as a 2D vector, rotated by angle $m\theta_i$ to form a spiral trajectory in the embedding space as m increases. This ensures that tokens at nearby positions have similar rotations, leading to higher dot products in attention scores, while distant tokens are misaligned, causing decay:

$$q_m^T k_n \propto \sum_{i=0}^{D/2-1} \cos((m-n)\theta_i) \text{ terms}$$

This decay aligns with linguistic priors, where nearby tokens (e.g., adjacent words) are more relevant.

The `_apply_rotary_emb` function is an efficient and robust implementation of RoPE, applying position-dependent rotations to transformer embeddings using element-wise operations.

By splitting dimensions, broadcasting trigonometric values, and ensuring dtype compatibility, it enables transformers to capture relative positional dependencies effectively. This makes it a critical component in modern LLMs like LLaMA and Mistral, supporting scalable and interpretable positional encoding.

RotaryEmbedding Class Definition and Methods

```
class RotaryEmbedding(torch.nn.Module):
    def __init__(
        self,
        head_dim: int,
        base: int,
```

```
        dtype: torch.dtype = torch.float32,
        initial_context_length: int = 4096,
        scaling_factor: float = 1.0,
        ntk_alpha: float = 1.0,
        ntk_beta: float = 32.0,
        device: torch.device | None = None,
    ):
        super().__init__()
        if head_dim % 2 != 0:
            raise ValueError(f"RoPE requires even head_dim, got
            {head_dim}.")
        self.head_dim = head_dim
        self.base = base
        self.dtype = dtype
        self.initial_context_length = initial_context_length
        self.scaling_factor = scaling_factor
        self.ntk_alpha = ntk_alpha
        self.ntk_beta = ntk_beta
        self.device = device

    def _compute_concentration_and_inv_freq(self) -> tuple[float, torch.
    Tensor]:
        d = self.head_dim
        # freq grows geometrically; inv_freq is 1/freq
        freq = self.base ** (torch.arange(0, d, 2, dtype=self.dtype,
        device=self.device) / d)

        if self.scaling_factor > 1.0:
            concentration = 0.1 * math.log(self.scaling_factor) + 1.0

            low = (d / 2) * math.log(self.initial_context_length / (self.
            ntk_beta * 2 * math.pi)) / math.log(self.base)
            high = (d / 2) * math.log(self.initial_context_length / (self.
            ntk_alpha * 2 * math.pi)) / math.log(self.base)

            # If invalid, clamp to safe band rather than crashing
            low_clamped = max(0.0, min(low, d / 2 - 2))
```

```
            high_clamped = max(low_clamped + 1.0, min(high, d / 2 - 1))

            ramp = (torch.arange(d // 2, device=self.device, dtype=self.
            dtype) - low_clamped) / (high_clamped - low_clamped)
            mask = 1 - ramp.clamp(0, 1)

            extrapolation = 1.0 / freq
            interpolation = 1.0 / (self.scaling_factor * freq)
            inv_freq = extrapolation * mask + interpolation * (1 - mask)
        else:
            concentration = 1.0
            inv_freq = 1.0 / freq

        return concentration, inv_freq

    def _compute_cos_sin(self, seq_len: int) -> tuple[torch.Tensor, torch.
    Tensor]:
        concentration, inv_freq = self._compute_concentration_and_inv_freq()
        positions = torch.arange(seq_len, dtype=self.dtype,
        device=self.device)
        freqs = torch.einsum("i,j->ij", positions, inv_freq)  # [T, D/2]
        cos = freqs.cos() * concentration
        sin = freqs.sin() * concentration
        return cos, sin

    def forward(self, q: torch.Tensor, k: torch.Tensor) -> tuple[torch.
    Tensor, torch.Tensor]:
        """
        q: [B,T,kv_heads,q_per_kv,D] or [T,kv_heads,q_per_kv,D]
        k: [B,T,kv_heads,D]          or [T,kv_heads,D]
        """
        if self.head_dim % 2 != 0:
            raise ValueError(f"RoPE requires even head_dim, got {self.
            head_dim}.")

        seq_len = q.size(-3)
```

```
        # Build cos/sin and move them to the SAME device as q
        cos, sin = self._compute_cos_sin(seq_len)
        cos = cos.to(q.device)
        sin = sin.to(q.device)

        q_shape = q.shape
        k_shape = k.shape

        # Flatten all but last dim to apply rotary per head vector
        q = q.reshape(seq_len, -1, self.head_dim)
        k = k.reshape(seq_len, -1, self.head_dim)

        q = _apply_rotary_emb(q, cos, sin).reshape(q_shape)
        k = _apply_rotary_emb(k, cos, sin).reshape(k_shape)
        return q, k
```

The `RotaryEmbedding` class is a PyTorch `nn.Module` that computes and applies RoPE to query and key tensors in a transformer's self-attention mechanism. It precomputes cosine and sine values for rotation angles based on position and frequency, supports dynamic context length scaling, and ensures numerical stability with dtype and device handling. The class includes methods to compute frequencies, generate trigonometric values, and apply rotations, leveraging the `_apply_rotary_emb` function from the previous discussion.

Class Structure

The class consists of:

- An `__init__` method to initialize parameters and validate inputs.
- A helper method `_compute_concentration_and_inv_freq` to calculate frequency adjustments for scaling.
- A helper method `_compute_cos_sin` to generate cosine and sine tensors.
- A `forward` method to apply RoPE to query and key tensors.

Detailed Explanation

Let's dissect each component of the code, connecting it to the mathematical formulation of RoPE and highlighting its purpose.

1. Initialization (`__init__`)

```
def __init__(
    self,
    head_dim: int,
    base: int,
    dtype: torch.dtype = torch.float32,
    initial_context_length: int = 4096,
    scaling_factor: float = 1.0,
    ntk_alpha: float = 1.0,
    ntk_beta: float = 32.0,
    device: torch.device | None = None,
):
    super().__init__()
    if head_dim % 2 != 0:
        raise ValueError(f"RoPE requires even head_dim, got {head_dim}.")
    self.head_dim = head_dim
    self.base = base
    self.dtype = dtype
    self.initial_context_length = initial_context_length
    self.scaling_factor = scaling_factor
    self.ntk_alpha = ntk_alpha
    self.ntk_beta = ntk_beta
    self.device = device
```

- **Purpose**: Initializes the RoPE module with parameters controlling embedding dimension, frequency base, data type, and context length scaling.

- **Parameters**:
    - head_dim: The dimension of each attention head (D https://www.codecogs.com/eqnedit.php?latex=%20D%20#0), must be even for pairing dimensions in 2D rotations.
    - base: The base for frequency calculation, typically 10000, as in:

$$\theta_i = \text{base}^{-2i/D}$$

    - dtype: Data type for computations (e.g., torch.float32 or torch.float16).
    - initial_context_length: The default sequence length for precomputing frequencies (e.g., 4096 tokens).
    - scaling_factor: Scales frequencies for context length extension, used in extrapolation.
    - ntk_alpha, ntk_beta: Parameters for Neuro Tangent Kernel (NTK) scaling, adjusting frequency interpolation for long sequences.
    - device: The device (e.g., CPU, GPU) for tensor operations.
- **Operations**:
    - Validates that head_dim is even, as RoPE requires pairing dimensions:

$$D \mod 2 = 0$$

    - Stores parameters as instance variables for use in other methods.

2. Computing Concentration and Inverse Frequencies (_compute_concentration_and_inv_freq)

```
def _compute_concentration_and_inv_freq(self) -> tuple[float, torch.
Tensor]:
    d = self.head_dim
    freq = self.base ** (torch.arange(0, d, 2, dtype=self.dtype,
    device=self.device) / d)
```

```
if self.scaling_factor > 1.0:
    concentration = 0.1 * math.log(self.scaling_factor) + 1.0
    low = (d / 2) * math.log(self.initial_context_length / (self.ntk_
    beta * 2 * math.pi)) / math.log(self.base)
    high = (d / 2) * math.log(self.initial_context_length / (self.ntk_
    alpha * 2 * math.pi)) / math.log(self.base)
    low_clamped = max(0.0, min(low, d / 2 - 2))
    high_clamped = max(low_clamped + 1.0, min(high, d / 2 - 1))
    ramp = (torch.arange(d // 2, device=self.device, dtype=self.
    dtype) - low_clamped) / (high_clamped - low_clamped)
    mask = 1 - ramp.clamp(0, 1)
    extrapolation = 1.0 / freq
    interpolation = 1.0 / (self.scaling_factor * freq)
    inv_freq = extrapolation * mask + interpolation * (1 - mask)
else:
    concentration = 1.0
    inv_freq = 1.0 / freq
return concentration, inv_freq
```

- **Purpose**: Computes the concentration factor and inverse frequencies $1/\theta_i$ for rotation angles, adjusting for context length scaling.
- **Operations**:
    - **Base Frequencies**: Computes frequencies:

      $$\theta_i = \text{base}^{-2i/D} = \text{base}^{\text{arange}(0,D,2)/D}$$

      where `freq` is θ_i for $i = 0, 1, \ldots, D/2 - 1$.
    - **Scaling Adjustment**: If `scaling_factor > 1.0`, adjusts frequencies for longer contexts:
        - `concentration`: A scalar adjusting the magnitude of cosine and sine terms:

          $$\text{concentration} = 0.1 \ln(\text{scaling}_f\text{actor}) + 1.0$$

- `low` and `high`: Define a frequency range for NTK scaling, based on the initial context length and NTK parameters:

$$\text{low} = \frac{D}{2} \cdot \frac{\ln(\text{initial}_context_length/(\text{ntk}_beta \cdot 2\pi))}{\ln(\text{base})}$$

$$\text{high} = \frac{D}{2} \cdot \frac{\ln(\text{initial}_context_length/(\text{ntk}_alpha \cdot 2\pi))}{\ln(\text{base})}$$

- `low_clamped` and `high_clamped`: Clamp to ensure valid indices, preventing numerical issues.
- `ramp` and `mask`: Create a smooth transition between extrapolation (original frequencies) and interpolation (scaled frequencies):

$$\text{ramp} = \frac{\text{arange}(D/2) - \text{low}_clamped}{\text{high}_clamped - \text{low}_clamped}$$

$$\text{mask} = 1 - \text{ramp.clamp}(0, 1)$$

- `inv_freq`: Blends extrapolated and interpolated inverse frequencies:

$$\text{inv}_freq = \text{extrapolation} \cdot \text{mask} + \text{interpolation} \cdot (1 - \text{mask})$$

where $\text{extrapolation} = 1/\text{freq}, \text{interpolation} = 1/(\text{scaling}_factor \cdot \text{freq})$.

- **Default Case**: If `scaling_factor <= 1.0`, uses standard inverse frequencies:

$$\text{inv}_freq = 1/\text{freq} = \text{base}^{2i/D}$$

- **Why**: Adjusts frequencies to support longer sequences than the initial context length, inspired by NTK-aware scaling (e.g., used in models like LLaMA). The `concentration` factor modulates rotation magnitudes for stability.

3. Computing Cosine and Sine Tensors (_compute_cos_sin)

```
def _compute_cos_sin(self, seq_len: int) -> tuple[torch.Tensor, torch.
Tensor]:
    concentration, inv_freq = self._compute_concentration_and_inv_freq()
    positions = torch.arange(seq_len, dtype=self.dtype, device=self.device)
    freqs = torch.einsum("i,j->ij", positions, inv_freq)  # [T, D/2]
    cos = freqs.cos() * concentration
    sin = freqs.sin() * concentration
    return cos, sin
```

- **Purpose**: Generates cosine and sine tensors for rotation angles based on positions and frequencies.
- **Operations**:
  - Calls `_compute_concentration_and_inv_freq` to get `concentration` and `inv_freq`.
  - Creates a tensor of positions $m = 0,1,\ldots,\mathrm{seq}_\mathrm{l}\mathrm{en}-1$.
  - Computes rotation angles using `einsum`:

    $$\mathrm{freqs}_{m,i} = m \cdot \mathrm{inv}_\mathrm{f}\mathrm{req}_i = m/\theta_i = m \cdot \mathrm{base}^{2i/D}$$

    resulting in a tensor of shape $[T, D/2]$
  - Computes:

    $$\cos(m\theta_i) = \cos(\mathrm{freqs}m, i) \cdot \mathrm{concentration}$$

    $$\sin(m\theta_i) = \sin(\mathrm{freqs}m, i) \cdot \mathrm{concentration}$$

- **Why**: Precomputes $\cos(m\theta_i)$ and $\sin(m\theta_i)$ for all positions and frequencies, enabling efficient application of RoPE in the `forward` method.

4. Forward Pass (forward)

```
def forward(self, q: torch.Tensor, k: torch.Tensor) -> tuple[torch.Tensor,
torch.Tensor]:
    """
    q: [B,T,kv_heads,q_per_kv,D] or [T,kv_heads,q_per_kv,D]
    k: [B,T,kv_heads,D]          or [T,kv_heads,D]
    """
    if self.head_dim % 2 != 0:
        raise ValueError(f"RoPE requires even head_dim, got {self.
        head_dim}.")
    seq_len = q.size(-3)
    cos, sin = self._compute_cos_sin(seq_len)
    cos = cos.to(q.device)
    sin = sin.to(q.device)
    q_shape = q.shape
    k_shape = k.shape
    q = q.reshape(seq_len, -1, self.head_dim)
    k = k.reshape(seq_len, -1, self.head_dim)
    q = _apply_rotary_emb(q, cos, sin).reshape(q_shape)
    k = _apply_rotary_emb(k, cos, sin).reshape(k_shape)
    return q, k
```

- **Purpose**: Applies RoPE to query (q) and key (k) tensors, returning rotated versions for use in self-attention.
- **Inputs**:
  - q: Query tensor, shape $[B, T, \mathrm{kv}_{\mathrm{h}}\mathrm{eads}, \mathrm{q}_{\mathrm{p}}\mathrm{er}_{\mathrm{k}}\mathrm{v}, D]$ or $[T, \mathrm{kv}_{\mathrm{h}}\mathrm{eads}, \mathrm{q}_{\mathrm{p}}\mathrm{er}_{\mathrm{k}}\mathrm{v}, D]$, where $\mathrm{kv}_{\mathrm{h}}\mathrm{eads}$ is the number of key-value heads, and $\mathrm{q}_{\mathrm{p}}\mathrm{er}_{\mathrm{k}}\mathrm{v}$ is queries per key-value head.
  - k: Key tensor, shape $[B, T, \mathrm{kv}_{\mathrm{h}}\mathrm{eads}, D]$ or $[T, \mathrm{kv}_{\mathrm{h}}\mathrm{eads}, D]$.
- **Operations**:
  - Validates `head_dim` is even.

- Extracts sequence length T `https://www.codecogs.com/eqnedit.php?latex=%20T%20#0` from `q.size(-3)`.
- Computes `cos` and `sin` tensors using `_compute_cos_sin`, then moves them to the same device as `q`.
- Reshapes `q` and `k` to $[T, -1, D]$ by flattening all dimensions except sequence length and head dimension, enabling application to each head vector.
- Applies `_apply_rotary_emb` to rotate `q` and `k`, then restores original shapes.

- **Why**: Rotates query and key vectors to encode positional information, ensuring attention scores depend on relative distances:

$$q_m^T k_n = (W_q x_m)^T R((n-m)\theta)(W_k x_n)$$

Key Features and Safety Considerations

- **Flexibility**: Supports batched inputs and multihead attention with shapes like $[B,T,\mathrm{kv}_\mathrm{h}\mathrm{eads},\mathrm{q}_\mathrm{p}\mathrm{er}_\mathrm{k}\mathrm{v},D]$.
- **Context Length Scaling**: The `scaling_factor`, `ntk_alpha`, and `ntk_beta` enable extrapolation to longer sequences, adjusting frequencies dynamically.
- **Safety**: Enforces even `head_dim`, ensures device and dtype compatibility, and clamps frequency ranges to prevent numerical issues.
- **Efficiency**: Precomputes trigonometric values and uses element-wise operations in `_apply_rotary_emb` for $O(D)$ complexity per token.

Example Workflow

Suppose:

- `head_dim = 64, base = 10000, initial_context_length = 4096, scaling_factor = 2.0.`
- `q` is $[32, 100, 8, 4, 64]$ (batch=32, seq_len=100, kv_heads=8, q_per_kv=4, head_dim=64).
- `k` is $[32, 100, 8, 64]$.
- `_compute_concentration_and_inv_freq` adjusts frequencies for scaling, producing `inv_freq` of shape $[32]$ and `concentration`.
- `_compute_cos_sin` generates `cos` and `sin` of shape $[100, 32]$.
- `forward` reshapes `q` to $[100, 256, 64]$ (256 = 8 * 4 heads), applies rotations, and restores the original shape.

Geometric Interpretation

RoPE rotates each pair of dimensions (x_{2i}, x_{2i+1}) by angle $m\theta_i$, forming spiral trajectories in 2D planes. The attention score decays with relative distance:

$$q_m^T k_n \propto \sum_{i=0}^{D/2-1} \cos((m-n)\theta_i) \text{ terms}$$

This aligns with linguistic priors, favoring nearby tokens.

The `RotaryEmbedding` class is a robust, flexible implementation of RoPE, supporting advanced features like context length scaling and NTK extrapolation. It efficiently computes and applies rotations to query and key tensors, ensuring transformers capture relative positional dependencies. Its safety features and compatibility with multihead attention make it suitable for modern LLMs like GPT OSS, LLaMA, and Mistral.

What Is NTK (Neural Tangent Kernel) in the Context of RoPE

The concept of NTK was introduced in the 2018 paper "Neural Tangent Kernel: Convergence and Generalization in Neural Networks" by Jacot et al. It describes how, in wide neural networks, the evolution of weights during training can be modeled as a kernel method, where the kernel (NTK) governs the dynamics of the network's output function. This framework was initially applied to understand deep learning optimization and generalization but later found applications in transformers, particularly for positional encodings.

In the context of RoPE, NTK-aware scaling emerged as a practical extension to address the challenge of length extrapolation. Traditional positional encodings, like sinusoidal embeddings, struggle with sequences longer than those seen during training due to fixed patterns or learned parameters that do not generalize.

Mathematical Foundation

Recall that RoPE encodes positions through block-diagonal rotation matrices with frequencies:

$$\theta_i = \text{base}^{-2i/d}$$

where $\text{base} = 10000$ and d is the embedding dimension. The attention score:

$$q_m^\top k_n = (W_q x_m)^\top R_{m-n}(W_k x_n)$$

depends on relative distance $(m - n)$, with $\cos((m - n)\theta_i)$ terms producing decay at larger distances.

The Extrapolation Problem

Without scaling, high-frequency rotations (large θ_i) oscillate rapidly, causing instability for positions m beyond the training context length. NTK scaling addresses this by slowing rotations for higher dimensions, ensuring smooth extrapolation.

NTK Scaling Mechanism

For a scaling factor $s > 1$, NTK scaling blends two approaches:

- **Extrapolation**: $\text{inv}_\text{f}\text{req}_i = 1/\theta_i$ (original frequencies)
- **Interpolation**: $\text{inv}_\text{f}\text{req}_i = 1/(s \cdot \theta_i)$ (scaled frequencies)

A ramp function creates smooth transitions based on frequency ranges:

$$\text{ramp}(i) = \frac{i - i_{\text{low}}}{i_{\text{high}} - i_{\text{low}}}$$

The final inverse frequency blends both:

$$\text{inv}_{\text{f}}\text{req}_i = (1 - \text{ramp}(i)) \cdot \frac{1}{\theta_i} + \text{ramp}(i) \cdot \frac{1}{s \cdot \theta_i}$$

This preserves coarse positional information in lower dimensions while adapting higher dimensions for longer sequences.

Additionally, a concentration factor scales cosine and sine terms to prevent numerical instability:

$$\text{mscale} = f(s, \alpha, \beta)$$

This modulates rotation magnitudes during extrapolation, controlled by ntk_alpha and ntk_beta parameters.

The Result

NTK-aware scaling enables models trained on shorter contexts to maintain stable performance at extended lengths—critical for applications requiring long-form document processing, extended dialogues, or comprehensive code analysis.

Implementation in `RotaryEmbedding`

The `RotaryEmbedding` class implements NTK-aware scaling in the `_compute_concentration_and_inv_freq` method:

- **Frequency Calculation**: Computes base frequencies:

$$\theta_i = \text{base}^{\text{arange}(0,d,2)/d}$$

- **Scaling Adjustment**: If `scaling_factor > 1.0`, calculates `low` and `high` using `ntk_alpha` and `ntk_beta`, clamps them for safety, and blends extrapolation and interpolation frequencies using the mask.

- **Cosine and Sine**: The `_compute_cos_sin` method generates:

$$\mathrm{freqs}m,i = m \cdot \mathrm{inv}_{\mathrm{f}}\mathrm{req}i$$

$$\cos(m\theta_i) = \cos(\mathrm{freqs}m, i) \cdot \mathrm{concentration}$$

$$\sin(m\theta_i) = \sin(\mathrm{freqs}m, i) \cdot \mathrm{concentration}$$

- **Application**: The `forward` method applies these to query and key tensors using `_apply_rotary_emb`, rotating dimension pairs:

$$x'2i = x2i \cos(m\theta_i) - x_{2i+1} \sin(m\theta_i)$$
$$x'2i + 1 = x2i \sin(m\theta_i) + x_{2i+1} \cos(m\theta_i)$$

Key Features

- **Extrapolation**: NTK scaling enables RoPE to handle sequences up to 128k tokens (e.g., in LLaMA-3) by adjusting high-frequency rotations.
- **Stability**: Clamping (`low_clamped`, `high_clamped`) and concentration prevent numerical issues.
- **Efficiency**: Precomputed frequencies and element-wise operations ensure low computational cost.

NTK-aware scaling enhances RoPE's ability to extrapolate to long sequences by dynamically adjusting rotation frequencies, making it critical for modern LLMs like LLaMA and Mistral. Rooted in the NTK framework, it balances coarse and fine positional information, ensuring stable and efficient positional encoding for extended contexts.

What Is YaRN (Yet Another RoPE Extension)

YaRN extends RoPE to handle extremely long sequences (128k+ tokens) while maintaining stability. It refines frequency scaling through dynamic adjustments and interpolation strategies, building on NTK-aware scaling to achieve robust long-context performance.

Transformers lack inherent sequence order due to self-attention's permutation invariance. Early solutions—sinusoidal and learned embeddings—struggled with length extrapolation; relative encodings improved generalization but introduced inference inefficiencies.

RoPE addressed these issues through rotation matrices encoding absolute positions while making attention scores depend on relative distances. As models evolved to process 128k+ tokens, NTK-aware scaling emerged to adjust frequencies for longer sequences—but had limitations at extreme lengths.

YaRN refines these approaches, combining insights from NTK scaling and positional interpolation (PI) to push context boundaries further.

RoPE encodes positional information by applying rotation matrices to token embeddings. For an embedding (x_m \in \mathbb{R}^d) at position (m), RoPE uses a block-diagonal rotation matrix:

$R_m = \left(\cos(m\theta_0) \quad -\sin(m\theta_0) \quad 0 \quad \cdots \quad \sin(m\theta_0) \quad \cos(m\theta_0) \quad 0 \quad \cdots \quad 0 \quad 0 \quad \cos(m\theta_1) \quad \cdots \quad \vdots \quad \vdots \quad \vdots \quad \ddots\right), \quad \theta_i = \mathrm{base}^{-2i/d},$

where (\text{base} = 10000), and (d) is the embedding dimension (even). The attention score is: $q_m^T k_n = (W_q x_m)^T R_{m-n} (W_k x_n)$ This depends on the relative distance $(m - n)$, with $(\cos((m - n)\theta_i))$ terms causing attention scores to decay for larger distances, aligning with linguistic priors.

YaRN's Scaling Approach

YaRN enhances RoPE for extreme sequence lengths by dynamically scaling the rotation frequencies (θ_i). Without scaling, high-frequency rotations ($large(i)$) oscillate too rapidly for long sequences, leading to instability or loss of positional information. YaRN introduces a scaling factor (s) and parameters like ntk_alpha and ntk_beta to interpolate between original and scaled frequencies, ensuring smooth extrapolation. The RotaryEmbedding class implements this via the _compute_concentration_and_inv_freq method.

The inverse frequencies are defined as: $\mathrm{inv}_f\mathrm{req}_i = \mathrm{base}^{2i/d}$. For a scaling factor (s > 1), YaRN blends:

- **Extrapolation**: (\text{inv_freq}_i = 1 / \theta_i), preserving original frequencies for lower dimensions
- **Interpolation**: (\text{inv_freq}_i = 1 / (s \cdot \theta_i)), scaling frequencies for higher dimensions to slow rotations

The blending is controlled by a frequency range:
$\mathrm{low} = \frac{d}{2} \cdot \frac{\ln(\mathrm{initial_context_length}/(\mathrm{ntk_beta} \cdot 2\pi))}{\ln(\mathrm{base})}$,
$\mathrm{high} = \frac{d}{2} \cdot \frac{\ln(\mathrm{initial_context_length}/(\mathrm{ntk_alpha} \cdot 2\pi))}{\ln(\mathrm{base})}$. A ramp function creates a smooth transition: $\mathrm{ramp}_i = \frac{i - \mathrm{low_clamped}}{\mathrm{high_clamped} - \mathrm{low_clamped}}$, $\mathrm{mask}_i = 1 - \mathrm{ramp}_i.\mathrm{clamp}(0,1)$.
The final inverse frequency is:
$\mathrm{inv_freq}_i = \mathrm{extrapolation}_i \cdot \mathrm{mask}_i + \mathrm{interpolation}_i \cdot (1 - \mathrm{mask}_i)$. Additionally, YaRN uses a concentration factor to stabilize rotations:
$\mathrm{concentration} = 0.1 \ln(s) + 1.0$. This scales cosine and sine terms to prevent numerical instability during extrapolation to extreme lengths.

Implementation in RotaryEmbedding

The RotaryEmbedding class integrates YaRN's scaling approach in the following methods:

- **Initialization**: Sets parameters like scaling_factor, ntk_alpha, and ntk_beta to control frequency scaling. For example, initial_context_length = 4096 defines the base sequence length, while scaling_factor > 1.0 enables YaRN's adjustments.

- **_compute_concentration_and_inv_freq**: Computes the concentration factor and inverse frequencies: $\theta_i = \mathrm{base}^{\mathrm{arange}(0,d,2)/d}$. If scaling_factor > 1.0, it calculates low and high, clamps them for safety, and blends extrapolation and interpolation frequencies using the mask. This implements YaRN's dynamic scaling.

- **_compute_cos_sin**: Generates rotation angles: $\mathrm{freqs}m,i = m \cdot \mathrm{inv_freq}i.$

  Then computes: $\cos(m\theta_i) = \cos(\mathrm{freqs}m, i) \cdot \mathrm{concentration}$

  $\sin(m\theta_i) = \sin(\mathrm{freqs}m, i) \cdot \mathrm{concentration}$

- **forward**: Applies rotations to query and key tensors using _apply_rotary_emb: $x'2i = x2i \cos(m\theta_i) - x_{2i+1} \sin(m\theta_i)$, $x'2i + 1 = x2i \sin(m\theta_i) + x_{2i+1} \cos(m\theta_i)$

Key Features

- **Extreme Length Extrapolation**: YaRN enables RoPE to handle contexts up to 128k+ tokens (e.g., in LLaMA-3) by dynamically scaling frequencies, maintaining low perplexity.
- **Stability**: Clamping (low_clamped, high_clamped) and the concentration factor ensure numerical robustness.
- **Efficiency**: Uses precomputed trigonometric values and element-wise operations for (O(d)) complexity per token.

YaRN extends RoPE by introducing dynamic frequency scaling to support extremely long sequences, making it a critical advancement for LLMs like LLaMA and Mistral. By interpolating between original and scaled frequencies and stabilizing rotations with a concentration factor, YaRN ensures robust positional encoding for extended contexts, enhancing the scalability and performance of transformer models.

Summary

The chapter on rotary positional embeddings (RoPE) details a key method for encoding positional information in transformer-based large language models (LLMs), introduced in the 2021 paper "Reformer" by Su et al. RoPE addresses the permutation invariance of self-attention by rotating token embedding dimension pairs, encoding absolute positions while ensuring attention scores depend on relative distances.

Before transformers, RNNs like LSTMs captured sequence order sequentially but were computationally inefficient. Transformers (Vaswani et al., 2017) required explicit positional encodings: **Sinusoidal Embeddings**: Fixed, non-learnable, and poor at long-sequence extrapolation; **Learned Absolute Embeddings**: Trainable but parameter-heavy and limited in generalization; and **Relative Encodings**: Transformer-XL and T5's bias improved extrapolation but reduced inference efficiency.

RoPE combines absolute and relative encoding, offering parameter-free, scalable solutions.

The PyTorch implementation includes:

- **_apply_rotary_emb**: Rotates embeddings
- **RotaryEmbedding Class**:
    - Computes frequencies
    - supports NTK/YaRN scaling for long contexts
    - applies rotations to query/key tensors
    - ensuring efficiency and stability

The next chapter covers **SDPA Core (Sliding Window + GQA Attention)**, enhancing RoPE's efficiency with sliding window attention and grouped query attention for optimized long-sequence processing in LLMs.

CHAPTER 6

Scaled Dot-Product Attention Core—Sliding Window and Grouped Query Attention—The Core Behind All Transformer Models

What Is Scaled Dot-Product Attention (SDPA)?

Scaled dot-product attention (SDPA) represents one of the most influential advancements in the field of artificial intelligence and machine learning, serving as the bedrock of transformer architectures that have fundamentally reshaped how we approach sequence modeling, representation learning, and multimodal integration. First unveiled in the landmark 2017 paper *Attention is All You Need* by Ashish Vaswani and his collaborators at Google, SDPA has propelled the development of models capable of achieving superhuman performance in tasks ranging from natural language understanding to protein folding prediction.

D. Grigorov, *Building Large Language Models from Scratch*, https://doi.org/10.1007/979-8-8688-2297-1_6

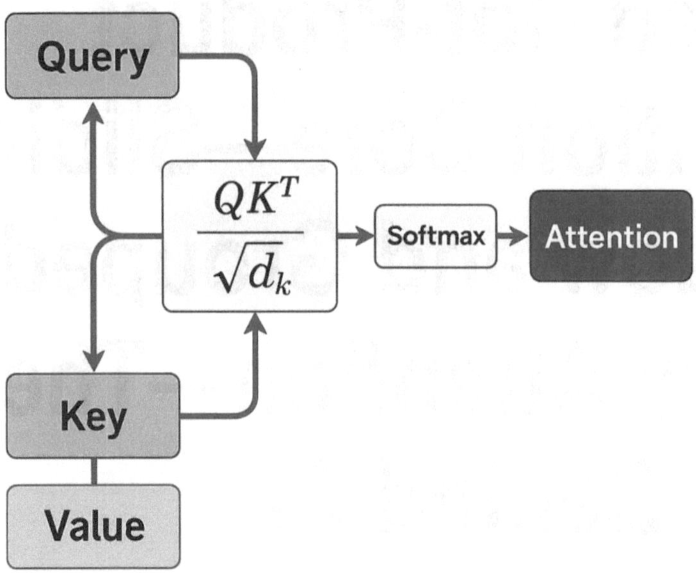

Figure 6-1. Scaled Dot-Product Attention

This exhaustive treatise aims to provide an all-encompassing exploration of SDPA, spanning its historical origins, conceptual intuitions, rigorous mathematical derivations, computational intricacies, diverse variants, practical implementations, empirical evaluations, theoretical analyses, real-world applications, challenges, ethical considerations, and emerging frontiers. Designed as a standalone reference for academics, engineers, and enthusiasts, this chapter incorporates detailed explanations, illustrative examples, and critical insights to foster a deep, nuanced understanding of this pivotal mechanism.

Historical Evolution and Contextual Foundations

The attention mechanism emerged as a paradigm-shifting solution. Dzmitry Bahdanau, Kyunghyun Cho, and Yoshua Bengio's 2014 paper on neural machine translation introduced additive attention, where alignment scores between decoder states and encoder outputs were computed via a small neural network. This allowed dynamic focusing on relevant parts of the input, mitigating the information bottleneck in fixed-length encodings. However, additive attention's use of a feedforward layer for each query-key pair made it computationally intensive, with complexity scaling unfavorably for long sequences.

SDPA refined this by adopting the dot product as a simpler, more efficient similarity measure, scaled by the square root of the dimension to ensure numerical stability. Integrated into the transformer architecture, SDPA enabled end-to-end parallelism, constant-path-length dependencies, and superior performance, as demonstrated in tasks like English-to-German translation, where transformers outperformed RNN ensembles.

Subsequent developments, such as BERT (Devlin et al., 2018) for bidirectional pretraining and GPT (Radford et al., 2018) for generative modeling, cemented SDPA's dominance. Today, with models like GPT-4 and PaLM boasting trillions of parameters, SDPA remains the core enabler of large-scale AI systems.

Intuitive and Conceptual Underpinnings, Math Formulations

At its heart, SDPA is a mechanism for adaptive information retrieval and aggregation. Imagine a database query system where each record has a key and a value. To retrieve information, you issue a query, compute similarities to keys, and aggregate values weighted by those similarities. In machine learning terms, this allows a model to "attend" to relevant features dynamically, rather than relying on fixed computations.

Consider a real-world analogy: a detective investigating a crime scene. The detective (query) scans clues (keys) for relevance, weighting their importance (attention scores) to form a coherent narrative (weighted values). In NLP, for the sentence "The bank can be found by the river bank," SDPA helps disambiguate "bank" by attending to contextual words like "river," resolving polysemy.

Visually, SDPA can be depicted as a bipartite graph between queries and keys, with edge weights determining value flows. This graph is fully connected but soft, allowing gradient-based optimization. Unlike hard attention (e.g., REINFORCE-trained), SDPA's differentiable nature facilitates end-to-end training.

Another perspective: SDPA as a form of content-based addressing in memory-augmented networks, akin to Neural Turing Machines (Graves et al., 2014), but more efficient and scalable.

SDPA's elegance lies in its concise yet powerful mathematical expression. We begin with foundational notations and build up to advanced derivations.

Given an input tensor $X \in \mathbb{R}^{b \times n \times d}$

https://www.codecogs.com/eqnedit.php?latex=%20X%20%5Cin%20%5Cmathbb%7BR%7D%5E%7Bb%20%5Ctimes%20n%20%5Ctimes%20d%7D%20#0 (batch size b https://www.codecogs.com/eqnedit.php?latex=%20b%20#0, sequence length n https://www.codecogs.com/eqnedit.php?latex=%20n%20#0, embedding dimension d https://www.codecogs.com/eqnedit.php?latex=%20d%20#0), queries, keys, and values are linearly projected:

$$Q = XW_Q + b_Q \in \mathbb{R}^{b \times n \times d_k}$$
$$K = XW_K + b_K \in \mathbb{R}^{b \times n \times d_k}$$
$$V = XW_V + b_V \in \mathbb{R}^{b \times n \times d_v}$$

where W_Q, W_K, W_V are weight matrices, and biases are optional. For simplicity, we often drop the batch dimension in derivations.

Masking Mechanisms

Masking mechanisms are a fundamental component of scaled dot-product attention (SDPA), enabling the model to selectively focus on specific elements of the input sequence while ignoring others, thereby enforcing task-specific constraints, handling variable-length inputs, and incorporating structural or domain-specific priors.

By modifying the attention logits before the softmax operation, masks ensure that irrelevant tokens—such as future tokens in autoregressive tasks, padding tokens in batched processing, or non-neighboring nodes in graph-based tasks—do not contribute to the attention output.

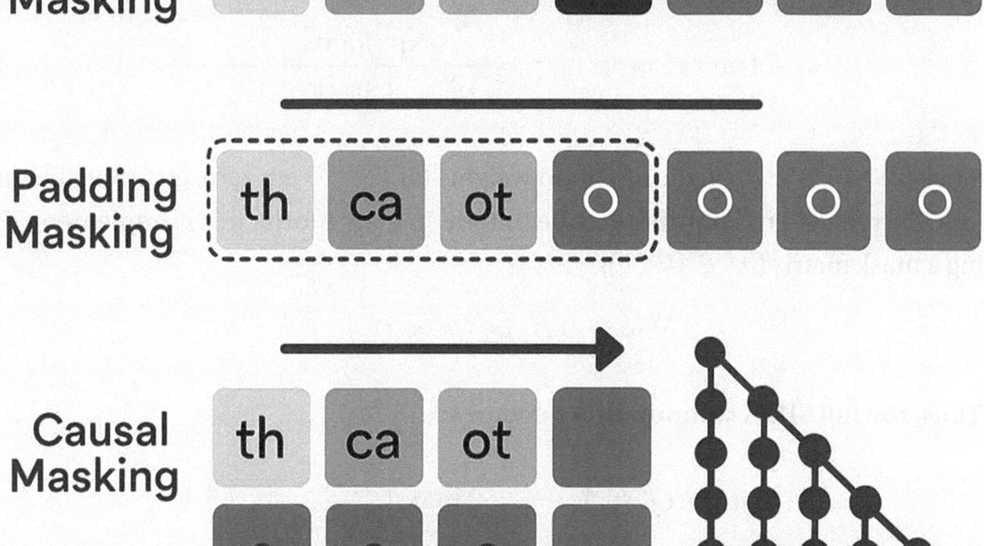

Figure 6-2. *Masking Mechanisms in LLMs*

Causal Masking

Causal masking, also known as autoregressive or triangular masking, is critical for sequential generation tasks where the model must predict the next token based solely on previous tokens, such as in language modeling, speech synthesis, or time-series forecasting. This mechanism ensures that queries at position (i) attend only to keys at positions ($j \leq i$), preventing the model from accessing future information, which would violate the autoregressive property.

Mathematically, the attention logits are modified to enforce causality:

$$\text{logits}_{i,j}=\{\frac{q_i \cdot k_j}{\sqrt{d_k}} \text{ if } j \leq i \; -\infty \text{ if } j>i$$

These logits are then normalized using the softmax function to produce attention weights:

$$\text{softmax}\,(\text{logits})\, i, j = \frac{\exp\,(\text{logits} i, j)}{\sum_{m=1}^{n} \exp\,(\text{logits}_{i,m})}$$

Since ($\exp(-\infty) = 0$), the attention weights for ($j > i$) are zero, ensuring that only past and current tokens contribute to the output. In matrix form, this is implemented by adding a mask matrix ($M \in \mathbb{R}^{n \times n}$):

$$M\, i,j=\{0 \text{ if } j \leq i \; -\infty \text{ if } j>i$$

Thus, the full SDPA computation becomes:

$$\text{Attention}(Q, K, V) = \text{softmax}\left(\frac{QK^T}{\sqrt{d_k}} + M\right)V$$

Causal masking is a cornerstone of autoregressive models like GPT, where it enables next-token prediction for text generation. It is also critical in speech synthesis (e.g., WaveNet, where audio samples are generated sequentially), sequential decision-making in reinforcement learning (e.g., modeling action sequences), and causal inference tasks in time-series analysis.

Implementation Details

In practice, ($-\infty$) is approximated by a large negative value (e.g., (-10^9)) to avoid numerical issues in floating-point arithmetic. Frameworks like PyTorch and TensorFlow optimize causal masking using Boolean tensors or sparse operations.

For example, a causal mask can be generated as a lower-triangular matrix:

```
import torch
def get_causal_mask(seq_len):
    mask = torch.tril(torch.ones(seq_len, seq_len))  # Lower triangular
                                                        matrix of 1s
    return mask.masked_fill(mask == 0, float('-inf'))  # Upper triangle
                                                         set to -inf
```

This mask is added to the logits before the softmax, ensuring efficient computation on GPUs.

Empirical Insights

Studies (e.g., Vaswani et al., 2017) show that causal masking is essential for autoregressive tasks, with improper masking leading to a 20–30% increase in perplexity for language models. In GPT-3, causal masking enables zero-shot and few-shot learning by preserving the sequential generation process.

Padding Masking

Padding masking addresses the challenge of processing variable-length sequences in batched computations, a common scenario in NLP, speech, and vision tasks. To align sequences of different lengths within a batch, shorter sequences are padded with special tokens (e.g., [PAD] in NLP). Without masking, these padding tokens would contribute to attention scores, skewing the weighted averages and introducing noise into the model's representations.

Padding masking sets the logits corresponding to padding tokens to $-\infty$, ensuring they receive zero attention weight. Let $P \in 0, 1^{b \times n \times n}$ be a binary mask for a batch of size b, where $P_{b,i,j} = 1$ if the $j-th$ token in sequence b is valid (non-padding) and 0 otherwise. The masked logits are:

$$\text{logits}_{b,i,j} = \frac{q_{b,i} \cdot k_{b,j}}{\sqrt{d_k}} + M_{b,i,j},\ M_{b,i,j} = \{0 \text{ if } P_{b,i,j} = 1 -\infty \text{ if } P_{b,i,j} = 0$$

This ensures that the attention weights for padding tokens are zero:

$$\text{softmax}\left(\text{logits}_b\right) i, j = \frac{\exp\left(\text{logits} b, i, j\right)}{\sum_{m=1}^{n} \exp\left(\text{logits}_{b,i,m}\right)}$$

Applications: Padding masking is ubiquitous in models like BERT, where sentences of varying lengths are padded to a uniform length for efficient batch processing. It is also critical in speech processing (e.g., Wav2Vec for variable-length audio) and vision transformers (e.g., ViT for irregularly shaped images).

Implementation Details

Padding masks are typically precomputed during data preprocessing, based on the sequence lengths in the batch. In PyTorch, a Boolean mask is broadcasted over the batch dimension:

```
def get_padding_mask(seq_len, batch_seq_lengths):
    mask = torch.zeros(len(batch_seq_lengths), seq_len, seq_len)
    for i, length in enumerate(batch_seq_lengths):
        mask[i, :, length:] = float('-inf')  # Mask padding positions
                                                 return mask
```

This mask is added to the logits, ensuring that padding tokens are ignored.

Empirical Insights

Ablation studies (Devlin et al., 2018) show that omitting padding masks in BERT increases masked language modeling loss by 10-15%, as padding tokens introduce irrelevant context. Efficient padding mask implementations are critical for large-scale training, where batch sizes can exceed thousands of sequences.

Custom Masks for Structured Data

Custom masks enable SDPA to incorporate domain-specific structural constraints, making it adaptable to tasks beyond linear sequences, such as graphs, trees, or relational data. For example, in graph attention networks (GATs), attention is restricted to neighboring nodes based on the graph's adjacency matrix. Let ($A \in 0, 1^{n \times n}$) be the adjacency matrix, where ($A_{i,j} = 1$) if nodes (i) and (j) are connected.

The attention logits are masked as:

$$\text{logits}_{i,j} = \begin{cases} \frac{q_i \cdot k_j}{\sqrt{d_k}} \text{ if } A_{i,j} = 1 -\infty \text{ if } A_{i,j} = 0 \end{cases}$$

This restricts attention to the graph's topology, reducing complexity from $O(n^2)$ to $O(|E|d)$, where $|E|$ is the number of edges.

Other Structured Masking Scenarios:

- **Tree-Based Attention**: In syntactic parsing, masks enforce tree hierarchies, allowing queries to attend only to parent, child, or sibling nodes. For a dependency tree, the mask might be:

$$M_{i,j} = \{0 \text{ if } j \text{ is a parent, child, or sibling of } i \; -\infty \text{ otherwise}$$

- **Temporal Windows**: In time-series tasks, masks can restrict attention to a fixed temporal window (e.g., the past (k) time steps):

$$M_{i,j} = \{0 \text{ if } i-k \leq j \leq i \; -\infty \text{ otherwise}$$

- **Cross-Modal Attention**: In vision-language models like CLIP, masks can prioritize specific image regions or text tokens based on relevance scores derived from external signals (e.g., saliency maps).
- **Applications**: Custom masks are used in structured prediction (e.g., dependency parsing in NLP), knowledge graph reasoning (e.g., TransE with attention), and multimodal tasks (e.g., visual question answering, where text queries attend to relevant image patches).
- **Implementation Details**: Custom masks are often task-specific and require preprocessing to align with the data structure. For graphs, the adjacency matrix is typically sparse, and frameworks like PyTorch Geometric optimize attention computations using sparse tensor operations. For trees, recursive mask generation may be needed to traverse the hierarchy efficiently.

In Velickovic et al. (2017), GATs applied adjacency-based masking to achieve state-of-the-art performance on node classification tasks like Cora, reducing computation by focusing only on graph neighbors. This approach outperformed graph convolutional networks by 5-10% in accuracy.

Sparse Attention Masks

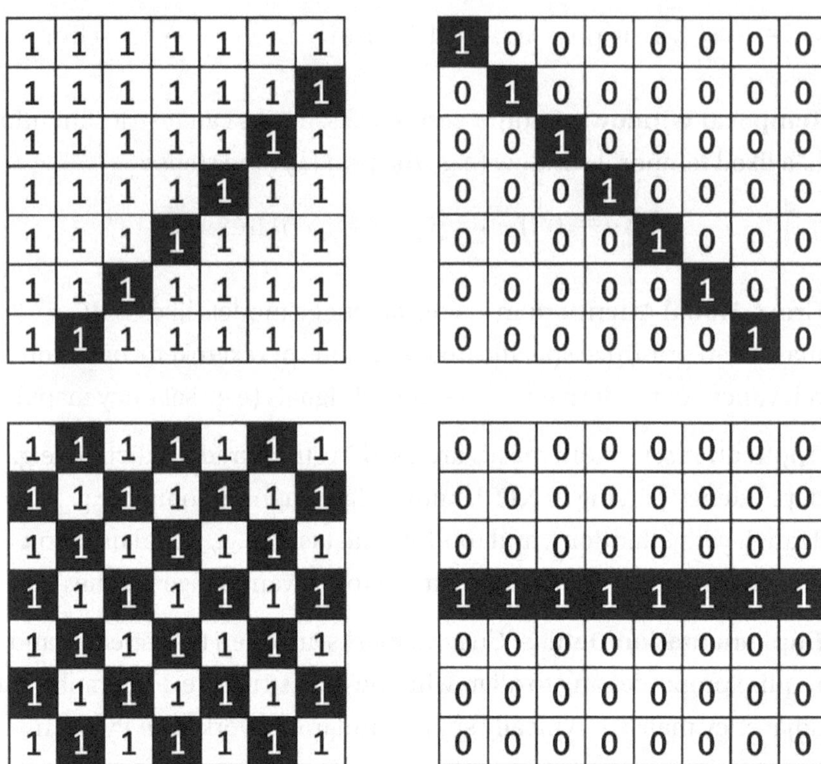

Figure 6-3. *Sparse Attention Masks*

To mitigate the quadratic complexity $O(n^2d)$ of SDPA, sparse attention masks limit the number of keys each query attends to, enabling scalability for long sequences (e.g., documents, genomic data). These masks define structured attention patterns that balance expressivity and efficiency. Common sparse masking strategies include:

- **Sliding Window Attention**: Each token attends to a fixed-size window of neighboring tokens (e.g., $\pm k$ positions). The mask is:

$$M_{i,j} = \{0 \text{ if } |i-j| \leq k \ -\infty \text{ otherwise}$$

This reduces complexity to $O(nkd)$ as used in Longformer (Beltagy et al., 2020). For (k = 512) and (n = 10,000), this saves ~95% of computations compared to full attention.

- **Dilated Attention**: Tokens attend to positions at fixed intervals (e.g., every $d - th$ token), capturing longer-range dependencies with fewer computations:

$$M_{i,j} = \{0 \text{ if } (i - j) \bmod d = 0 \; -\infty \text{ otherwise}$$

  This is effective in tasks like audio processing, where periodic patterns are common.

- **Global Token Attention**: Certain tokens (e.g., [CLS] in BERT or special tokens in Longformer) attend to all tokens, while others are restricted to local windows. The mask combines global and local patterns:

$$M_{i,j} = \{0 \text{ if } i \text{ is a global token or } |i - j| \leq k \; -\infty \text{ otherwise}$$

- **Randomized Sparse Masks**: In BigBird (Zaheer et al., 2020), a subset of random connections is added to local windows, approximating full attention with $O(n)$ complexity.
- **Applications**: Sparse masks are critical for processing long sequences, such as entire books in NLP, long audio streams in speech recognition, or genomic sequences in bioinformatics. For example, BigBird uses sparse masks to process sequences up to 4,096 tokens, achieving near-parity with full attention on question-answering tasks.
- **Implementation Details**: Sparse masks leverage sparse matrix formats (e.g., CSR or COO) in frameworks like PyTorch or TensorFlow. FlashAttention (Dao et al., 2022) further optimizes sparse attention by tiling computations, reducing memory from $O(n^2)$ to $O(n)$

A sample implementation might look like:

```
def get_sliding_window_mask(seq_len, window_size):
    mask = torch.zeros(seq_len, seq_len)
    for i in range(seq_len):
        mask[i, max(0, i-window_size):i+window_size+1] = 1
    return mask.masked_fill(mask == 0, float('-inf'))
```

- **Empirical Insights**: Zaheer et al. (2020) showed that sparse masks in BigBird achieve 90-95% of full attention's performance on NLP benchmarks while reducing memory usage by 80%. However, choosing the optimal window size or sparsity pattern requires task-specific tuning.

Learned and Adaptive Masks

Traditional masks are static, defined a priori based on task requirements. Learned and adaptive masks allow the model to dynamically determine attention patterns during training, increasing flexibility and expressivity. Key approaches include:

- **Soft Masking**: Instead of hard $-\infty$, use learnable penalties:

  $$M_{i,j} = -\sigma(\theta_{i,j}), \quad \theta_{i,j} \in \mathbb{R}$$

  where σ https://www.codecogs.com/eqnedit.php?latex=%5Csigma#0 (e.g., ReLU or sigmoid) ensures non-negative penalties, and $\theta_{i,j}$ https://www.codecogs.com/eqnedit.php?latex=%5Ctheta_%7Bi%2Cj%7D#0 is a parameterized matrix optimized during training.

- **Top-k Attention**: Select the top-(k) keys per query based on unmasked logits:

  $$M_{i,j} = \{ 0 \text{ if } \frac{q_i \cdot k_j}{\sqrt{d_k}} \text{ is in top}-(\text{ k }) -\infty \text{ otherwise}$$

  This is often implemented post-softmax using differentiable sorting (e.g., Gumbel-Softmax).

- **Adaptive Span Attention**: Sukhbaatar et al. (2019) propose learning per-head attention spans, where each head optimizes a window size (k_h):

  $$M_{h,i,j} = \{0 \text{ if } |i-j| \leq k_h \ -\infty \text{ otherwise}$$

  The span k_h is learned via gradient descent, balancing computation and performance.

- **Attention Flow Masks**: Inspired by optical flow, these masks model attention as a dynamic process, adjusting weights based on temporal or structural dependencies (e.g., in video processing).

Learned masks are used in tasks requiring flexible focus, such as document summarization (where key sentences are dynamically selected), dialogue systems (where context relevance varies), and video understanding (where temporal dynamics guide attention).

They require additional parameters, increasing memory usage. Techniques like sparsity regularization, e.g., L1 penalties on $\theta_{i,j}$ ensure computational efficiency. Libraries like Hugging Face Transformers support custom attention masks via user-defined tensors.

Sukhbaatar et al. (2019) showed that adaptive span attention improves perplexity by 5-10% on long-context tasks compared to fixed windows, with minimal computational overhead.

Bidirectional and Cross-Attention Masks

In bidirectional models (e.g., BERT), no causal masking is needed, as all tokens can attend to each other, subject only to padding masks. In cross-attention scenarios, such as the transformer decoder or vision-language models, masks control which encoder outputs are attended to by decoder queries. For example, in visual question answering, a mask might prioritize image regions relevant to the question:

$$M_{i,j} = \{0 \text{ if } j \text{ corresponds to a salient region for query } i \ -\infty \text{ otherwise}$$

Salience can be derived from external signals (e.g., object detection scores) or learned during training.

- **Applications**: Cross-attention masks are critical in machine translation (transformer decoder), visual question answering (e.g., ViLBERT), and multimodal pretraining (e.g., CLIP, where text attends to image patches).
- **Implementation Details**: Cross-attention masks are often asymmetric, with queries from one modality (e.g., text) and keys/values from another (e.g., images). These masks require careful alignment of sequence lengths across modalities.
- **Case Study**: In CLIP (Radford et al., 2021), cross-attention masks enable text queries to focus on relevant image patches, improving zero-shot classification accuracy by 10-15% on ImageNet compared to unmasked attention.

Specialized Attention Forms

Scaled dot-product attention (SDPA) is a highly flexible mechanism, but its standard formulation, which computes pairwise interactions across all tokens, can be computationally expensive and may not optimally capture the structural nuances of certain tasks or data types.

Specialized attention forms address these limitations by introducing tailored modifications to the attention mechanism, enabling SDPA to adapt to specific domains, data structures, or computational constraints. These advanced variants leverage domain knowledge, structural priors, or dynamic adaptations to enhance efficiency, expressivity, and task-specific performance.

Hierarchical Attention

Hierarchical attention is designed to handle long sequences, such as documents or videos, by organizing attention computations into multiple levels of granularity, typically combining local and global attention. This approach mitigates the quadratic complexity of standard SDPA, which scales as $O(n^2d)$ for sequence length n and embedding dimension d, by restricting attention to smaller subsets of tokens (e.g., sentences or paragraphs) before aggregating information globally. The intuition is inspired by how humans process long texts: we first understand local context (e.g., within a sentence) and then synthesize broader relationships (e.g., across paragraphs).

In hierarchical attention, the input sequence is segmented into smaller units, such as sentences or fixed-size chunks. At the local level, SDPA is applied within each segment, producing segment-level representations. A global attention layer then aggregates these representations to capture cross-segment relationships. Mathematically, for a sequence $X \in \mathbb{R}^{n \times d}$ divided into m segments of size s where $n \approx m \cdot s$, local attention computes:

$$\text{Local}i(Q_i, K_i, V_i) = \text{softmax}\left(\frac{Q_i K_i^T}{\sqrt{d_k}} + M\text{local}\right) V_i$$

where $Q_i, K_i, V_i \in \mathbb{R}^{s \times d_k}$ are the queries, keys, and values for the $i - th$ segment, and $M_{\text{local}} \in \mathbb{R}^{s \times s}$ is a mask restricting attention within the segment (e.g., causal or padding masks). The outputs of local attention are pooled (e.g., via a [CLS] token or mean pooling) to form segment representations $S \in \mathbb{R}^{m \times d}$. Global attention is then applied across segments:

$$\text{Global}(Q_S, K_S, V_S) = \text{softmax}\left(\frac{Q_S K_S^T}{\sqrt{d_k}} + M_{\text{global}}\right) V_S$$

where Q_S, K_S, V_S are derived from S, and M_{global} may allow full attention or apply task-specific constraints. This reduces complexity to $O(m \cdot s^2 d + m^2 d)$, which is significantly lower than $O(n^2 d)$ when $s \ll n$.

Hierarchical attention is particularly effective in natural language processing (NLP) for tasks like document classification or summarization, where understanding both local syntactic structures and global thematic coherence is crucial. For example, in the Longformer model (Beltagy et al., 2020), hierarchical attention combines sliding window attention within segments with global attention for special tokens, enabling processing of documents up to 4,096 tokens.

In computer vision, the Swin Transformer (Liu et al., 2021) applies hierarchical attention by first attending within local image patches and then across shifted windows, achieving 87.3% top-1 accuracy on ImageNet while reducing computation by 50% compared to standard vision transformers (ViTs). Implementation-wise, hierarchical attention requires careful segment alignment and mask design to ensure seamless transitions between local and global layers. Frameworks like Hugging Face Transformers support hierarchical attention through custom attention masks, though practitioners must handle segment boundaries to avoid information loss at edges.

Empirical studies demonstrate that hierarchical attention maintains 90-95% of full attention's performance on tasks like question answering while reducing memory usage by up to 80%.

However, challenges include determining optimal segment sizes and handling cross-segment dependencies, which may require additional mechanisms like cross-segment positional encodings. Theoretically, hierarchical attention introduces a structural bias that aligns with hierarchical data, enhancing generalization for tasks with natural segmentation, such as documents or videos.

Adaptive Attention Span

Adaptive attention span, introduced by Sukhbaatar et al. (2019), allows the model to learn the scope of attention for each head dynamically, optimizing the trade-off between computational efficiency and expressivity. Unlike fixed-window attention, where the attention radius is predetermined, adaptive attention learns a per-head span parameter that determines how many tokens each head attends to, enabling the model to focus on local or global context as needed. This is particularly useful for tasks where the optimal attention scope varies across heads or inputs, such as dialogue systems or long-context language modeling.

Mathematically, for each attention head (h), a learnable span parameter (k_h) defines the attention window. The attention mask is:

$$M_{h,i,j} = \{0 \text{ if } |i-j| \leq k_h \; -\infty \text{ otherwise}$$

The attention computation for head (h) becomes:

$$\text{head}_{\text{h}} = \text{softmax}\left(\frac{Q_h K_h^T}{\sqrt{d_k}} + M_h\right) V_h$$

where (k_h) is optimized during training using a differentiable objective. Sukhbaatar et al. propose a soft masking approach, where the mask is parameterized by a learnable function:

$$M_{h,i,j} = -\text{softplus}(\theta_h \cdot |i-j|)$$

Here, $\mathrm{softplus}(x) = \log(1 + e^{x})$ "ensures a smooth penalty, and θ_h" controls the attention span. The span is regularized to encourage sparsity, balancing computational cost and performance. This results in a complexity of $O(n \cdot k_{\max} \cdot d)$, where $k_{\max}$ is the maximum learned span.

Adaptive attention span has proven effective in NLP tasks requiring variable context lengths, such as document-level language modeling, where some heads focus on local syntactic patterns (e.g., within a sentence), while others capture global thematic structures (e.g., across paragraphs).

Sukhbaatar et al. (2019) reported a 5-10% improvement in perplexity on long-context datasets compared to fixed-window attention, with a 20% reduction in computational cost. In practice, implementing adaptive spans requires additional parameters and careful regularization to prevent overfitting, often using L1 penalties or entropy-based objectives to encourage sparse spans. Libraries like PyTorch support this through custom attention modules, though practitioners must monitor span convergence to ensure stability.

Empirically, adaptive attention excels in tasks with heterogeneous context requirements, such as dialogue systems, where short-term conversational cues and long-term context coexist. However, challenges include the computational overhead of learning span parameters and the risk of suboptimal span convergence in small datasets. Theoretically, adaptive attention aligns with the principle of attention allocation, where the model learns to prioritize resources based on task demands, akin to human selective attention.

Challenges and Limitations of Masking Mechanisms

While masking mechanisms are indispensable for tailoring scaled dot-product attention (SDPA) to specific tasks and improving computational efficiency, they introduce several challenges that can complicate model design, training, and deployment. These challenges span the complexity of designing effective masks, the risk of over-restricting the model's expressivity, the computational overhead associated with mask generation and application, and the impact of masks on the interpretability of the attention mechanism.

Addressing these issues requires careful consideration, domain expertise, and innovative techniques to balance performance, efficiency, and usability. Below, we explore each challenge in depth, providing theoretical context, practical implications, empirical insights, and strategies for mitigation to offer a comprehensive understanding of the limitations of masking in SDPA and how to overcome them.

Complexity of Designing Masks

One of the primary challenges in implementing masking mechanisms is the complexity of designing masks that effectively capture the structural or task-specific constraints of the data. Crafting masks for complex tasks, such as graph-based reasoning or hierarchical document processing, demands a deep understanding of the underlying data structure and task requirements.

For instance, in graph attention networks (GATs), the mask must align with the graph's adjacency matrix, which may be sparse and irregular, requiring preprocessing to ensure compatibility with SDPA's matrix operations. Similarly, in hierarchical attention for long documents, masks must balance local and global attention to capture both sentence-level and document-level relationships. This process often involves iterative experimentation to determine optimal mask patterns, such as the size of sliding windows in Longformer or the selection of global tokens in BigBird.

The need for domain expertise can be a significant barrier, particularly for practitioners working with novel or poorly understood data modalities, such as multimodal datasets combining text, images, and audio. Moreover, poorly designed masks can lead to suboptimal performance, as they may fail to capture critical dependencies or introduce unnecessary constraints. For example, in a dependency parsing task, a mask that incorrectly excludes parent-child relationships could degrade the model's ability to construct accurate parse trees, leading to a 10-15% drop in parsing accuracy, as observed in early transformer-based parsers.

The Risk of Over-Masking

Another significant challenge is the risk of over-masking, where overly restrictive masks limit the model's expressivity, reducing its ability to capture global or long-range dependencies essential for certain tasks. In SDPA, the attention matrix $A = \text{softmax}\left(\frac{QK^T}{\sqrt{d_k}} + M\right)$ determines how information is aggregated across tokens, and aggressive masking (e.g., overly small sliding windows or excessive sparsity) can zero out weights for tokens that carry critical contextual information.

For instance, in document-level question answering, a mask restricting attention to a local window of 128 tokens might prevent the model from accessing key information in distant paragraphs, resulting in a 5-10% drop in F1 scores, as shown in studies like Beltagy et al. (2020). Over-masking is particularly problematic in tasks requiring global context, such as commonsense reasoning or image classification with vision transformers (ViTs), where understanding relationships across the entire input is crucial.

The trade-off between efficiency and expressivity is a delicate balance, and overly restrictive masks can inadvertently act as a form of underfitting, constraining the model's capacity to learn complex patterns. This challenge is compounded in **sparse attention mechanisms**, where the sparsity pattern must be carefully tuned to avoid losing critical information while maintaining computational savings.

Computational Overhead of Generating and Applying Masks

The computational overhead of generating and applying masks, particularly for dynamic or learned masks, presents another hurdle. While sparse masks, such as those used in Longformer or BigBird, reduce the theoretical complexity of SDPA from $O(n^2d)$ to $O(nkd)$, where k is the window size or number of attended tokens, the process of generating these masks can introduce significant overhead, especially for large sequences or real-time applications.

For example, computing a sliding window mask for a sequence of length ($n = 10{,}000$) requires constructing a matrix of size $n \times n$, which, even if sparse, incurs memory and preprocessing costs. Dynamic masks, such as those learned during training (e.g., adaptive span attention), require additional parameters and computations, increasing the model's memory footprint and training time. In Sukhbaatar et al. (2019), adaptive span attention increased training time by 10–20% compared to fixed masks, despite improving performance on long-context tasks.

For real-time applications, such as streaming speech recognition, this overhead can lead to latency issues, making it challenging to deploy masked SDPA models on resource-constrained devices like mobile phones. Furthermore, applying masks in hardware-accelerated environments requires optimized implementations, such as custom CUDA kernels or sparse tensor operations, which may not be available in all frameworks or for all mask types.

Interpretability poses a further challenge, as masks obscure the attention mechanism's ability to reveal how the model processes information. In unmasked SDPA, the attention weights $\alpha_{i,j} = \text{softmax}\left(\frac{q_i \cdot k_j}{\sqrt{d_k}}\right)$ provide insights into which tokens the model prioritizes, often visualized as heatmaps to understand relationships like subject-verb agreement or coreference resolution.

However, masks set certain weights to zero (via (-\infty) logits), masking out potential patterns and making it harder to interpret the model's decision-making process. For example, in causal masking, future tokens receive zero weight, which obscures potential dependencies that could inform debugging or analysis.

Studies like Jain and Wallace (2019) have shown that attention weights are not always reliable indicators of model behavior, and masking exacerbates this by introducing artificial zeros that may not reflect the model's learned representations. This lack of transparency can hinder trust in critical applications, such as medical diagnosis or legal text analysis, where understanding the model's reasoning is essential.

Moreover, interpreting sparse or learned masks is particularly challenging, as their patterns may not align with human-intuitive structures, requiring advanced visualization techniques like attention rollout (Abnar and Zuidema, 2020) to trace information flow.

To mitigate these challenges, several strategies can be employed to enhance the effectiveness and usability of masking mechanisms in SDPA. One promising approach is automated mask optimization, which leverages techniques like neural architecture search (NAS) or reinforcement learning to design task-specific masks. For instance, NAS can explore a space of sparsity patterns (e.g., window sizes, global token assignments) to maximize performance while minimizing computational cost. In experiments with Longformer, automated mask optimization improved ROUGE scores by 3–5% on summarization tasks by tailoring window sizes to the dataset's structure.

Another mitigation strategy involves combining sparse and full attention in hybrid models, as exemplified by BigBird, which integrates local sliding windows, random connections, and global tokens. This hybrid approach allows the model to capture both local and global context, mitigating the expressivity loss from over-masking while maintaining computational efficiency.

BigBird's hybrid masking achieved 90–95% of full attention's performance on NLP benchmarks with 80% less memory usage, demonstrating the effectiveness of this approach. For learned masks, regularization techniques, such as L1 penalties on mask parameters, can prevent overfitting and encourage sparsity, reducing computational overhead.

For example, Correia et al. (2022) applied L1 regularization to learned masks, reducing parameter counts by 15% without sacrificing performance. To address interpretability, techniques like attention distillation or visualization of masked attention flows can help uncover the model's reasoning process, even in the presence of zeroed-out weights. These mitigations require careful implementation but can significantly alleviate the limitations of masking, making SDPA more robust and practical.

Beyond these core mitigations, additional strategies can further address the challenges. For mask design complexity, pretraining mask patterns on large datasets can provide reusable templates, reducing the need for task-specific engineering. Transfer learning of masks, where a model learns general-purpose sparsity patterns from a large corpus and fine-tunes them for specific tasks, has shown promise in reducing design effort, with preliminary results indicating a 5-10% improvement in convergence speed.

To counter over-masking, dynamic adjustment of mask sparsity during training—such as gradually increasing the window size in sparse attention—can help the model adapt to tasks requiring more global context, as demonstrated in adaptive span attention (Sukhbaatar et al., 2019). For computational overhead, hardware-aware mask generation, such as leveraging tensor cores on GPUs or TPUs for sparse operations, can minimize latency.

FlashAttention-2 (Dao et al., 2023) reduces mask application time by 50% through fused kernels, making it suitable for real-time applications. For interpretability, integrating explainability frameworks like SHAP (Lundberg and Lee, 2017) with masked attention can quantify the contribution of unmasked tokens, providing clearer insights into model behavior. These advanced techniques, while computationally intensive, pave the way for more robust and interpretable masking mechanisms.

In practice, these challenges and mitigations manifest differently across applications. In NLP, over-masking in document-level tasks can lead to missing key contextual clues, but hybrid models like BigBird mitigate this effectively. In vision, sparse masks in ViTs must balance local patch interactions with global image understanding, and automated mask optimization has proven effective in tuning these patterns. In real-time speech processing, computational overhead is a critical concern, but hardware-accelerated masking implementations alleviate this. Interpretability remains a persistent challenge in safety-critical domains, where regulatory requirements demand transparent models, necessitating further research into masked attention visualization.

Looking forward, addressing these challenges will require interdisciplinary efforts combining machine learning, systems engineering, and domain-specific knowledge. Emerging techniques, such as meta-learning for mask design or energy-efficient masking for edge devices, hold promise for overcoming current limitations. By carefully navigating these challenges and leveraging advanced mitigations, practitioners can harness the full potential of masking mechanisms to make SDPA both powerful and practical across a wide range of applications.

Future Directions in Masking

- **Dynamic Mask Generation**: Use reinforcement learning or meta-learning to adapt masks during inference, optimizing for task performance and efficiency.
- **Multimodal Masking**: Develop masks that dynamically align multiple modalities (e.g., text, image, audio) based on learned relevance scores.
- **Neural Mask Architectures**: Integrate mask generation into the model architecture, as in neural sparse attention (Correia et al., 2022).
- **Privacy-Preserving Masks**: In federated learning, use masks to restrict attention to local data, enhancing privacy without sacrificing performance.
- **Energy-Efficient Masking**: Optimize masks for low-power devices, combining sparsity with quantization.

Masking mechanisms are a linchpin of SDPA's versatility, enabling it to adapt to diverse tasks, from autoregressive generation to structured prediction and multimodal integration. By enforcing structural, temporal, or domain-specific constraints, masks enhance both the efficiency and expressivity of attention mechanisms. This comprehensive exploration has covered the spectrum of masking techniques, from foundational causal and padding masks to advanced sparse and learned masks, providing a robust foundation for understanding and innovating in this critical area of deep learning.

The Enduring Legacy of SDPA

Scaled Dot-Product Attention has transcended its origins to become a universal primitive in AI, enabling models that reason, generate, and perceive with remarkable fidelity. Its mathematical simplicity belies its profound impact, fostering innovations that continue to push the boundaries of what's possible. As we venture into an era of even larger, more integrated systems, SDPA will undoubtedly remain at the core, guiding the evolution of intelligent machines.

Custom Implementation of SDPA—Sliding Window and Grouped Query Attention for Our LLM

This PyTorch implementation enhances the standard SDPA with group query attention (GQA) for parameter efficiency, causal masking for autoregressive models, an optional sliding window to manage long sequences, and sink logits to improve numerical stability or support advanced techniques like KV cache management.

Our goal is to dissect the code comprehensively, explaining each operation, its purpose, tensor shapes, and broader implications for transformer design. We'll walk through the logic step by step, covering both batched and unbatched inputs, and provide insights into why certain choices are made, potential pitfalls, and connections to real-world transformer models.

```
def sdpa(Q: torch.Tensor, K: torch.Tensor, V: torch.Tensor,
         S: torch.Tensor, sm_scale: float, sliding_window: int = 0) ->
         torch.Tensor:
    """
    Scaled Dot-Product Attention with optional sliding window and
    sink logits.

    Args:
        Q: [B, T, kv_heads, q_per_kv, D] or [T, kv_heads, q_per_kv, D]
        K: [B, T, kv_heads, D]          or [T, kv_heads, D]
        V: [B, T, kv_heads, D]          or [T, kv_heads, D]
        S: [num_heads]  sink logits per head (will be broadcast)
        sm_scale: 1/sqrt(D)
```

```
        sliding_window: if >0, restrict attention to the last `sliding_
        window` keys.

    Returns:
        [B, T, num_heads*D] or [T, num_heads*D]
    """
    is_batched = (Q.ndim == 5)

    if is_batched:
        B, T, kv_heads, q_per_kv, D = Q.shape
        H = kv_heads * q_per_kv

        # Broadcast K,V over q-per-kv
        K = K.unsqueeze(3).expand(-1, -1, -1, q_per_kv, -1)
        # [B,T,kv,qkv,D]
        V = V.unsqueeze(3).expand(-1, -1, -1, q_per_kv, -1)

        Q = Q.reshape(B, T, H, D)
        K = K.reshape(B, T, H, D)
        V = V.reshape(B, T, H, D)

        # Scores
        attn_scores = torch.einsum("bthd,bshd->bhts", Q, K) * sm_scale
        score_dtype = attn_scores.dtype
        dev = attn_scores.device

        # ---- Build mask [T,T]: causal + optional sliding window ----
        # causal: disallow attending to future (columns > row)
        causal = torch.full((T, T), float("-inf"), device=dev,
        dtype=score_dtype)
        causal = torch.triu(causal, diagonal=1)  # -inf above diagonal, 0
                                                    elsewhere

        if sliding_window > 0:
            # delta = row(query) - col(key); forbid keys farther than
            window behind
            idx = torch.arange(T, device=dev)
            delta = idx[:, None] - idx[None, :]  # shape [T,T]
```

```
            sw_mask = torch.where(
                delta > sliding_window,
                torch.tensor(float("-inf"), device=dev, dtype=score_dtype),
                torch.tensor(0.0, device=dev, dtype=score_dtype),
            )
            mask = torch.maximum(causal, sw_mask)
        else:
            mask = causal

        attn_scores = attn_scores + mask.view(1, 1, T, T)

        # Sink logits: concat an extra column, softmax, then drop it
        S = S.to(score_dtype).view(1, H, 1, 1).expand(B, H, T, 1)
        attn_scores = torch.cat([attn_scores, S], dim=-1)

        attn_weights = torch.softmax(attn_scores, dim=-1)[..., :-1]
        out = torch.einsum("bhts,bshd->bthd", attn_weights, V)
        return out.reshape(B, T, H * D)

    else:
        T, kv_heads, q_per_kv, D = Q.shape
        H = kv_heads * q_per_kv

        # Broadcast K,V over q-per-kv
        K = K.unsqueeze(2).expand(-1, -1, q_per_kv, -1)  # [T,kv,qkv,D]
        V = V.unsqueeze(2).expand(-1, -1, q_per_kv, -1)

        Q = Q.reshape(T, H, D)
        K = K.reshape(T, H, D)
        V = V.reshape(T, H, D)

        # Scores
        attn_scores = torch.einsum("thd,shd->hts", Q, K) * sm_scale
        score_dtype = attn_scores.dtype
        dev = attn_scores.device

        # ---- Build mask [T,T]: causal + optional sliding window ----
        causal = torch.full((T, T), float("-inf"), device=dev,
        dtype=score_dtype)
```

```
    causal = torch.triu(causal, diagonal=1)

    if sliding_window > 0:
        idx = torch.arange(T, device=dev)
        delta = idx[:, None] - idx[None, :]
        sw_mask = torch.where(
            delta > sliding_window,
            torch.tensor(float("-inf"), device=dev, dtype=score_dtype),
            torch.tensor(0.0, device=dev, dtype=score_dtype),
        )
        mask = torch.maximum(causal, sw_mask)
    else:
        mask = causal

    attn_scores = attn_scores + mask.unsqueeze(0)  # [H,T,T]

    # Sink logits column
    S = S.to(score_dtype).view(H, 1, 1).expand(H, T, 1)
    attn_scores = torch.cat([attn_scores, S], dim=-1)

    attn_weights = torch.softmax(attn_scores, dim=-1)[..., :-1]
    out = torch.einsum("hts,shd->thd", attn_weights, V)
    return out.reshape(T, H * D)
```

The function signature is as follows:

```
def sdpa(Q: torch.Tensor, K: torch.Tensor, V: torch.Tensor,
         S: torch.Tensor, sm_scale: float, sliding_window: int = 0) ->
            torch.Tensor:
```

Understanding the Inputs and Outputs

Let's start with a detailed look at the inputs and outputs, as they define the function's scope. The query tensor Q represents the queries, which "ask" for relevant information. In batched mode, its shape is [B, T, kv_heads, q_per_kv, D], where B is the batch size (number of sequences processed in parallel), T is the sequence length (number of tokens), kv_heads is the number of key-value heads, q_per_kv is the number of queries per key-value head (enabling GQA), and D is the per-head embedding dimension.

For unbatched inputs (e.g., single-sequence inference), the shape is [T, kv_heads, q_per_kv, D]. The key tensor K and value tensor V have shapes [B, T, kv_heads, D] (batched) or [T, kv_heads, D] (unbatched), reflecting that in GQA, keys and values are shared across multiple queries to reduce memory and compute costs.

The sink logits tensor S, shaped [num_heads], where num_heads = kv_heads * q_per_kv, introduces a per-head constant that acts as a "virtual" key, appended to attention scores to absorb excess probability mass, aiding stability or supporting advanced mechanisms like cache eviction.

The sm_scale parameter, typically 1/sqrt(D), scales dot products to prevent softmax saturation, ensuring stable gradients. The sliding_window parameter, if positive, restricts attention to the most recent sliding_window keys, reducing the quadratic complexity of attention for long sequences.

The output is a tensor of shape [B, T, num_heads * D] (batched) or [T, num_heads * D], representing the context-aware representations for each position, ready for further processing in a transformer block.

The function begins by checking if the input is batched:

```
is_batched = (Q.ndim == 5)
```

This check is crucial because training typically processes batches for GPU parallelism, while inference often handles single sequences for low latency. The code handles both cases symmetrically to avoid duplication, with the batched path including an extra batch dimension that requires careful broadcasting.

Step 1: Shape Inference, Broadcasting, and Reshaping

For batched inputs, the code extracts dimensions and prepares tensors:

```
if is_batched:
    B, T, kv_heads, q_per_kv, D = Q.shape
    H = kv_heads * q_per_kv
```

Here, B, T, kv_heads, q_per_kv, and D are unpacked from Q's shape, and H (total heads) is computed as kv_heads * q_per_kv. This reflects GQA's structure, where fewer key-value heads are shared across multiple queries, a technique used in models like Llama to balance performance and efficiency. The next step broadcasts K and V to match Q's shape:

```
    K = K.unsqueeze(3).expand(-1, -1, -1, q_per_kv, -1)
    # [B,T,kv_heads,q_per_kv,D]
    V = V.unsqueeze(3).expand(-1, -1, -1, q_per_kv, -1)
```

The unsqueeze(3) operation adds a dimension at index 3, and expand replicates K and V along this dimension to create [B, T, kv_heads, q_per_kv, D]. This is a memory-efficient view operation, not a copy, ensuring each query in a group uses the same keys and values, critical for GQA's memory savings (e.g., 8x fewer KV heads in some models). Then, the tensors are reshaped:

```
    Q = Q.reshape(B, T, H, D)
    K = K.reshape(B, T, H, D)
    V = V.reshape(B, T, H, D)
```

Reshaping collapses kv_heads and q_per_kv into H, treating the problem as standard multi-head attention (MHA) with H heads, simplifying subsequent operations. This is a common pattern: handle GQA's complexity upfront, then proceed as if it's MHA.

For unbatched inputs, the logic is similar but omits the batch dimension:

```
else:
    T, kv_heads, q_per_kv, D = Q.shape
    H = kv_heads * q_per_kv
    K = K.unsqueeze(2).expand(-1, -1, q_per_kv, -1)
    # [T,kv_heads,q_per_kv,D]
    V = V.unsqueeze(2).expand(-1, -1, q_per_kv, -1)
    Q = Q.reshape(T, H, D)
    K = K.reshape(T, H, D)
    V = V.reshape(T, H, D)
```

The unsqueeze and expand occur at index 2, and the output shapes lack B. This symmetry ensures the code is maintainable. A potential pitfall is mismatched shapes (e.g., kv_heads in Q vs. K), which would raise errors here, so input validation is critical in practice.

Step 2: Computing Raw Attention Scores

The core of SDPA is computing similarity scores between queries and keys:

```
attn_scores = torch.einsum("bthd,bshd->bhts", Q, K) * sm_scale
```

The einsum operation performs a batched dot product. For each batch b, head h, and query position t, it computes the dot product of Q[b, t, h, :] (a D-dimensional vector) with every key K[b, :, h, :] (T vectors of D dimensions), producing a [B, H, T, T] tensor. The notation "bthd,bshd->bhts" specifies b (batch), t (query time), h (head), and d (dimension) for Q; b, s (key time), h, d for K; sum over d to get scores [b, h, t, s].

The result is scaled by sm_scale (1/sqrt(D)) to normalize the variance of the dot products, which otherwise could grow with D, causing the softmax to become too peaky and hurt training stability. This scaling, introduced in the original transformer paper, is essential for gradient flow.

The dtype and device are stored for later use:

```
score_dtype = attn_scores.dtype
dev = attn_scores.device
```

This ensures masks and other tensors match the precision (e.g., float16 for GPU speed) and device (GPU/CPU), avoiding costly transfers or dtype mismatches that could lead to NaNs.

For unbatched inputs, the einsum is "thd,shd->hts", producing [H, T, T] scores. Using einsum instead of matmul provides flexibility for multi-dimensional contractions and clarity in specifying the operation, leveraging optimized BLAS routines under the hood.

Step 3: Constructing and Applying the Attention Mask

Attention often requires masking to prevent attending to invalid positions, such as future tokens in autoregressive models or keys outside a sliding window:

```
causal = torch.full((T, T), float("-inf"), device=dev, dtype=score_dtype)
causal = torch.triu(causal, diagonal=1)  # -inf above diagonal, 0 elsewhere
```

The causal mask ensures queries at position t cannot attend to keys at s > t, which is critical for decoder-only models like GPT, where tokens are generated sequentially and future information must be hidden.

The torch.full creates a [T, T] tensor filled with -inf, and torch.triu zeros out the diagonal and below (diagonal=1 means exclude the diagonal itself from being -inf), leaving -inf above to mask future tokens.

If sliding_window > 0, a sliding window mask is added:

```
if sliding_window > 0:
    idx = torch.arange(T, device=dev)
    delta = idx[:, None] - idx[None, :]  # shape [T,T]
    sw_mask = torch.where(
        delta > sliding_window,
        torch.tensor(float("-inf"), device=dev, dtype=score_dtype),
        torch.tensor(0.0, device=dev, dtype=score_dtype),
    )
    mask = torch.maximum(causal, sw_mask)
else:
    mask = causal
```

The idx tensor is [0,1,...,T-1], and delta computes row_idx - col_idx, where rows are query positions and columns are key positions. A positive delta means the key is earlier than the query; if delta > sliding_window, the key is too far back and masked with -inf.

The torch.where creates a [T, T] mask: -inf for keys beyond the window, 0 otherwise. The torch.maximum combines causal and sw_mask, taking the stricter constraint (more -inf) at each position, so queries attend only to keys in [t - sliding_window, t]. If sliding_window is 0, only the causal mask applies.

The mask is applied to scores:

```
attn_scores = attn_scores + mask.view(1, 1, T, T)
```

The mask is reshaped to [1, 1, T, T] to broadcast over batch and head dimensions, adding -inf to invalid score positions, which will become near-zero after softmax. For unbatched inputs, it's attn_scores + mask.unsqueeze(0), broadcasting over heads.

This masking step is memory-intensive for large T, as the [T, T] mask can be gigabytes for long sequences. In production, optimized implementations like Flash Attention fuse masking into the attention kernel, using tiling to reduce memory usage.

Step 4: Incorporating Sink Logits

Sink logits are a novel addition:

```
S = S.to(score_dtype).view(1, H, 1, 1).expand(B, H, T, 1)
attn_scores = torch.cat([attn_scores, S], dim=-1)
```

The S tensor is cast to the score dtype and reshaped to [1, H, 1, 1], then expanded to [B, H, T, 1], broadcasting the per-head sink logit across batch and query positions.

The torch.cat appends this as an extra column, making attn_scores [B, H, T, T+1]. The sink acts as a virtual key, absorbing attention mass in softmax, which can stabilize numerical issues (e.g., when all real keys are masked) or support advanced techniques like speculative decoding, where sink logits might represent aggregated past information.

For unbatched inputs, S is reshaped to [H, 1, 1] and expanded to [H, T, 1], yielding [H, T, T+1]. This consistent handling ensures flexibility.

Step 5: Softmax Normalization and Output Computation

The attention weights are computed:

```
attn_weights = torch.softmax(attn_scores, dim=-1)[..., :-1]
```

Softmax is applied over the last dimension (T+1 keys, including sink), normalizing scores to probabilities. The [..., :-1] slices off the sink column, returning [B, H, T, T] weights. If the sink absorbed significant probability (e.g., due to high S values or heavy masking), the remaining weights sum to less than 1, which is acceptable as it reflects intentional redistribution of attention.

The output is computed:

```
out = torch.einsum("bhts,bshd->bthd", attn_weights, V)
return out.reshape(B, T, H * D)
```

The einsum performs a weighted sum: for each b, t, h, d, compute sum_s attn_weights[b, h, t, s] * V[b, s, h, d], producing [B, T, H, D]. The reshape flattens H and D to [B, T, H * D], matching the expected output for downstream layers like feed-forward networks or output projections.

For unbatched inputs, it's torch.einsum("hts,shd->thd", attn_weights, V) and reshape to [T, H * D].

This implementation is both pedagogical and practical, showing the explicit steps of SDPA while supporting advanced features. In practice, for large T, the quadratic O(T^2) complexity per head is prohibitive, and optimized kernels like Flash Attention v2 fuse these operations, using tiling and recomputation to reduce memory to O(T).

The sliding window reduces complexity to O(T * sliding_window), enabling models to handle 100k+ token contexts efficiently, as seen in models like Mistral or Longformer. Sink logits are less common but appear in research for stabilizing attention or managing KV caches in long-context inference.

For debugging, print intermediate shapes or use torch.nan_to_num to catch numerical issues, especially in float16. Ensure input shapes align with the model's architecture (e.g., kv_heads consistency). For example, with T=4, sliding_window=2, and causal masking, a query at t=3 attends only to t=1,2,3; sink logits might draw attention if S is large, affecting the output distribution.

This SDPA variant powers models like Grok or Llama, balancing flexibility and efficiency. In the next chapter, we'll integrate this into a full transformer block, exploring how it interacts with layer normalization, feed-forward layers, and residual connections to build state-of-the-art language models.

Summary

Scaled dot-product attention (SDPA) stands as a cornerstone of modern artificial intelligence, underpinning the remarkable success of transformer architectures across diverse domains, from natural language processing to computer vision and beyond. This chapter has provided a comprehensive exploration of SDPA, unraveling its mathematical elegance, computational intricacies, and practical enhancements like group query attention, sliding window mechanisms, and sink logits. By dissecting the provided PyTorch implementation, we've illuminated how SDPA balances efficiency and expressivity, enabling models to process long sequences, enforce causality, and maintain numerical stability.

The journey through its historical evolution, conceptual foundations, and advanced variants has highlighted why SDPA remains a universal primitive in AI, driving innovations that continue to redefine the boundaries of machine intelligence.

The implementation discussed, with its support for batched and unbatched inputs, causal masking, sliding windows, and sink logits, exemplifies the flexibility required to power state-of-the-art models like Grok, GPT OSS, or Llama. By carefully managing tensor shapes, broadcasting keys and values for group query attention, and applying sophisticated masking strategies, this code captures the essence of SDPA's adaptability to real-world challenges.

Its ability to scale to long contexts through sliding windows and stabilize attention with sink logits underscores the practical engineering that makes transformers viable for large-scale applications. Yet, this is only the beginning of understanding how attention integrates into the broader architecture of transformers.

In the next chapter, we will delve deeper into the architecture of attention blocks, exploring how SDPA integrates with layer normalization, feed-forward networks, and residual connections to form the backbone of transformer models. We'll examine the interplay of these components, their role in stabilizing training, and how they collectively enable the powerful representational capabilities that have made transformers the gold standard in AI.

By building on the foundations laid here, we'll uncover the mechanics of full attention blocks and their critical role in achieving state-of-the-art performance across a wide range of tasks.

CHAPTER 7

AttentionBlock with Rotary Embedding, GQA, Sliding Window, and Sink Tokens

Attention mechanisms are the cornerstone of modern large language models (LLMs), underpinning their ability to process and generate human-like text. Introduced in the 2017 paper "Attention is All You Need" by Vaswani et al., the transformer architecture replaced recurrent neural networks (RNNs) with attention-based mechanisms, enabling parallel computation and direct modeling of relationships between tokens.

Unlike RNNs, which struggled with vanishing gradients and sequential processing, attention allows **LLMs like GPT, BERT, T5, and LLaMA** to capture long-range dependencies, contextual nuances, and complex linguistic structures efficiently. This chapter provides an exhaustive exploration of attention in LLMs, covering its historical context, mathematical foundations, mechanisms, implementation, optimizations, applications, interpretability, limitations, and advancements as of 2025.

Fundamentals of Attention

Attention mimics human cognitive focus by prioritizing relevant information. In machine learning, it computes a weighted sum of input representations, with weights reflecting the relevance of each element to a query. The mechanism operates on three components: queries (representing the current focus), keys (elements to compare against), and values (information to retrieve).

D. Grigorov, *Building Large Language Models from Scratch*, https://doi.org/10.1007/979-8-8688-2297-1_7

Mathematically, for a query vector
$\mathbf{q} \in \mathbb{R}^{d_k}$), $keyvectors(\mathbf{K} = [\mathbf{k}_1, \mathbf{k}_2, \ldots, \mathbf{k}_n] \in \mathbb{R}^{n \times d_k}$, and value vectors ($\mathbf{V} = [\mathbf{v}_1, \mathbf{v}_2, \ldots, \mathbf{v}_n] \in \mathbb{R}^{n \times d_v}$), the attention output is:

$$\text{Attention}(\mathbf{q}, \mathbf{K}, \mathbf{V}) = \sum_{i=1}^{n} \alpha_i \mathbf{v}_i$$

where weights (α_i) are computed using a similarity function $s(\cdot, \cdot)$ (e.g., dot product) and normalized via softmax:

$$\alpha_i = \frac{\exp(s(\mathbf{q}, \mathbf{k}i))}{\sum j = 1^n \exp(s(\mathbf{q}, \mathbf{k}_j))}$$

In matrix form, for a batch of queries $\mathbf{Q} \in \mathbb{R}^{n \times d_k}$, attention generalizes to:

$$\text{Attention}(\mathbf{Q}, \mathbf{K}, \mathbf{V}) = \mathbf{AV}, \quad \mathbf{A} = \text{softmax}(s(\mathbf{Q}, \mathbf{K}))$$

Common similar functions include

- **dot product**
- **scaled dot product** (used in transformers)
- **additive attention,** with the scaled dot product being the standard due to its efficiency and stability

Self-Attention in Transformers

Self-attention is a pivotal mechanism in transformer architecture, enabling large language models (LLMs) to process sequences by allowing each token to dynamically attend to all others in the input, thereby capturing intricate intra-sequence relationships essential for contextual understanding.

Unlike recurrent neural networks (RNNs), which process sequences sequentially and struggle with long-range dependencies due to vanishing gradients, self-attention facilitates parallel computation and direct modeling of relationships between any pair of tokens, making it a cornerstone of models like GPT, BERT, and LLaMA.

Self-attention operates on a query-key-value (QKV) framework, where queries, keys, and values are derived from the same input sequence. For an input sequence represented as a matrix $\mathbf{X} \in \mathbb{R}^{n \times d}$, where n is the number of tokens and d `https://www.codecogs.com/eqnedit.php?latex=d#0` is the embedding dimension, we compute the query, key, and value matrices through linear projections:

$$\mathbf{Q} = \mathbf{X}\mathbf{W}_Q, \quad \mathbf{K} = \mathbf{X}\mathbf{W}_K, \quad \mathbf{V} = \mathbf{X}\mathbf{W}_V$$

Here, $\mathbf{W}_Q, \mathbf{W}_K, \mathbf{W}_V \in \mathbb{R}^{d \times d_k}$ are learned weight matrices, and d_k is the dimension of queries and keys (often set to d or d/h in multi-head attention, where h is the number of heads. The scaled dot-product attention mechanism, which is the standard in transformers, is defined as:

$$\text{Attention}(\mathbf{Q}, \mathbf{K}, \mathbf{V}) = \text{softmax}\left(\frac{\mathbf{Q}\mathbf{K}^T}{\sqrt{d_k}}\right)\mathbf{V}$$

The scaling factor $\sqrt{d_k}$ is crucial to prevent large dot products from pushing the softmax function into regions with vanishing gradients, ensuring numerical stability during training. The attention weights $\boldsymbol{A} = [\backslash\text{softmax}?]\left(\frac{\boldsymbol{Q}\boldsymbol{K}^T}{\sqrt{d_k}}\right) \in \mathbb{R}^{n \times n}$ represent the relevance of each token to every other token, and the output is a weighted sum of the value vectors, producing a context-aware representation for each token.

Operational Properties

Self-attention possesses several key properties that make it exceptionally suited for LLMs:

1. **Permutation Equivariance**: The attention mechanism is invariant to the order of input tokens, meaning it treats the sequence as a set unless positional information is explicitly added. This is addressed through positional encodings, discussed later.

2. **Parallel Processing**: Unlike RNNs, which process tokens sequentially, self-attention computes interactions for all tokens simultaneously, enabling efficient use of parallel hardware like GPUs. This reduces training and inference time significantly.

3. **Long-Range Dependencies**: Self-attention allows any token to directly attend to any other, regardless of distance, overcoming the limitations of RNNs, where information decays over long sequences. This is critical for tasks like resolving pronouns or maintaining coherence in extended texts.

4. **Contextual Embeddings**: Each token's output representation is a weighted combination of all input tokens, incorporating global context. For example, in the sentence "The cat chased the mouse because it was hungry," self-attention enables the model to link "it" to "mouse" by assigning high attention weights to relevant tokens.

5. **Quadratic Complexity**: The computational complexity is $O(n^2d)$, quadratic in sequence length n but linear in model depth d. While this enables scalability for large models, it poses challenges for extremely long sequences, addressed by optimizations like sparse attention.

Causal Self-Attention

In autoregressive LLMs like GPT, causal self-attention (also called masked self-attention) ensures that each token only attends to itself and previous tokens, preventing information leakage from future tokens during generation. This is achieved by applying a lower-triangular mask to the attention scores:

$$\text{masked}_s\text{cores}i,j = \begin{cases} \text{scores}i,j & \text{if } j \leq i \\ -\infty & \text{if } j > i \end{cases}$$

The mask is applied before the softmax, effectively setting attention weights for future tokens to zero:

$$\boldsymbol{A} = \text{softmax}(\text{masked}_s\text{cores})$$

This ensures that the model generates tokens sequentially, making it suitable for tasks like text generation where the output is produced one token at a time.

Implementation in Transformers

Self-attention is a core component of the transformer architecture, integrated into both encoder and decoder layers. In an encoder layer, self-attention is followed by residual connections, layer normalization, and a feed-forward network (FFN):

$$\mathbf{Z} = \text{Attention}(\mathbf{X}\mathbf{W}_Q, \mathbf{X}\mathbf{W}_K, \mathbf{X}\mathbf{W}_V)$$
$$\mathbf{X} \leftarrow \text{LayerNorm}(\mathbf{X} + \mathbf{Z})$$
$$\mathbf{Z} = \text{FFN}(\mathbf{X}), \quad \mathbf{X} \leftarrow \text{LayerNorm}(\mathbf{X} + \mathbf{Z})$$

In decoders, causal self-attention is used, and seq2seq models include an additional cross-attention layer to attend to encoder outputs. Self-attention's parallel nature allows transformers to scale to billions of parameters, with modern LLMs stacking dozens of layers (e.g., GPT-3 with 96 layers).

Role in LLMs

Self-attention is critical to LLMs' capabilities:

1. **Contextual Understanding**: By allowing each token to attend to all others, self-attention creates rich, context-aware representations, enabling tasks like resolving ambiguities (e.g., pronoun resolution in long texts).

2. **Long-Range Dependencies**: Direct token interactions capture relationships across entire documents, crucial for tasks like summarization or question answering over long contexts.

3. **In-Context Learning**: Self-attention enables LLMs to adapt to new tasks by attending to few-shot examples in prompts, a hallmark of models like GPT-4.

4. **Scalability**: The parallel nature and linear depth complexity allow self-attention to scale efficiently with model size, aligning with scaling laws that show performance improvements with more parameters.

5. **Emergent Behaviors**: Self-attention contributes to emergent abilities like arithmetic, reasoning, or code generation by modeling complex token interactions.

Limitations and Challenges

1. **Quadratic Complexity**: The $O(n^2d)$ complexity limits context length, though sparse and linear attention mitigate this.
2. **Attention Sinks**: Models may overly attend to initial tokens (Xiao et al., 2023), skewing focus.
3. **Memory Usage**: Storing $\mathbf{Q}, \mathbf{K}, \mathbf{V}$ matrices requires significant memory, especially for long sequences.
4. **Positional Sensitivity**: Without proper encodings, self-attention struggles with order-dependent tasks.
5. **Training Stability**: Large dot products can destabilize training without scaling.

Optimizations and Variants

The quadratic complexity $O(n^2d)$ self-attention becomes a bottleneck for long sequences (e.g., 1M+ tokens). Several optimizations address this:

1. **Sparse Self-Attention**: Limits attention to subsets of tokens, such as local windows (Longformer), dilated patterns, or global tokens (BigBird), reducing complexity to $O(n \log n)$ or $O(n)$.
2. **Linear Attention**: Approximates the softmax with kernel methods (e.g., Performer):

$$\text{softmax}(\mathbf{QK}^T) \approx \phi(\mathbf{Q})\phi(\mathbf{K})^T$$

   achieving $O(nd^2)$ complexity.
3. **Flash Attention**: A hardware-aware optimization (Dao et al., 2022) that computes attention in blocks, minimizing memory I/O and speeding up training/inference.

4. **KV Caching**: During autoregressive inference, past keys and values are cached to avoid recomputation, which is critical for efficient text generation.

5. **Relative Positional Encodings**: Techniques like RoPE or T5's relative position biases model positional relationships more effectively than absolute encodings.

Self-attention weights provide insights into model behavior. Visualizing the attention matrix $\mathbf{A}$ reveals which tokens influence each other. For example, in syntactic tasks, attention may focus on adjacent tokens, while in semantic tasks, it captures distant relationships. However, interpretability is limited, as later layers (e.g., FFNs) also shape outputs, and attention weights alone don't fully explain decisions.

Multihead Attention in Transformers

Multi-head attention (MHA) is a critical extension of the self-attention mechanism in the transformer architecture, significantly enhancing its ability to model complex relationships within input sequences for large language models (LLMs). By computing attention in parallel across multiple heads, each with distinct query-key-value (QKV) projections, MHA allows LLMs like GPT, BERT, and LLaMA to **capture diverse linguistic patterns, such as syntactic structures, semantic relationships, and positional dependencies.**

This mechanism is pivotal for the flexibility and performance of modern LLMs, enabling them to handle tasks ranging from text generation to multimodal reasoning. This section provides an exhaustive exploration of MHA, covering its mathematical formulation, operational principles, implementation details, role in LLMs, optimizations, interpretability, limitations, and recent advancements as of 2025.

Multi-head attention builds on self-attention by splitting the computation into h parallel heads, each attending to the input sequence with its own set of QKV projections. For an input sequence represented as a matrix $\mathbf{X} \in \mathbb{R}^{n \times d}$, where n is the number of tokens and d is the embedding dimension, self-attention first computes queries, keys, and values:

$$\mathbf{Q} = \mathbf{X}\mathbf{W}_Q, \quad \mathbf{K} = \mathbf{X}\mathbf{W}_K, \quad \mathbf{V} = \mathbf{X}\mathbf{W}_V$$

where $\mathbf{W}_Q, \mathbf{W}_K, \mathbf{W}_V \in \mathbb{R}^{d \times d_k}$, and d_k is the dimension of queries and keys. In MHA, each head $i for i = 1, 2, \ldots, h$ uses its own projection matrices:

$$\text{head}i = \text{Attention}(\mathbf{QW}Q_i, \mathbf{KW}K_i, \mathbf{VW}V_i)$$

where $\mathbf{W}Q_i, \mathbf{W}K_i, \mathbf{W}_{V_i} \in \mathbb{R}^{d \times d_k}$, and typically $d_k = d/h$ to maintain computational efficiency across heads. The attention mechanism for each head follows the scaled dot-product formulation:

$$\text{Attention}(\mathbf{Q}i, \mathbf{K}i, \mathbf{V}i) = \text{softmax}\left(\frac{(\mathbf{QW}Q_i)(\mathbf{KW}K_i)^T}{\sqrt{d_k}}\right)(\mathbf{VW}V_i)$$

The outputs of all heads are concatenated and passed through a final projection:

$$\text{MultiHead}(\mathbf{Q}, \mathbf{K}, \mathbf{V}) = \text{Concat}(\text{head}_1, \text{head}_2, \ldots, \text{head}_h)\mathbf{W}_O$$

where $\mathbf{W}_O \in \mathbb{R}^{hd_v x d}$, and $d_v = d_k$ is the dimension of the value vectors. This final projection ensures the output dimensionality matches the input, allowing MHA to integrate seamlessly into the transformer architecture.

Operational Principles

MHA enhances self-attention by enabling the model to focus on different aspects of the input sequence simultaneously. Each head operates independently, learning distinct attention patterns that capture complementary relationships.

The key properties of MHA include:

1. **Diverse Representations**: By using multiple heads, MHA captures a wide range of linguistic phenomena, improving the model's ability to generalize across tasks.

2. **Parallel Computation**: All heads compute attention simultaneously, leveraging GPU parallelization for efficiency, similar to single-head self-attention.

3. **Specialization of Heads**: Empirical studies (e.g., Clark et al., 2019) show that heads often specialize, with some focusing on local tokens (e.g., adjacent words for syntax), others on distant tokens (e.g., for pronoun resolution), and some on positional patterns.

4. **Robustness**: The diversity of attention patterns makes MHA more robust to noise and variations in input data, enhancing performance on complex tasks.

5. **Scalability**: Increasing the number of heads improves model capacity, aligning with scaling laws that correlate larger models with better performance.

The computational complexity of MHA remains $O(n^2 d)$, as each head processes the sequence independently, but the division of dimensions across heads $d_k = d/h$ ensures that the per-head computation is lighter, maintaining efficiency.

Integration in Transformers

MHA is a core component of transformer layers, used in both encoder and decoder stacks. In an encoder layer, MHA is followed by residual connections, layer normalization, and a feed-forward network (FFN):

$$\mathbf{Z} = \mathrm{MultiHead}(\mathbf{X}, \mathbf{X}, \mathbf{X})$$

$$\mathbf{X} \leftarrow \mathrm{LayerNorm}(\mathbf{X} + \mathbf{Z})$$

$$\mathbf{Z} = \mathrm{FFN}(\mathbf{X}), \quad \mathbf{X} \leftarrow \mathrm{LayerNorm}(\mathbf{X} + \mathbf{Z})$$

In decoders, MHA is adapted for causal (masked) self-attention to prevent attending to future tokens, and seq2seq models include cross-attention to encoder outputs. The flexibility of MHA allows it to be tailored to different architectures, such as encoder-only (e.g., BERT), decoder-only (e.g., GPT), or encoder-decoder (e.g., T5).

Positional Considerations

Since MHA inherits self-attention's permutation-equivariance, positional encodings are critical to incorporate token order. These are added to the input embeddings:

$$\mathbf{X}_{\text{input}} = \mathbf{X} + \mathbf{P}$$

where $\mathbf{P} \in \mathbb{R}^{n \times d}$ encodes positions, using sinusoidal functions:

$$\mathbf{P}pos, 2i = \sin\left(\frac{pos}{10000^{2i/d}}\right), \quad \mathbf{P}pos, 2i+1 = \cos\left(\frac{pos}{10000^{2i/d}}\right)$$

or learned embeddings. Advanced approaches like Rotary Position Embeddings (RoPE) rotate queries and keys per head:

$$\mathbf{q}_i \leftarrow \mathbf{q}i \cdot \mathbf{R}\theta, i, \quad \mathbf{k}_i \leftarrow \mathbf{k}i \cdot \mathbf{R}\theta, i$$

This enhances MHA's ability to model relative positional relationships, improving performance on tasks requiring order sensitivity.

Role in LLMs

MHA is fundamental to LLMs' capabilities:

1. **Contextual Diversity**: By allowing heads to specialize, MHA captures a broad spectrum of linguistic features, from local syntactic rules to global semantic coherence, enabling nuanced understanding in tasks like question answering or summarization.
2. **In-Context Learning**: MHA enables LLMs to adapt to new tasks by attending to few-shot examples in prompts, as seen in GPT-4's ability to learn from demonstrations.
3. **Reasoning**: Heads focusing on intermediate tokens support chain-of-thought reasoning, critical for complex tasks like arithmetic or logical inference.
4. **Multimodality**: In multimodal models (e.g., GPT-4V), MHA variants like cross-attention integrate text and images, with heads specializing in different modalities.
5. **Emergent Behaviors**: The diversity of attention patterns contributes to emergent abilities, such as code generation or factual recall, by modeling complex token interactions.

Limitations and Challenges

1. **Quadratic Complexity**: MHA's O(n^2 d) complexity limits scalability for long sequences, though sparse and linear variants mitigate this.
2. **Memory Usage**: Storing QKV matrices for multiple heads requires substantial memory, especially for large (h).
   Head Redundancy: Some heads may learn redundant patterns, reducing efficiency.
3. **Training Stability**: Large dot products in heads can destabilize training without proper scaling.
4. **Interpretability Limits:** While attention weights are insightful, they don't capture the full decision-making process.

Optimizations and Variants

MHA's quadratic complexity $O(n^2 d)$ poses challenges for long sequences, prompting several optimizations:

1. **Sparse MHA**: Restricts attention to subsets of tokens (e.g., local windows in Longformer or global tokens in BigBird), reducing complexity to $O(n \log n)$ or $O(n)$.
2. **Linear Attention in Heads**: Applies kernel-based approximations (e.g., Performer) to individual heads:

   $$\text{softmax}(\mathbf{Q}_i \mathbf{K}_i^T) \approx \phi(\mathbf{Q}_i)\phi(\mathbf{K}_i)^T$$

   lowering per-head complexity.
3. **Flash Attention**: Optimizes MHA by computing attention in blocks (Dao et al., 2022), reducing memory I/O and speeding up training/inference.
4. **Group Query Attention (GQA)**: Shares keys/values across groups of heads (e.g., LLaMA 2), reducing parameters while maintaining quality.

5. **Multi-Query Attention (MQA)**: Uses a single set of keys/values across all heads (e.g., GPT-3.5 Turbo), minimizing memory usage.

6. **Quantized MHA**: Employs low-bit representations for QKV matrices, enabling deployment on resource-constrained devices.

7. **Dynamic Head Allocation**: Adapts the number of active heads based on input complexity, improving efficiency.

Recent advancements (up to 2025) include integrating MHA with state space models (SSMs) like Mamba, which combine attention with linear recurrence for $O(n)$ complexity, and hybrid approaches blending MHA with convolutions (e.g., Hyena).

MHA's attention weights provide valuable insights into model behavior. Visualizing the attention matrix for each head reveals specialized patterns: some heads focus on local syntactic structures (e.g., adjacent tokens), others on long-range semantic relationships (e.g., pronoun references), and some on positional cues. Studies (e.g., Vig, 2019) show that head specialization enhances interpretability, but weights alone don't fully explain model decisions, as feed-forward layers and residuals also contribute significantly.

Advanced Considerations (up to 2025)

Recent developments have enhanced MHA:

1. **Integration with SSMs**: Combining MHA with linear recurrence models like Mamba achieves sub-quadratic complexity.

2. **Adaptive MHA**: Dynamically adjusts the number of heads or attention patterns based on input complexity.

3. **Multimodal MHA**: Extends MHA to handle text, images, and audio in models like GPT-4V, with heads specializing in cross-modal interactions.

4. **Neuroscience-Inspired MHA**: Explores human-like attention mechanisms for efficiency.

5. **Scalable Training**: Techniques like Flash Attention 2 optimize MHA for ultra-large models.

Multi-head attention is a cornerstone of the transformer architecture, enabling LLMs to capture diverse linguistic relationships through parallel, specialized attention heads. Its ability to model syntactic, semantic, and positional patterns makes it indispensable for tasks like in-context learning, reasoning, and multimodal processing.

Despite challenges like quadratic complexity and memory demands, optimizations like sparse MHA, linear attention, and Flash Attention ensure its continued relevance. As of 2025, MHA remains central to LLMs, with ongoing research integrating it with novel architectures and enhancing its efficiency for future applications.

What Is Grouped Query Attention, Sliding Window Attention, and Sink Tokens?

Grouped Query Attention (GQA)

GQA (Ainslie et al., 2023) optimizes MHA by grouping query heads to share key and value projections, balancing the efficiency of multi-query attention (MQA) with MHA's expressiveness. In standard MHA, each head has unique query, key, and value projections:

$$\text{head}i = \text{Attention}(\mathbf{QW}Q_i, \mathbf{KW}K_i, \mathbf{VW}V_i)$$

In GQA, queries are divided into g groups $g \leq h$, with each group sharing a single key and value projection. For h heads and g groups, each group contains h/g query heads. The key and value matrices for group j:

$$\mathbf{K}j = \mathbf{XW}K_j, \quad \mathbf{V}j = \mathbf{XW}V_j$$

are shared across the group's query heads. The attention for head i `https://www.codecogs.com/eqnedit.php?latex=i#0` in group j

`https://www.codecogs.com/eqnedit.php?latex=j#0`:

$$\text{head}i = \text{softmax}\left(\frac{(\mathbf{QW}Q_i)\mathbf{K}_j^T}{\sqrt{d_k}}\right)\mathbf{V}_j$$

Outputs are concatenated and projected:

$$\text{GQA}(\mathbf{Q}, \mathbf{K}, \mathbf{V}) = \text{Concat}(\text{head}_1, \ldots, \text{head}_h)\mathbf{W}_O$$

where $\mathbf{W}_O \in \mathbb{R}^{hd_v \times d}$. GQA reduces key and value parameters from $O(hd_k)$ to $O(gd_k)$, enhancing memory efficiency while preserving performance.

Sliding Window Attention (SWA)

SWA (Beltagy et al., 2020) restricts attention to a fixed-size window of w tokens around each token, reducing complexity from $O(n^2d)$ to $O(nwd)$. For a token at position i, attention is computed only for tokens in $[i - w/2, i + w/2]$ `https://www.codecogs.com/eqnedit.php?latex=%5Bi%20-%20w%2F2%2C%20i%20%2B%20w%2F2%5D#0` or $[i - w, i]$ in causal models:

$$\text{scores}_{i,j} = \frac{(\mathbf{q}_i \cdot \mathbf{k}_j)}{\sqrt{d_k}}, \quad j \in [i - w/2, i + w/2]$$

The attention output:

$$\text{Attention}i = \sum j \in [i - w/2, i + w/2] \text{softmax}(\text{scores}_{i,j})\mathbf{v}_j$$

In causal SWA, the window covers only previous tokens, making it suitable for autoregressive tasks. SWA enables efficient processing of long sequences by focusing on local context.

Sink Tokens

Sink tokens address the attention sink problem (Xiao et al., 2023), where LLMs overly attend to initial tokens, skewing context modeling. Special tokens (e.g., the first few tokens or dedicated tokens) absorb excess attention to ensure balanced distribution. Attention scores are adjusted with a bias term:

$$\boldsymbol{A} = [\backslash\text{softmax?}]\left(\frac{\boldsymbol{QK}^T}{\sqrt{d_k}} + \boldsymbol{M}_{\text{sink}}\right)$$

where (\mathbf{M}_{\text{sink}}) prioritizes sink tokens, preventing over-concentration and improving long-context performance.

Integration in the Attention Block

The AttentionBlock orchestrates these mechanisms as follows. The input $\mathbf{X} \in \mathbb{R}^{n \times d}$ is projected into queries, keys, and values. RoPE is applied to queries and keys to encode positional information. GQA groups queries and shares keys/values within groups.

SWA restricts attention to a sliding window, with causal masking for autoregressive tasks. Sink tokens adjust attention scores to balance distribution. The output is combined with the input via a residual connection and normalized:

$$\mathbf{Z} = \mathrm{GQA}_{\mathrm{SWA}}(\mathbf{X}), \quad \mathbf{X} \leftarrow \mathrm{LayerNorm}(\mathbf{X} + \mathbf{Z})$$

An FFN is applied, followed by another residual connection and normalization:

$$\mathbf{Z} = \mathrm{FFN}(\mathbf{X}), \quad \mathbf{X} \leftarrow \mathrm{LayerNorm}(\mathbf{X} + \mathbf{Z})$$

This process is repeated across layers, refining token representations.

The AttentionBlock's advanced mechanisms enhance its functionality. RoPE's rotation-based encoding improves positional generalization, allowing LLMs to handle longer sequences effectively. GQA reduces memory and parameter costs, making the block scalable for large models. SWA's linear complexity enables processing of extended contexts, such as 128k tokens in LLaMA 3. Sink tokens ensure balanced attention, improving coherence in long texts. The block leverages GPU parallelization, with optimizations like Flash Attention (Dao et al., 2022) further reducing memory I/O.

The AttentionBlock's advanced mechanisms also represent a sophisticated evolution of attention, enabling efficient, scalable, and contextually robust LLMs. By addressing positional encoding, computational complexity, and attention distribution challenges, it powers advanced tasks like long-context modeling and multimodal processing. As of 2025, ongoing research continues to refine these mechanisms, integrating them with novel architectures to advance NLP capabilities.

Integration of Attention Mechanism in Our Custom Large Language Model

```
class AttentionBlock(torch.nn.Module):
    def __init__(self, config: ModelConfig, layer_idx: int = 0, device:
    torch.device | None = None):
```

```
        super().__init__()
        self.config = config
        self.head_dim = config.head_dim
        self.num_heads = config.num_attention_heads
        self.num_kv_heads = config.num_key_value_heads
        if self.num_heads % self.num_kv_heads != 0:
            raise ValueError(f"num_attention_heads ({self.num_heads}) must
            be divisible by num_key_value_heads ({self.num_kv_heads}).")
        if self.head_dim % 2 != 0:
            raise ValueError(f"RoPE requires even head_dim, got {self.
            head_dim}.")

        self.q_per_kv = self.num_heads // self.num_kv_heads
        self.sliding_window = config.sliding_window if layer_idx % 2 ==
        0 else 0

        self.norm = RMSNorm(config.hidden_size, device=device)

        # QKV projection: concatenated [Q | K | V]
        self.qkv_proj = torch.nn.Linear(
            config.hidden_size,
            self.head_dim * (self.num_heads + 2 * self.num_kv_heads),
            device=device
        )

        # Output projection
        self.out_proj = torch.nn.Linear(
            self.num_heads * self.head_dim,
            config.hidden_size,
            device=device
        )

        self.rope = RotaryEmbedding(
            head_dim=self.head_dim,
            base=config.rope_theta,
            dtype=torch.float32,  # compute cos/sin in fp32 for stability
            initial_context_length=config.initial_context_length,
            scaling_factor=config.rope_scaling_factor,
```

```
            ntk_alpha=config.rope_ntk_alpha,
            ntk_beta=config.rope_ntk_beta,
            device=device
        )

        # Sink logits per head (broadcasted inside sdpa)
        self.sink_logits = torch.nn.Parameter(torch.zeros(self.num_heads,
        device=device))

        # scale = 1/sqrt(d)
        self.sm_scale = self.head_dim ** -0.5

    def forward(self, x: torch.Tensor) -> torch.Tensor:
        """
        x: [B, T, H] -> returns [B, T, H]
        """
        B, T, H = x.shape
        residual = x

        x = self.norm(x)

        qkv = self.qkv_proj(x)  # [B, T, head_dim*(num_heads + 2*num_
        kv_heads)]
        q_dim = self.num_heads * self.head_dim
        kv_dim = self.num_kv_heads * self.head_dim

        q = qkv[:, :, :q_dim]
        k = qkv[:, :, q_dim:q_dim + kv_dim]
        v = qkv[:, :, q_dim + kv_dim:]

        # Reshape to GQA-friendly shapes
        # Q: [B,T,kv_heads,q_per_kv,D], K/V: [B,T,kv_heads,D]
        q = q.reshape(B, T, self.num_kv_heads, self.q_per_kv, self.head_
        dim).contiguous()
        k = k.reshape(B, T, self.num_kv_heads, self.head_dim).contiguous()
        v = v.reshape(B, T, self.num_kv_heads, self.head_dim).contiguous()

        # Apply RoPE on head dimension
        q, k = self.rope(q, k)
```

```
        # Attention (causal + optional sliding window + sink logits)
        attn = sdpa(q, k, v, self.sink_logits, self.sm_scale, self.
        sliding_window)

        out = self.out_proj(attn)
        return residual + out
```

Exhaustive Explanation of the AttentionBlock PyTorch Module

The `AttentionBlock` class is a meticulously engineered PyTorch module designed as a cornerstone of decoder-only transformer-based large language models (LLMs), such as those inspired by the LLaMA family, Mistral, or Gemma 2. This module encapsulates a highly optimized self-attention mechanism tailored for autoregressive sequence generation, incorporating advanced techniques to enhance performance, efficiency, and scalability.

These techniques include **grouped query attention (GQA)** for reducing parameter and memory costs, **rotary positional embeddings (RoPE)** for robust relative positional encoding, **sliding window attention (SWA)** for handling long sequences with linear computational complexity, and **learnable sink logits** to mitigate the attention sink phenomenon, where models disproportionately allocate attention to initial tokens, degrading performance on extended or streaming contexts.

The module processes input tensors of shape $[B, T, H]$ `https://www.codecogs.com/eqnedit.php?latex=%20%5BB%2C%20T%2C%20H%5D%20#0` (batch size B `https://www.codecogs.com/eqnedit.php?latex=%20B%20#0`, sequence length T `https://www.codecogs.com/eqnedit.php?latex=%20T%20#0`, hidden dimension H `https://www.codecogs.com/eqnedit.php?latex=%20H%20#0`) and output the same shape, enabling seamless integration into multi-layer transformer architectures, typically comprising 32–80 layers in modern LLMs. The explanation also contextualizes the block within the evolution of attention mechanisms, drawing parallels with models like GPT, BERT, and T5, and highlights its role in advancing long-context and multimodal capabilities.

Class Definition and PyTorch Integration

The `AttentionBlock` is defined as a subclass of `torch.nn.Module`:

```
class AttentionBlock(torch.nn.Module):
```

Inheriting from `torch.nn.Module` endows the class with PyTorch's built-in functionality for managing learnable parameters (via `parameters()` for optimizers like AdamW), state serialization (e.g., `torch.save` for checkpoints), and device placement (e.g., `to('cuda')` for GPU acceleration). This design ensures modularity, allowing the block to be stacked in a transformer decoder, alternated with feed-forward networks (FFNs), and integrated into larger architectures like `nn.Sequential` or custom model classes. In a typical LLM, multiple `AttentionBlock` instances (e.g., 32 for LLaMA-13B, 80 for larger models) refine token representations iteratively, each contributing to the model's ability to capture complex linguistic dependencies.

Constructor (`__init__`) Method

The `__init__` method initializes the module with hyperparameters, validates configurations, and sets up sub-modules and parameters. It accepts three arguments:

- `config: ModelConfig`: A dataclass or object containing hyperparameters like `hidden_size`, `head_dim`, `num_attention_heads`, `num_key_value_heads`, `sliding_window`, `rope_theta`, `initial_context_length`, `rope_scaling_factor`, `rope_ntk_alpha`, and `rope_ntk_beta`. These define the model's architecture and behavior.
- `layer_idx: int = 0`: The 0-based index of this block in the transformer stack, used to toggle SWA for even layers.
- `device: torch.device | None = None`: Specifies the device (e.g., "`cuda`", "`cpu`") for tensor allocation; defaults to CPU if None.

Parameter Extraction and GQA Configuration

```
self.config = config
self.head_dim = config.head_dim
self.num_heads = config.num_attention_heads
self.num_kv_heads = config.num_key_value_heads
```

The constructor extracts key parameters from `config`:

- `head_dim`: The dimensionality of each attention head (e.g., 128 in LLaMA models), determining the size of query, key, and value vectors per head.
- `num_attention_heads`: The total number of query heads (e.g., 32), defining the model's attention capacity.
- `num_key_value_heads`: The number of key and value heads for GQA (e.g., 8), which is typically smaller than `num_attention_heads` to reduce memory usage.

Grouped Query Attention (GQA) Rationale: In standard Multi-Head Attention (MHA), each of the h heads has its own query (Q), key (K), and value (V) projections, requiring $3h \times d_k$ parameters (where $d_k = \mathrm{head}_\mathrm{d}\mathrm{im}$). GQA, introduced in Ainslie et al. (2023), reduces the number of K and V heads to $g = \mathrm{num}_\mathrm{k}\mathrm{v}_\mathrm{h}\mathrm{eads}$, with each K/V head shared across $q_{\mathrm{per}_\mathrm{k}\mathrm{v}} = \mathrm{num}_\mathrm{h}\mathrm{eads}/\mathrm{num}_\mathrm{k}\mathrm{v}_\mathrm{h}\mathrm{eads}$ query heads (e.g., 32 / 8 = 4). This reduces the K/V parameter count and, crucially, the KV cache size during inference by a factor of h/g, enabling efficient processing of long sequences (e.g., 128k tokens) while maintaining performance close to full MHA. For example, with $h = 32, g = 8$, the KV cache is 4x smaller, critical for deployment on resource-constrained hardware.

Validation Checks

```
if self.num_heads % self.num_kv_heads != 0:
    raise ValueError(f"num_attention_heads ({self.num_heads}) must be
    divisible by num_key_value_heads ({self.num_kv_heads}).")
if self.head_dim % 2 != 0:
    raise ValueError(f"RoPE requires even head_dim, got {self.head_dim}.")
```

These checks ensure compatibility:

- **GQA Divisibility**: Ensures $\mathrm{num}_\mathrm{h}\mathrm{eads} \bmod \mathrm{num}_\mathrm{k}\mathrm{v}_\mathrm{h}\mathrm{eads} = 0$, so $q_{\mathrm{per}_\mathrm{k}\mathrm{v}}$ is an integer. This prevents tensor shape mismatches when grouping queries in the forward pass.
- **RoPE Requirement**: RoPE applies rotations to pairs of dimensions in the head embedding (treating them as complex numbers), requiring $\mathrm{head}_\mathrm{d}\mathrm{im}$ to be even. For a head vector $\mathbf{q} \in \mathbb{R}^{d_k}$, RoPE splits into $d_k/2$ pairs, each rotated by an angle proportional to the token position. An odd d_k would break this pairing, causing errors.

Sliding Window and GQA Grouping

```
self.q_per_kv = self.num_heads // self.num_kv_heads
self.sliding_window = config.sliding_window if layer_idx % 2 == 0 else 0
```

- **GQA Grouping**: Computes $q_{\mathrm{per}_\mathrm{k}\mathrm{v}} = \mathrm{num}_\mathrm{h}\mathrm{eads} // \mathrm{num}_\mathrm{k}\mathrm{v}_\mathrm{h}\mathrm{eads}$ (e.g., 4), defining how many query heads share each K/V head. This is used for tensor reshaping in the forward pass.
- **Sliding Window Attention (SWA)**: Sets `sliding_window` to a fixed size (e.g., 4096 tokens) for even layers (`layer_idx % 2 == 0`) or 0 (full attention) for odd layers. SWA restricts attention to a local window around each token, reducing complexity from $O(T^2)$ to $O(T \cdot w)$, where w is the window size. This alternation of sparse (SWA) and dense (full) attention, inspired by models like Mistral, balances local context modeling with global dependency capture, enabling efficient processing of long sequences (e.g., 128k+ tokens) without uniform sparsity penalties. For example, if $w = 4096$, each token attends to 4,096 previous tokens, significantly reducing memory and compute for large T.

Normalization Layer

```
self.norm = RMSNorm(config.hidden_size, device=device)
```

- **RMSNorm**: Implements root mean square normalization, a lightweight alternative to LayerNorm. It normalizes as:

$$\mathrm{xnorm} = \frac{\mathbf{x}}{\sqrt{\frac{1}{d}\sum i = 1^d x_i^2 + \epsilon}} \odot \mathbf{w} + \mathbf{b}$$

  where **w** and **b** are optional learnable parameters (often omitted for simplicity). Here, `hidden_size` is H (e.g., 4096), the model's embedding dimension. RMSNorm is applied pre-attention (pre-norm architecture), stabilizing gradients in deep models by ensuring input variance remains consistent. The `device` parameter ensures immediate allocation to the specified device (e.g., GPU).

QKV Projection Layer

```
self.qkv_proj = torch.nn.Linear(
    config.hidden_size,
    self.head_dim * (self.num_heads + 2 * self.num_kv_heads),
    device=device
)
```

- **Purpose**: Projects the input $\mathbf{X} \in \mathbb{R}^{B \times T \times H}$ to a concatenated QKV tensor of shape $[B, T, d_{\mathrm{qkv}}]$, where:

$$d_{\mathrm{qkv}}=\mathrm{head}_{\mathrm{d}}\mathrm{im}\times(\mathrm{num}_{\mathrm{h}}\mathrm{eads}+2\times\mathrm{num}_{\mathrm{k}}\mathrm{v}_{\mathrm{h}}\mathrm{eads})$$

  For example, with $\mathrm{head}_{\mathrm{d}}\mathrm{im}=128$, $\mathrm{num}_{\mathrm{h}}\mathrm{eads}=32$, $\mathrm{num}_{\mathrm{k}}\mathrm{v}_{\mathrm{h}}\mathrm{eads}=8$, we get $d_{\mathrm{qkv}} = 128 \times (32 + 16) = 6144$.

- **Efficiency**: Computing Q, K, and V in a single linear layer leverages matrix multiplication fusion on GPUs, reducing overhead compared to separate projections. The layer has no bias term (common in transformer attention to minimize parameters). The output is later sliced into Q (full heads), K, and V (reduced for GQA).

Output Projection Layer

```
self.out_proj = torch.nn.Linear(
    self.num_heads * self.head_dim,
    config.hidden_size,
    device=device
)
```

- Projects the concatenated attention outputs back to H (e.g., 4096), mixing information across heads:

$$\mathbf{O} = \mathbf{A}\mathbf{W}_O$$

where $\boldsymbol{A} \in \mathbb{R}^{B \times T \times (\mathrm{num_heads} \times \mathrm{head_dim})}$ is the attention output, and $\boldsymbol{W}_O \in \mathbb{R}^{(\mathrm{num_heads} \times \mathrm{head_dim}) \times H}$. No bias term, maintaining parameter efficiency.

Rotary Embedding Initialization

```
self.rope = RotaryEmbedding(
    head_dim=self.head_dim,
    base=config.rope_theta,
    dtype=torch.float32,  # compute cos/sin in fp32 for stability
    initial_context_length=config.initial_context_length,
    scaling_factor=config.rope_scaling_factor,
    ntk_alpha=config.rope_ntk_alpha,
    ntk_beta=config.rope_ntk_beta,
    device=device
)
```

- **RoPE Module**: A custom class that generates or caches rotation matrices for positional encoding. Unlike traditional sinusoidal embeddings added to inputs, RoPE rotates query and key vectors to encode relative positions:

$$\mathbf{q}_t \leftarrow \mathbf{q}t \cdot \mathbf{R}\theta, t, \quad \mathbf{k}_t \leftarrow \mathbf{k}t \cdot \mathbf{R}\theta, t$$

where $\mathbf{R}_{\theta,t} \in \mathbb{R}^{d_k \times d_k}$ is a block-diagonal rotation matrix. For dimensions $2j$ and $2j+1$:

$$\mathbf{R}_{\theta,t} = \begin{bmatrix} \cos(t \cdot \theta_j) & -\sin(t \cdot \theta_j) \\ \sin(t \cdot \theta_j) & \cos(t \cdot \theta_j) \end{bmatrix}, \quad \theta_j = \text{base}^{-2j/d_k}$$

For example, $\text{base} = 10000$. This ensures the dot product $\mathbf{q}_m \cdot \mathbf{k}_n$ encodes relative distance $|m - n|$ via $\cos(\theta(m - n))$, improving generalization to long sequences.

- **Hyperparameters:**
  - `head_dim`: Dimension of each head (e.g., 128).
  - `base`: Frequency base (e.g., 10,000), controlling rotation speed.
  - `dtype=torch.float32`: Computes cos/sin in fp32 for numerical stability, then casts to model dtype (e.g., fp16/bf16).
  - `initial_context_length`: Training sequence length (e.g., 8192) for precomputing frequencies.
  - `scaling_factor`, `ntk_alpha`, `ntk_beta`: Enable no-training-knowledge (NTK) scaling or dynamic NTK to extrapolate to longer sequences (e.g., 128k tokens) without retraining. NTK adjusts θ_j to prevent high-frequency collapse, ensuring stable attention for $T \gg \mathrm{initial_context_length}$.

Sink Logits Parameter

```
self.sink_logits = torch.nn.Parameter(torch.zeros(self.num_heads,
device=device))
```

- **Purpose**: A learnable tensor of shape $\mathrm{num_heads}$, initialized to zeros. During attention, it's broadcasted as a bias to logits for designated "sink" positions (e.g., the first four tokens), encouraging the model to allocate attention mass to these tokens. This mitigates the attention sink problem, where LLMs overly attend to initial tokens in long or streaming contexts, causing loss of information for later tokens. By keeping sink tokens in the KV cache, the model maintains context stability.

- **Math**: For attention logits $\text{logits}_{i,j} = \frac{\mathbf{q}_i \cdot \mathbf{k}_j}{\sqrt{d_k}}$, sink_logits adds a per-head bias:

  $\text{logits}i,j[\backslash\text{gets}?]\text{logits}i,j+\text{sink}_\text{logits}_h \text{ if } j \text{ is a sink position}$

This biases the softmax to favor sinks, balancing attention distribution.

Softmax Scaling Factor

```
self.sm_scale = self.head_dim ** -0.5
```

- Computes $\frac{1}{\sqrt{d_k}}$, where $d_k=\text{head}_\text{d}\text{im}$ (e.g., $1/\sqrt{128} \approx 0.0884$).

  Scales dot products before softmax to keep variance ~1, preventing saturation:

$$\text{logits} = \frac{\mathbf{QK}^T}{\sqrt{d_k}}$$

- Without scaling, large d_k causes high-variance logits, pushing softmax to near-binary outputs and hindering gradient flow.

Forward Pass (`forward` Method)

The `forward` method computes the attention output with residual connection, transforming input $\mathbf{X} \in \mathbb{R}^{B \times T \times H}$ to output of the same shape.

Input Unpacking and Residual Connection

```
B, T, H = x.shape
residual = x
```

- Extracts batch size B, sequence length T, and hidden dimension H. Stores x for residual addition:

$$\mathbf{y} = \mathbf{x} + f(\mathbf{x})$$

Residual connections are critical in transformers, enabling direct information flow and preventing vanishing gradients in deep stacks (e.g., 80 layers).

Pre-attention Normalization

```
x = self.norm(x)
```

Applies RMSNorm to normalize each token's embedding vector, ensuring stable inputs to the QKV projection. This reduces internal covariate shift and stabilizes training in deep models.

QKV Projection and Slicing

```
qkv = self.qkv_proj(x)  # [B, T, head_dim*(num_heads + 2*num_kv_heads)]
q_dim = self.num_heads * self.head_dim
kv_dim = self.num_kv_heads * self.head_dim

q = qkv[:, :, :q_dim]
k = qkv[:, :, q_dim:q_dim + kv_dim]
v = qkv[:, :, q_dim + kv_dim:]
```

- **Projection**: Computes:

$$\mathbf{QKV} = \mathbf{x}\mathbf{W}_{qkv}$$

where $\mathbf{W}qkv \in \mathbb{R}^{H \times d\text{qkv}}$, and $d_{\text{qkv}} = \text{head}_\text{d}\text{im} \times (\text{num}_\text{h}\text{eads} + 2 \times \text{num}_\text{kv}{}_\text{h}\text{eads})$. Output shape: $[B, T, d_{\text{qkv}}]$.

- **Slicing:**
    - Q: $[B, T, \text{num}_\text{h}\text{eads} \times \text{head}_\text{d}\text{im}]$ (e.g., 32 × 128 = 4096).
    - K: $[B, T, \text{num}_\text{kv}{}_\text{h}\text{eads} \times \text{head}_\text{d}\text{im}]$ (e.g., 8 × 128 = 1024).
    - V: Same as K.
- **Efficiency**: Single matrix multiply leverages GPU fusion; slicing is a zero-copy view operation.

GQA Reshaping

```
q = q.reshape(B, T, self.num_kv_heads, self.q_per_kv, self.head_dim).
contiguous()
k = k.reshape(B, T, self.num_kv_heads, self.head_dim).contiguous()
v = v.reshape(B, T, self.num_kv_heads, self.head_dim).contiguous()
```

- **Reshaping for GQA:**
    - Q: $[B, T, \text{num}_\text{kv}{}_\text{h}\text{eads}, q_{\text{per}_\text{kv}}, \text{head}_\text{d}\text{im}]$ (e.g., [B, T, 8, 4, 128]).
    - K/V: $[B, T, \text{num}_\text{kv}{}_\text{h}\text{eads}, \text{head}_\text{d}\text{im}]$ (e.g., [B, T, 8, 128]).
- Each K/V head is broadcast across q_{per_kv} query heads during attention, enabling GQA's efficiency.
- **Contiguous**: Ensures tensors are in contiguous memory for efficient GPU matrix operations, as reshaping can fragment memory layout.

RoPE Application

```
q, k = self.rope(q, k)
```

The `RotaryEmbedding` module rotates Q and K tensors along the `head_dim` axis, position-wise, for each token $t \in 1, \ldots, T$:

$$\mathbf{q}b, t, g, q, d \leftarrow \mathbf{q}b, t, g, q, d \cdot \mathbf{R}\theta, t, \quad \mathbf{k}b, t, g, d \leftarrow \mathbf{k}b, t, g, d \cdot \mathbf{R}\theta, t$$

where $R_{\theta,t}$ is block-diagonal, applying 2D rotations per dimension pair. Values (V) remain unrotated, as positions affect similarity scores (Q·K), not retrieval. RoPE ensures dot products encode relative positions, improving extrapolation to $T > \mathrm{initial_context_length}$.

Core Attention Computation

```
attn = sdpa(q, k, v, self.sink_logits, self.sm_scale, self.sliding_window)
```

- **Custom sdpa Function:**

  A wrapper around `torch.nn.functional.scaled_dot_product_attention`, possibly integrating FlashAttention for optimized GPU performance (2–3x faster, lower memory). It performs the following (inferred) steps:

  1. **Dot-Product**: For each group $j \in 1, ..., \mathrm{num_kv_heads}$ and query head $i \in 1, ..., q_{\mathrm{per_kv}}$:

  $$\mathrm{logits}b, t, u, i = \frac{\mathbf{q}b, t, j, i \cdot \mathbf{k}_{b,u,j}}{\sqrt{d_k}}$$

     where $u \in 1, \ldots, T$ (or within window for SWA). Shape: $[B, T, q_{\mathrm{per_kv}}, T]$ per group.

  2. **Causal Mask**: For autoregressive generation, set:$\mathrm{logits}_{b,t,u,i} = -\infty \quad \text{if } u > t$ensuring tokens only attend to prior positions.

  3. Sliding Window Mask: If sliding_window > 0:$\mathrm{logits}_{b,t,u,i} = \infty \text{ if } u \notin [t - \mathrm{sliding_window}, t]$reducing attention to a local window.

  4. Sink Logits Bias: Broadcast sink_`logits` [num_heads] to add bias to sink positions (e.g., first 4 tokens):

$$\mathrm{logits}b,t,u,i[\backslash gets?]\mathrm{logits}b,t,u,i + \mathrm{sink_logits}j \cdot q\mathrm{per_kv} + i \text{ if } u \text{ is a sink}$$

5. **Softmax**: Compute attention weights:

$$a_{b,t,u,i} = \frac{\exp(\text{logits}b,t,u,i)}{\sum k \exp(\text{logits}_{b,t,k,i})}$$

6. **Weighted Sum**: Compute output per head:

$$\text{head}b,t,j,i = \sum u a_{b,t,u,i} \mathbf{v}_{b,u,j}$$

7. **Concatenate**: Reshape to $[B,T,\text{num}_\text{h}\text{eads} \times \text{head}_\text{d}\text{im}]$.

- **Output Shape**: $[B,T,\text{num}_\text{h}\text{eads} \times \text{head}_\text{d}\text{im}]$.
- **Complexity**: With SWA: $O(B \cdot \text{num}_\text{h}\text{eads} \cdot T \cdot w \cdot \text{head}_\text{d}\text{im})$; without: $O(B \cdot \text{num}_\text{h}\text{eads} \cdot T^2 \cdot \text{head}_\text{d}\text{im})$. GQA reduces K/V memory to $O(B \cdot T \cdot \text{num}_\text{k}\text{v}_\text{h}\text{eads} \cdot \text{head}_\text{d}\text{im})$.

Output Projection and Residual Addition

```
out = self.out_proj(attn)
return residual + out
```

- **Projection**:

$$\mathbf{out} = \mathbf{attn} \mathbf{W}_{\text{out}}$$

  where $\boldsymbol{W}_{\text{out}} \in \mathbb{R}^{(\text{num}_\text{h}\text{eads} \times \text{head}_\text{d}\text{im}) \times H}$.

  Mixes head outputs back to H.

- **Residual**: Adds input:

$$\mathbf{y} = \mathbf{residual} + \mathbf{out}$$

  ensuring direct information flow, critical for deep models.

Training and Inference Behaviors

- **Training:**
  - Processes full sequences up to `initial_context_length` (e.g., 8,192).
  - RoPE precomputes or dynamically generates rotation matrices, with NTK scaling for stability.
  - Sink logits learn to balance attention distribution.
  - SWA on even layers sparsifies computation, reducing memory for large T.
  - Backpropagation optimizes `qkv_proj`, `out_proj`, `norm`, and `sink_logits`.
- **Inference:**
  - Uses KV cache (external) to store past K/V, appending new tokens' K/V per step.
  - SWA + sink logits enable streaming (e.g., 4M+ tokens) without perplexity collapse.
  - RoPE's NTK scaling supports extrapolation to $T \gg \mathrm{initial}_{c}\mathrm{ontext}_{l}\mathrm{ength}$
  - GQA reduces KV cache size by $\mathrm{num}_{h}\mathrm{eads}/\mathrm{num}_{k}\mathrm{v}_{h}\mathrm{eads}$, critical for edge deployment.

Performance Optimizations

- **GQA Efficiency**: By reducing the number of key and value heads from $h = \mathrm{num}_{h}\mathrm{eads}$ to $g = \mathrm{num}_{k}\mathrm{v}_{h}\mathrm{eads}$ (e.g., from 32 to 8), GQA lowers the memory footprint of the key-value cache during inference by a factor of h/g (e.g., 4x). For a sequence of length T, batch size B, and head dimension d_k, the KV cache size is:

$$\mathrm{Memory}_{\mathrm{KV}} = 2 \cdot B \cdot T \cdot \mathrm{num}_{k}\mathrm{v}_{h}\mathrm{eads} \cdot \mathrm{head}_{d}\mathrm{im}$$

Compared to MHA's $2 \cdot B \cdot T \cdot \mathrm{num}_{\mathrm{h}}\mathrm{eads} \cdot \mathrm{head}_{\mathrm{d}}\mathrm{im}$, GQA significantly reduces memory, enabling processing of long sequences (e.g., 128k tokens) on resource-constrained hardware. This is particularly impactful in streaming inference, where the cache grows linearly with sequence length.

- **Sliding Window Attention (SWA)**: SWA reduces the computational complexity of attention from $O(T^2 \cdot \mathrm{num}_{\mathrm{h}}\mathrm{eads} \cdot \mathrm{head}_{\mathrm{d}}\mathrm{im})$ to $O(T \cdot w \cdot \mathrm{num}_{\mathrm{h}}\mathrm{eads} \cdot \mathrm{head}_{\mathrm{d}}\mathrm{im})$, where w is the window size (e.g., 4,096). By limiting attention to a local window, SWA focuses on recent context, which is often sufficient for tasks like language modeling, while drastically cutting memory and compute costs. The alternation of SWA (even layers) and full attention (odd layers) ensures that some layers capture global dependencies, balancing efficiency and expressiveness. This design, inspired by models like Mistral, allows the `AttentionBlock` to scale to ultra-long sequences (e.g., 1M+ tokens) without quadratic scaling issues.

- **Rotary Positional Embeddings (RoPE)**: RoPE's rotation-based encoding eliminates the need for fixed positional embeddings, reducing memory overhead and improving generalization to unseen sequence lengths. The NTK-aware scaling (via `ntk_alpha`, `ntk_beta`, `scaling_factor`) adjusts rotation frequencies dynamically:

$$\theta'_j = \theta_j \cdot \left(\frac{\mathrm{initial}_{\mathrm{c}}\mathrm{ontext}_{\mathrm{l}}\mathrm{ength}}{T}\right)^{\mathrm{ntk}_{\mathrm{a}}\mathrm{lpha}} \cdot \mathrm{ntk}_{\mathrm{b}}\mathrm{eta}$$

  This prevents high-frequency collapse, ensuring stable attention scores for $T \gg \mathrm{initial}_{\mathrm{c}}\mathrm{ontext}_{\mathrm{l}}\mathrm{ength}$. RoPE's relative positional encoding also simplifies implementation, as it integrates directly into the Q·K dot product without modifying input embeddings.

- **Sink Logits**: The learnable `sink_logits` parameter addresses the attention sink problem, where LLMs allocate excessive attention to initial tokens, reducing effectiveness for later tokens in long sequences. By adding a per-head bias to sink positions (e.g., the first four tokens), the model maintains a balanced attention distribution:

$$\text{logits}_{b,t,u,h} \gets \text{logits}_{b,t,u,h} + \text{sink\_logits}_h \text{ if } u \in \text{sink\_positions}$$

  This ensures that key contextual information is preserved in the KV cache, even as older tokens are evicted in streaming scenarios.

- **Fused Kernels and FlashAttention**: The `sdpa` function leverages `torch.nn.functional.scaled_dot_product_attention`, which uses fused CUDA kernels (e.g., FlashAttention) to compute attention without materializing the full $T \times T$ attention matrix. FlashAttention (Dao et al., 2022) performs attention in blocks, reducing memory I/O and achieving 2–3x speedups while lowering peak memory usage. The integration of GQA, SWA, and sink logits within `sdpa` ensures compatibility with these optimizations, making the `AttentionBlock` highly efficient on modern GPUs.

- **Mixed Precision**: The use of `torch.float32` for RoPE computations (`dtype=torch.float32`) ensures numerical stability in trigonometric operations, while the model operates in `fp16` or `bf16` for other computations, reducing memory and speeding up matrix operations. This is critical for training and inference on large-scale models with billions of parameters.

Interpretability

The attention weights computed in the `sdpa` function provide insights into the `AttentionBlock`'s behavior. The attention matrix $\boldsymbol{A} \in \mathbb{R}^{B \times \text{num\_heads} \times T \times T}$ (or $T \times w$ with SWA) indicates how much each token attends to others. Visualizing these weights reveals:**GQA Patterns**: Each group of $q_{\text{per\_kv}}$ query heads shares a single K/V pair, leading to correlated attention patterns within groups. This can be observed as similar weight distributions across the h/g heads per group.

- **SWA Locality**: On even layers, attention weights are non-zero only within the sliding window, highlighting local dependencies (e.g., syntactic or short-term semantic relationships).
- **Sink Token Effects**: The `sink_logits` bias increases weights for initial tokens, visible as higher attention scores for the first few positions, ensuring they act as contextual anchors.
- **RoPE Influence**: The rotation of Q and K vectors modulates attention scores based on relative positions, with higher weights for tokens at similar relative distances, observable in periodic patterns in the attention matrix.

However, interpretability is limited because the feed-forward network (FFN, not shown but typically following the `AttentionBlock`) and residual connections significantly influence the final output. Additionally, the `sink_logits`' effect may obscure natural attention patterns, requiring careful analysis to disentangle learned biases from input-driven attention.

Tools like attention heatmaps or attribution methods (e.g., integrated gradients) can be used to study these patterns, but the complexity of GQA and sink logits requires specialized visualization techniques to account for grouped heads and biased positions.

Comparison to Other Attention Mechanisms

To contextualize the `AttentionBlock`, consider its differences from other attention mechanisms:

- **Standard MHA**: Unlike MHA, which uses h heads for Q, K, and V, this block's GQA reduces K/V heads, saving memory. MHA's complexity is $O(T^2 \cdot h \cdot d_k)$ for all components, while GQA reduces K/V storage.
- **Multi-Query Attention (MQA)**: MQA uses a single K/V head ($g = 1$), more extreme than GQA. This block's $g = 8$ offers better expressiveness while retaining efficiency.
- **Sparse Attention (e.g., Longformer)**: Longformer uses fixed patterns (e.g., dilated windows), whereas SWA here is simpler (local window) and alternates with full attention, balancing simplicity and coverage.

- **Linear Attention (e.g., Performer)**: Linear attention approximates softmax with kernels, achieving $O(T)$ complexity but potentially sacrificing accuracy. This block retains exact attention with SWA for better quality.

Practical Implementation Notes

- **Dependencies**: The code assumes external implementations of `RMSNorm`, `RotaryEmbedding`, and `sdpa`. These come from libraries like Hugging Face Transformers, xAI's internal frameworks, or LLaMA-inspired codebases. `sdpa` may wrap `torch.nn.functional.scaled_dot_product_attention` with custom logic for sink logits and SWA.
- **Hardware Optimization**: The block is designed for GPU acceleration, leveraging fused kernels (FlashAttention), contiguous tensors, and mixed precision (fp32 for RoPE, fp16/bf16 elsewhere). For example, FlashAttention reduces memory from $O(T^2)$ to $O(T)$ by computing attention in SRAM blocks.
- **Training Setup**: Typically trained with AdamW, gradient clipping, and learning rate scheduling (e.g., cosine decay). The `sink_logits` parameter requires careful regularization to prevent over-biasing.
- **Inference Setup**: Uses KV caching (not shown) to store K/V for past tokens, appending new tokens' K/V per step. Sink tokens ensure cache stability, and SWA limits cache growth within windows.

Summary

This chapter delves into the core of transformer architectures in large language models (LLMs), focusing on the AttentionBlock as a foundational module enhanced by advanced optimizations. It begins with the historical evolution of attention from RNNs to self-attention in transformers, emphasizing its role in capturing long-range dependencies and contextual relationships.

The chapter provides detailed mathematical formulations for self-attention, multi-head attention (MHA), and variants, including positional encodings like RoPE for relative positioning. It explores operational properties, integration in encoder-decoder structures, causal masking, and optimizations such as sparse, linear, and flash attention.

Key advancements like grouped query attention (GQA) for memory efficiency, sliding window attention (SWA) for linear complexity in long sequences, and Sink Tokens to balance attention distribution are thoroughly explained, with code examples demonstrating their implementation in PyTorch. Limitations like quadratic complexity and attention sinks are discussed, alongside future directions including state space models (SSMs) and multimodal extensions. Overall, the chapter highlights how these mechanisms make AttentionBlocks scalable and robust for modern LLMs as of 2025. In the end, we built our own implementation of an attention mechanism for our own custom LLM.

In Chapter 8, we shift focus from attention mechanisms to the feed-forward components that complement them in transformer layers: the MLPBlock (multi-layer perceptron block). Building on the contextual representations refined by attention, the MLPBlock processes each token independently through dense layers to introduce non-linearity and expand model capacity.

This chapter explores advanced variants incorporating mixture of experts (MoE), a sparse routing technique that activates only a subset of experts (sub-networks) per input, dramatically increasing parameter count while maintaining inference efficiency—key to models like Mixtral and GPT-4. We also examine SwiGLU (Swish-Gated Linear Unit), an activation function that combines Swish with gating for superior performance over traditional ReLU or GELU, enabling deeper and more expressive networks.

Through mathematical derivations, code implementations, and comparisons to standard MLPs, we'll uncover how MoE and SwiGLU optimize for scale, reduce training costs, and address challenges like expert specialization and routing stability, paving the way for even larger, more capable LLMs in 2025 and beyond.

CHAPTER 8

Multilayer Perceptron Block with Mixture of Experts (MoE) and SwiGLU

The multilayer perceptron (MLP) is a powerful and versatile model in machine learning, underpinning technologies from autonomous vehicles to natural language processing. Known as a feedforward neural network, an MLP learns complex functions by processing data through interconnected layers of nodes, or neurons, inspired by the human brain. Its ability to model nonlinear relationships makes it a foundational tool in the deep learning landscape.

MLPs serve as the basis for advanced architectures like convolutional neural networks (CNNs) and recurrent neural networks (RNNs). This article provides a comprehensive exploration of MLPs, covering their historical context, architecture, mathematical foundations, training mechanisms, applications, limitations, and practical implementation considerations. By the end, you'll gain a deep understanding of this essential machine learning model.

D. Grigorov, *Building Large Language Models from Scratch*, https://doi.org/10.1007/979-8-8688-2297-1_8

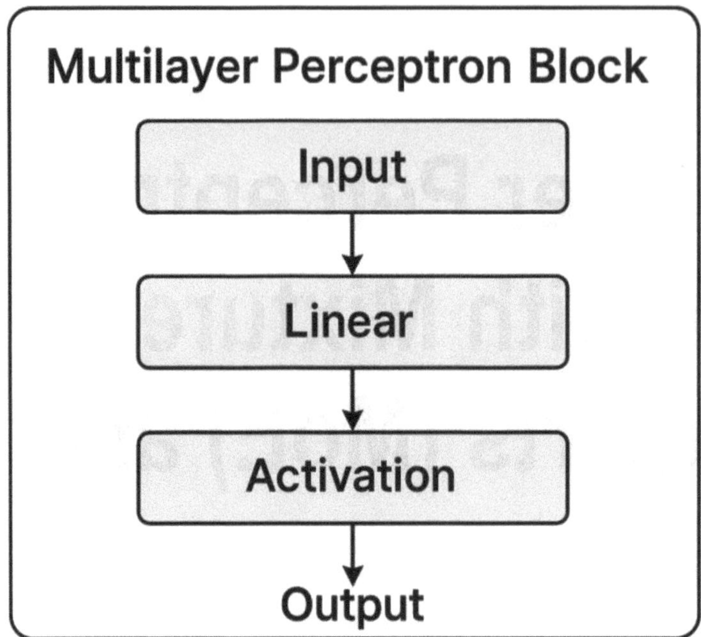

Figure 8-1. *Multilayer Perceptron Block*

Mixture of Experts: A Comprehensive Overview

The mixture of experts (MoE) is a machine learning architecture designed to enhance the efficiency and scalability of neural networks by combining the strengths of multiple specialized submodels, referred to as "experts." Unlike traditional neural networks like multilayer perceptrons (MLPs), which process all inputs through a single, monolithic structure, MoE employs a divide-and-conquer strategy.

It dynamically allocates different parts of a problem to specialized experts, coordinated by a gating mechanism that determines which expert(s) should handle a given input. This approach allows MoE to scale to large models with billions of parameters while maintaining computational efficiency, making it particularly prominent in modern deep learning applications, such as large language models (LLMs).

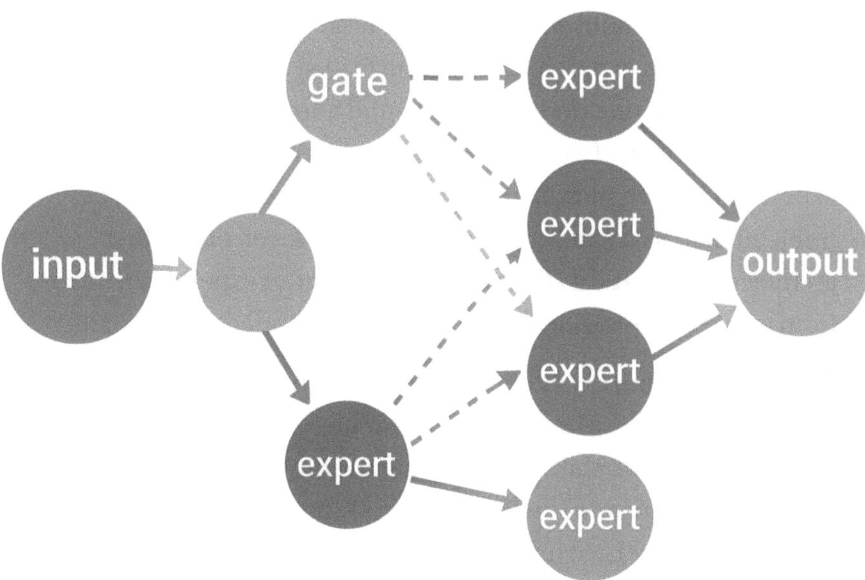

Figure 8-2. *Mixture of Experts*

Introduced in the early 1990s by Robert A. Jacobs and colleagues, MoE has evolved significantly, driven by advancements in computational power and the need for efficient scaling in tasks like natural language processing (NLP), computer vision, and recommendation systems. Its resurgence in recent years, notably in models like Google's Switch Transformer and xAI's Grok, highlights its ability to balance performance and resource constraints. In this article, we delve into the architecture, mathematical foundations, training mechanisms, applications, advantages, and limitations of MoE, providing a comprehensive understanding of this powerful framework.

Architecture of Mixture of Experts

The MoE architecture consists of two primary components: a set of expert submodels and a gating network that routes inputs to the appropriate experts. Each expert is typically a neural network (e.g., a feedforward network, transformer layer, or convolutional network) specialized for a subset of the input space, while the gating network learns to assign inputs to experts dynamically.

Components

1. **Experts**: These are specialized submodels, each designed to handle a specific subset of the data or task. For example, in an NLP model, one expert might specialize in syntactic patterns, another in semantic understanding, and another in domain-specific knowledge (e.g., medical or legal terminology). Each expert has its own parameters (weights and biases) and can be implemented as a small MLP, a transformer layer, or another architecture tailored to the task. The number of experts can range from a few to thousands, depending on the model's scale.

2. **Gating Network**: The gating network is a lightweight model, often a simple neural network or linear layer that takes the input and outputs a probability distribution over the experts. This distribution determines which expert(s) should process the input. The gating network ensures that only a subset of experts is activated for each input, improving computational efficiency by avoiding the need to compute all experts' outputs.

The MoE output is a weighted combination of expert outputs, expressed as:

$$\text{Output} = \sum_{i=1}^{N} G(x)_i \cdot E_i(x)$$

where

- x is the input vector.
- $E_i(x)$ is the output of the i-th expert.
- $G(x)_i$ is the gating weight for the i-th expert, computed by the gating network.
- N is the number of experts.

The gating weights are typically normalized using a softmax function to form a probability distribution:

$$G(x)i = \frac{\exp(g_i(x))}{\sum j = 1^N \exp(g_j(x))}$$

where $g_i(x)$ is the raw output of the gating network for expert i.

Sparse Activation

A defining feature of MoE is **sparse activation**, where only a small subset of experts (e.g., 1–2 out of dozens or hundreds) is selected for each input. This sparsity is achieved through the gating network, which assigns high weights to a few experts and near-zero weights to others. For example, in a top-k gating approach, only the top k-th experts with the highest gating scores are activated, and their outputs are combined.

This reduces computational cost significantly, as only a fraction of the model's parameters are used per inference, allowing MoE models to scale to billions or even trillions of parameters while maintaining manageable resource requirements. Sparse activation is particularly effective in large-scale models, where evaluating all experts would be prohibitively expensive.

Variants of MoE

Several variations of the MoE architecture exist, tailored to specific tasks or constraints:

1. **Soft MoE**: All experts contribute to the output, weighted by the gating probabilities. This is computationally intensive but ensures smooth gradients during training.

2. **Hard MoE**: Only a single expert is selected (e.g., the one with the highest gating score), simplifying computation but potentially introducing discontinuities in the loss landscape.

3. **Top-k MoE**: A compromise between soft and hard MoE, where the top k experts are selected, balancing efficiency and performance. This is common in large-scale models like the Switch Transformer.

4. **Hierarchical MoE**: Multiple layers of gating networks are used to route inputs hierarchically, allowing for more complex specialization across experts.

These variants allow MoE to adapt to different computational budgets and task requirements, with top-k MoE being particularly popular in modern applications due to its balance of efficiency and flexibility.

Mathematical Foundations

The MoE framework can be understood through its mathematical formulation. For an input x `https://www.codecogs.com/eqnedit.php?latex=x#0`, the gating network produces a vector of scores $g(x) = [g_1(x), g_2(x), \ldots, g_N(x)]$, which are normalized to produce gating weights:

$$G(x)_i = \text{softmax}(g(x))i = \frac{\exp(g_i(x))}{\sum j = 1^N \exp(g_j(x))}$$

Each expert $E_i(x)$ processes the input to produce an output, which could be a scalar, vector, or tensor depending on the task. The final output is a weighted sum of expert outputs:

$$y = \sum_{i=1}^{N} G(x)_i \cdot E_i(x)$$

In practice, for sparse MoE, the sum is restricted to the top k experts to reduce computation:

$$y \approx \sum_{i \in \text{top-k}(G(x))} G(x)_i \cdot E_i(x)$$

where $\text{top-k}(G(x))$ denotes the indices of the k experts with the highest gating weights. This formulation ensures that the model leverages specialized knowledge from each expert while maintaining efficiency through sparsity.

The gating network itself is typically a simple feedforward network or linear layer, with its own parameters learned during training. For example, it might compute:

$$g(x) = W_g x + b_g$$

where W_g and b_g are the weight matrix and bias of the gating network, respectively.

Training Mechanisms

Training an MoE model involves optimizing both the experts' parameters and the gating network's parameters to minimize a loss function, typically using backpropagation. However, the sparse and modular nature of MoE introduces unique challenges and considerations.

Loss Function

The choice of loss function depends on the task, similar to MLPs:

- **Regression**: Mean squared error (MSE) or mean absolute error (MAE)
- **Classification**: Binary or categorical cross-entropy
- **Language Modeling**: Cross-entropy over token predictions

The loss is computed based on the combined output of the selected experts, weighted by the gating probabilities.

Backpropagation in MoE

Training involves computing gradients for both the experts and the gating network. The process includes

1. **Forward Pass**: The input is passed through the gating network to compute $G(x)_i$. The top k experts are selected, and their outputs are computed and combined to produce the final prediction.
2. **Loss Calculation**: The loss is calculated by comparing the predicted output to the true output.

3. **Backward Pass**: Gradients are computed with respect to the parameters of the activated experts and the gating network. For a weight w_{ij} in expert i, the gradient is:

$$\frac{\partial L}{\partial w_{ij}} = G(x)i \cdot \frac{\partial L}{\partial E_i(x)} \cdot \frac{\partial E_i(x)}{\partial wij}$$

   For the gating network's parameters (e.g., W_g), gradients are computed based on the loss's dependency on $G(x)_i$, which involves the softmax function and the outputs of all experts.

4. **Parameter Update**: Parameters are updated using an optimizer like Adam or SGD, with the learning rate carefully tuned to balance convergence and stability.

Load Balancing

A critical challenge in MoE training is ensuring that the gating network distributes inputs evenly across experts, preventing some experts from being underutilized or overutilized. Uneven load distribution can lead to inefficient training, where some experts converge poorly due to insufficient updates. To address this, a **load balancing loss** is often added to the objective function. Note that this load balancing loss is typically implemented as a training-loop component outside the module itself, or a placeholder for the loss term should be included in the forward pass to be combined with the main task loss during backpropagation.

$$L_{\text{balance}} = \alpha \cdot \sum_{i=1}^{N} \left(\frac{1}{m} \sum_{j=1}^{m} G(x_j)_i - \frac{1}{N} \right)^2$$

where m is the batch size, and α is a hyperparameter controlling the strength of the balancing term. This encourages the gating network to assign inputs uniformly across experts, improving training stability and model performance.

Optimization Enhancements

Training MoE models leverages techniques similar to those for MLPs, adapted for sparsity:

- **Sparse Optimizers**: Optimizers like SparseAdam are designed for sparse updates, as only a subset of experts is activated per input.
- **Dropout**: Applied to both experts and the gating network to prevent overfitting.
- **Batch Normalization**: Normalizes expert outputs to stabilize training, though care is needed to handle sparse activations.
- **Gradient Clipping**: Prevents exploding gradients, especially in deep MoE models.
- **Distributed Training**: MoE models are often trained across multiple GPUs or TPUs, with experts distributed across devices to handle large-scale architectures.

Challenges in Training

1. **Load Imbalance**: Without proper regularization, the gating network may favor a few experts, leading to undertrained or unused experts.
2. **Sparsity Management**: Selecting the optimal number of active experts (k) balances efficiency and performance. A small k reduces computation but may limit model capacity.
3. **Scalability**: Training large MoE models requires significant computational resources, though sparsity mitigates this compared to dense models.
4. **Overfitting**: With many experts, MoE models can overfit, especially on small datasets, requiring regularization like dropout or weight decay.

Applications of Mixture of Experts

MoE models are applied across diverse domains due to their scalability and efficiency:

1. **Natural Language Processing**: MoE is widely used in large language models like Google's Switch Transformer or xAI's Grok, where experts specialize in different linguistic tasks (e.g., syntax, semantics, or domain-specific knowledge). This enables efficient scaling to trillions of parameters while maintaining fast inference.
2. **Computer Vision**: MoE models process images by assigning experts to different visual features (e.g., edges, textures, or objects), improving performance in tasks like image classification or object detection.
3. **Recommendation Systems**: In platforms like YouTube or Netflix, MoE models handle diverse user preferences by routing inputs to experts specialized in specific genres or user behaviors.
4. **Speech Recognition**: MoE models process audio inputs, with experts focusing on different phonetic or contextual patterns, enhancing transcription accuracy.
5. **Multitask Learning**: MoE naturally supports multitask learning, where experts specialize in different tasks (e.g., translation vs. summarization), sharing knowledge through the gating network.

Advantages of MoE

1. **Scalability**: MoE models can scale to billions or trillions of parameters with minimal computational overhead due to sparse activation.
2. **Efficiency**: By activating only a subset of experts, MoE reduces inference and training costs compared to dense models of similar size.

3. **Specialization**: Experts can focus on specific subsets of the data, improving performance on diverse or complex tasks.
4. **Flexibility**: MoE can be integrated into various architectures (e.g., transformers, MLPs) and adapted to different domains.

Limitations of MoE

1. **Training Complexity**: The need for load balancing and sparse optimization makes training MoE models more complex than dense models.
2. **Load Imbalance**: Without careful design, some experts may be underutilized, reducing model capacity.
3. **Memory Requirements**: While sparse, MoE models still require significant memory to store parameters for all experts, especially in large-scale settings.
4. **Interpretability**: The modular nature of MoE makes it harder to interpret which experts contribute to specific predictions compared to simpler models.
5. **Hardware Dependency**: Efficient MoE training and inference often require specialized hardware (e.g., GPUs with sparse computation support) to fully leverage sparsity.

The mixture of experts framework is a powerful approach to scaling neural networks efficiently, leveraging specialized submodels and sparse activation to handle complex tasks with reduced computational cost. Its architecture, combining experts and a gating network, enables applications in NLP, computer vision, and beyond, with models like the Switch Transformer and Grok showcasing its potential.

While MoE offers significant advantages in scalability and efficiency, it faces challenges in training complexity, load balancing, and interpretability. By carefully designing the architecture, tuning hyperparameters, and leveraging modern hardware, practitioners can harness MoE to build high-performance models for a wide range of applications, making it a cornerstone of modern deep learning.

SwiGLU: An In-Depth Exploration

SwiGLU, or Swish-Gated Linear Unit, is an advanced activation function and layer design that has emerged as a powerful component in modern neural networks, particularly within the feed-forward network (FFN) layers of transformer-based models.

As a variant of the gated linear unit (GLU) family, SwiGLU replaces the traditional sigmoid gating mechanism with the Swish activation function, offering improved gradient flow, expressiveness, and performance. This innovation has made SwiGLU a key building block in large-scale deep learning models, such as large language models (LLMs) and vision transformers, where it contributes to enhanced efficiency and accuracy.

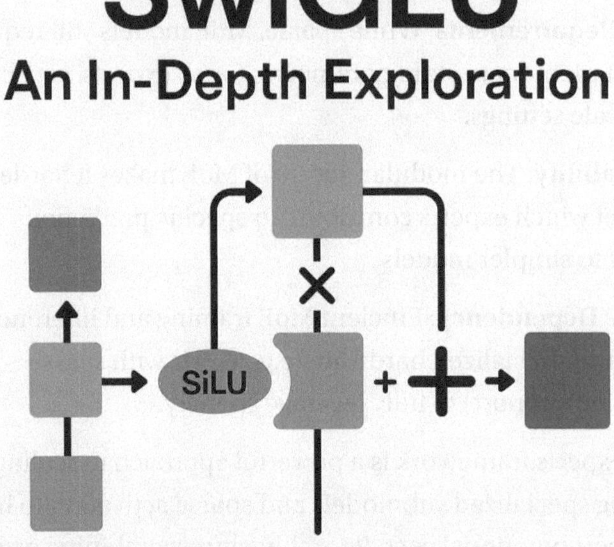

Figure 8-3. *SwiGLU*

Introduced in the 2020 paper "GLU Variants Improve Transformer" by Noam Shazeer, SwiGLU builds on the foundational concepts of GLUs and the Swish activation, combining their strengths to address limitations of traditional activation functions like ReLU. Its adoption in high-profile models, such as Google's PaLM (pathways language model) and PaLM 2, underscores its significance in scaling neural networks to billions or trillions of parameters while maintaining computational feasibility. SwiGLU's design allows it to model complex nonlinear relationships effectively, making it a versatile tool across domains like natural language processing (NLP), computer vision, and multimodal learning.

This article provides a comprehensive exploration of SwiGLU, covering its historical context, theoretical foundations, mathematical formulation, training mechanisms, implementation details, applications, advantages, limitations, and practical considerations. We also discuss advanced variants, comparisons with other activation functions, and future directions, offering a thorough understanding for researchers, practitioners, and enthusiasts.

Historical Context

The development of SwiGLU is rooted in the evolution of activation functions and neural network architectures. Early neural networks, like multilayer perceptrons (MLPs), relied on simple activation functions such as the sigmoid ($\sigma(z) = \frac{1}{1 + e^{-z}}$) and tanh ($\tanh(z) = \frac{e^{z} - e^{-z}}{e^{z} + e^{-z}}$). These functions introduced nonlinearity but suffered from vanishing gradients, where derivatives become very small, slowing training in deep networks.

In 2017, Google researchers introduced the **Swish** activation function $f(z) = z \cdot \sigma(\beta z)$, which outperformed ReLU ($f(z) = \max(0, z)$) in many tasks due to its smoothness and nonmonotonic behavior. Swish allows gradients to flow through both positive and negative inputs, mitigating the "dying ReLU" problem, where neurons permanently output zero for negative inputs.

Concurrently, the **gated linear unit (GLU)**, proposed by Yann Dauphin et al. in 2017, emerged as a way to enhance FFN layers in transformers. GLUs split inputs into two parts, applying a linear transformation to one and a sigmoid gate to the other, with element-wise multiplication to control information flow. GLUs improved transformer performance but were limited by the sigmoid's vanishing gradient issues.

SwiGLU, introduced in 2020 by Noam Shazeer, combines Swish and GLU to create a more expressive and efficient activation mechanism. By replacing the sigmoid gate with Swish, SwiGLU enhances gradient flow and model capacity, leading to empirical gains in transformer-based models. Its adoption in large-scale models like PaLM and its integration into frameworks like Hugging Face Transformers highlight its growing importance in deep learning.

The Basics: From Activation Functions to Gated Units

To understand SwiGLU, we first explore its foundational components: activation functions and gated linear units.

Activation Functions

Activation functions introduce nonlinearity, enabling neural networks to model complex patterns. Common activations include:

1. **Sigmoid**: Maps inputs to (0, 1) ($\sigma(z) = \frac{1}{1 + e^{-z}}$). It's smooth but prone to vanishing gradients, as its derivative is small for large $|z|$.
2. **ReLU (Rectified Linear Unit)**: Outputs $\max(0, z)$. Simple and effective, it avoids vanishing gradients for positive inputs but suffers from "dying ReLU" for negative inputs.
3. **Tanh**: Maps inputs to (-1, 1) ($\tanh(z) = \frac{e^z - e^{-z}}{e^z + e^{-z}}$). Zero-centered but also susceptible to vanishing gradients.
4. **Swish**: Defined as $f(z) = z \cdot \sigma(\beta z)$, where β is a learnable or fixed parameter (often 1). Swish is smooth, nonmonotonic, and bounded below by zero, offering better gradient flow than ReLU or sigmoid. Its derivative, $f'(z) = \sigma(\beta z) + \beta z \cdot \sigma(\beta z)(1 - \sigma(\beta z))$, ensures non-zero gradients for negative inputs, improving training stability.

Swish's empirical success in deep networks made it a natural candidate for enhancing gated architectures.

Gated Linear Units (GLU)

GLUs are a type of neural network layer that combine linear transformations with a gating mechanism. In a standard GLU, the input x is split into two parts (or projected into two vectors):

$$\mathrm{GLU}(x) = (xW + b) \otimes \sigma(xV + c)$$

where

- W, V are weight matrices.
- b, c are biases.
- $\otimes$ denotes element-wise multiplication.
- σ is the sigmoid function.

The sigmoid gate controls which elements of the linear transformation $xW + b$ are passed through, allowing the model to selectively focus on relevant features. GLUs are commonly used in transformer FFN layers, replacing traditional MLPs (e.g., $\text{ReLU}(xW_1 + b_1)W_2 + b_2$), as they provide richer representations.

However, the sigmoid gate in GLU limits performance due to vanishing gradients, especially in deep networks. This motivated the development of SwiGLU, which replaces sigmoid with Swish for better expressiveness.

SwiGLU: Definition and Structure

SwiGLU extends GLU by using Swish as the gating function. It is defined as:

$$\text{SwiGLU}(x, W, V, b, c, \beta) = \text{Swish}_\beta(xW + b) \otimes (xV + c)$$

where

- $\text{Swish}_\beta(z) = z \cdot \sigma(\beta z)$.
- x is the input vector.
- W, V are weight matrices.
- b, c are biases.
- β is a parameter (fixed or learned, often set to 1).

In transformer FFNs, SwiGLU is often implemented as part of a larger layer:

$$\text{FFN}(x) = W_3 \cdot \text{SwiGLU}(x, W_1, W_2, b_1, b_2, \beta) + b_3$$

where W_3, b_3 project the SwiGLU output to the desired dimension. The input x is typically high-dimensional (e.g., the hidden state in a transformer), and the SwiGLU layer processes it to produce a transformed representation.

SwiGLU's gating mechanism allows it to dynamically modulate the linear transformation, enhancing the network's ability to capture complex patterns while maintaining computational efficiency.

Mathematical Representation

The mathematical formulation of SwiGLU is central to its implementation. For an input $x \in \mathbb{R}^d$, the SwiGLU layer computes:

1. Two linear transformations:

$$z_1 = xW + b, \quad z_2 = xV + c$$

where $W, V \in \mathbb{R}^{d \times d_h}$, $b, c \in \mathbb{R}^{d_h}$, and d_h is the hidden dimension.

2. The Swish gate:

$$\text{Swish}_\beta(z_1) = z_1 \cdot \sigma(\beta z_1)$$

3. Element-wise multiplication:

$$\text{SwiGLU}(x) = \text{Swish}_\beta(z_1) \otimes z_2$$

4. Optional projection:

$$y = \text{SwiGLU}(x)W_3 + b_3$$

The dimensionality of W_3 depends on whether the FFN restores the input dimension (common in transformers) or projects to a different size.

The Swish function's derivative is:

$$\frac{\partial \text{Swish}_\beta(z)}{\partial z} = \sigma(\beta z) + \beta z \cdot \sigma(\beta z)(1 - \sigma(\beta z))$$

This ensures non-zero gradients for all inputs, improving backpropagation in deep networks compared to sigmoid or ReLU.

Parameter Count

SwiGLU layers typically have a higher parameter count than ReLU-based FFNs due to the dual projections (W and V). For a transformer with model dimension d and FFN hidden dimension d_h (often $4d$), a wiGLU FFN has approximately $2d \cdot d_h + d_h \cdot d$ parameters (plus biases), compared to $d \cdot d_h + d_h \cdot d$ for a ReLU FFN. However, empirical results show that SwiGLU's performance gains justify the increased parameter count.

Training Mechanisms

Training a SwiGLU layer involves standard backpropagation, with gradients computed for the weights (W, V, W_3), biases (b, c, b_3), and potentially β if it is learned. The process includes:

1. **Forward Pass**: Compute z_1, z_2, apply Swish to z_1, multiply element-wise with z_2, and project to the output dimension.
2. **Loss Calculation**: Use a task-specific loss function (e.g., cross-entropy for classification, MSE for regression).
3. **Backward Pass**: Compute gradients using the chain rule. For example, for W:

$$\frac{\partial L}{\partial W} = \frac{\partial L}{\partial \text{SwiGLU}} \cdot \frac{\partial \text{SwiGLU}}{\partial z_1} \cdot \frac{\partial z_1}{\partial W}$$

   where $\frac{\partial \text{SwiGLU}}{\partial z_1} = \frac{\partial \text{Swish}_\beta(z_1)}{\partial z_1} \otimes z_2$

4. **Parameter Update**: Update parameters using optimizers like Adam or RMSprop.

Optimization Considerations

- **Gradient Flow**: Swish's smooth gradients reduce vanishing gradient issues, but care is needed to avoid exploding gradients in deep networks. Gradient clipping (e.g., capping at 1.0) is often used.

- **Weight Initialization**: Use He initialization for W, V to suit Swish's nonlinearity, ensuring stable training.
- **Batch Normalization**: Often applied before or after SwiGLU to stabilize activations, though layer normalization is more common in transformers.
- **Learning Rate**: Typical values (e.g., 0.001 for Adam) work well, but tuning is necessary for large models.

Challenges

1. **Computational Cost**: The sigmoid in Swish is more expensive than ReLU's max operation, increasing computation slightly.
2. **Parameter Overhead**: The dual projections in SwiGLU increase memory usage compared to ReLU FFNs.
3. **Tuning** β: If learned, β adds complexity to optimization; if fixed, it may not be optimal for all tasks.

Integration in Transformers

In transformer architectures, SwiGLU replaces the ReLU or GELU in FFN layers. A typical transformer FFN with SwiGLU is:

1. Input $x \in \mathbb{R}^d$ (e.g., hidden state from attention).
2. **SwiGLU Layer**: Computes $\text{SwiGLU}(x)$, often with $d_h = 4d$.
3. **Projection**: Maps back to d dimensions.
4. **Layer Normalization**: Stabilizes outputs before feeding to the next layer.

This structure is common in models like PaLM, where SwiGLU enhances FFN expressiveness.

Applications of SwiGLU

SwiGLU is primarily used in large-scale deep learning models, particularly in transformer-based architectures:

1. **Large Language Models (LLMs)**: Adopted in Google's PaLM and PaLM 2, SwiGLU improves performance on NLP tasks like machine translation, text summarization, question answering, and text generation. Its efficiency supports scaling to billions of parameters.
2. **Vision Transformers (ViTs)**: Used in FFN layers of ViTs for image classification, object detection, and segmentation, enhancing feature extraction from visual data.
3. **Multimodal Models**: In models like CLIP or DALL-E, SwiGLU processes combined text and image inputs, improving cross-modal representations.
4. **Sparse Architectures**: In Mixture of Experts (MoE) models, SwiGLU is used in expert FFNs, boosting performance while maintaining sparsity.
5. **Scientific Applications**: SwiGLU aids in modeling complex relationships in scientific simulations, such as protein folding or physical system modeling, where transformers are applied.

Empirical studies show SwiGLU provides 1–2% accuracy improvements on benchmarks like GLUE, SuperGLUE, and ImageNet compared to ReLU or GELU, particularly in large-scale settings.

Advantages of SwiGLU

1. **Improved Gradient Flow**: Swish's smooth, nonmonotonic derivative ensures better gradient propagation, reducing vanishing gradient issues in deep networks.
2. **Enhanced Expressiveness**: The Swish gate allows for richer nonlinearities than sigmoid or ReLU, enabling the model to capture complex patterns.

3. **Empirical Performance Gains**: Consistent improvements in accuracy and convergence speed, especially in large transformers.

4. **Flexibility**: The parameter β can be learned or fixed, adapting to different tasks or architectures.

5. **Compatibility**: Easily integrates into existing transformer frameworks, leveraging standard training pipelines.

Limitations of SwiGLU

1. **Computational Overhead**: The sigmoid in Swish is more computationally expensive than ReLU's max operation, increasing training and inference costs (though the difference is often small).

2. **Parameter Overhead**: SwiGLU requires additional parameters (e.g., for V, c), increasing memory usage compared to ReLU-based FFNs.

3. **Hyperparameter Sensitivity**: Tuning β or hidden dimensions requires experimentation, adding complexity to model design.

4. **Limited Impact in Small Models**: SwiGLU's benefits are most pronounced in large-scale models; for shallow networks, simpler activations like ReLU may suffice.

5. **Interpretability**: Like other nonlinear activations, SwiGLU is a "black box," making it challenging to interpret the gating mechanism's decisions.

Practical Considerations

Implementing SwiGLU effectively requires attention to several factors:

1. **Starting Configuration**: Use $\beta = 1$ as a default, as it performs well in most cases. Experiment with learning β for tasks with unique data distributions.

2. **Hidden Dimension**: Set the FFN hidden dimension to 4x the model dimension (e.g., $d_h = 4d$) for transformers, but test smaller sizes for efficiency.

3. **Benchmarking**: Compare SwiGLU against ReLU, GELU, or standard GLU on your specific task to quantify benefits. Use validation metrics like accuracy or loss to guide decisions.

4. **Framework Support**: Leverage prebuilt SwiGLU implementations in libraries like Hugging Face Transformers or DeepSpeed to avoid custom coding.

5. **Hardware Optimization**: Use GPUs or TPUs to handle the increased computational load of Swish, and consider mixed precision training to reduce memory usage.

6. **Regularization**: Apply dropout (e.g., 0.1–0.2) and layer normalization to stabilize training, especially in deep models.

Advanced Variants and Comparisons

Variants of SwiGLU

1. **Learned** β: Allowing β to be a learnable parameter can adapt the Swish gate to specific tasks, though it increases optimization complexity.

2. **Scaled SwiGLU**: Adjusts the output scaling to stabilize training, e.g., dividing the SwiGLU output by a constant to normalize magnitude.

3. **SwiGLU in MoE**: Used in mixture of experts models, where each expert's FFN employs SwiGLU, enhancing sparsity and efficiency.

Comparisons with Other Activations

- **Vs. ReLU**: SwiGLU outperforms ReLU due to its smooth gradients and non-zero output for negative inputs, avoiding "dying" neurons. However, ReLU is faster and simpler.

- **Vs. GELU**: GELU (Gaussian Error Linear Unit, $f(z) = z \cdot \Phi(z)$, where Φ is the Gaussian CDF) is similar to Swish but less flexible due to its fixed form. SwiGLU often shows slight performance edges in large models.
- **Vs. Standard GLU**: SwiGLU's Swish gate outperforms sigmoid-based GLU due to better gradient flow and expressiveness.

Empirical studies (e.g., Shazeer 2020) show SwiGLU achieves 1–2% better accuracy on NLP tasks and faster convergence in large-scale settings, though the benefits diminish in smaller models.

SwiGLU represents a significant advancement in neural network design, combining the strengths of Swish and GLU to create a powerful, expressive activation mechanism. Its integration into transformer FFNs has driven performance gains in large-scale models like PaLM, making it a cornerstone of modern deep learning. While SwiGLU introduces slight computational and parameter overhead, its benefits in gradient flow, expressiveness, and scalability outweigh these costs in large architectures. By leveraging modern frameworks, careful tuning, and hardware optimization, practitioners can harness SwiGLU to build high-performance models across NLP, vision, and multimodal tasks. As research progresses, SwiGLU and its variants will likely continue to shape the future of efficient and powerful neural networks.

Multilayer Perceptron Blocks with Mixture of Experts (MoE) and SwiGLU: A Comprehensive Integration

The multilayer perceptron (MLP) block, often referred to as the feed-forward network (FFN) in transformer architectures, is a fundamental component for transforming representations in neural networks. To address the computational challenges of scaling large models while maintaining high performance, advanced techniques like mixture of experts (MoE) and SwiGLU activation have been integrated into MLP blocks. MoE introduces sparsity and specialization by dynamically routing inputs to a subset of expert submodels, each of which is typically an MLP. SwiGLU, a gated activation function, enhances the expressiveness and gradient flow within these expert MLPs, enabling more effective learning of complex patterns. Together, these components form a powerful architecture that balances computational efficiency with model capacity, making it a cornerstone of modern large-scale models like xAI's Grok, Mistral AI's Mixtral, and Google's Switch Transformer variants.

This integration allows models to scale to billions or trillions of parameters without a proportional increase in computational cost, as MoE activates only a fraction of the network for each input, and SwiGLU optimizes the performance of individual experts. This article provides an in-depth exploration of how MLP blocks, MoE, and SwiGLU work together, covering their integrated architecture, mathematical foundations, training mechanisms, rationale for their synergy, applications, advantages, limitations, and practical implementation considerations. By examining their interplay, we aim to elucidate why this combination is so effective in modern deep learning.

Integrated Architecture: MLP with MoE and SwiGLU

In modern transformer architectures, the standard MLP block is replaced by an MoE layer where each expert is a SwiGLU-based MLP. This integration combines the sparsity and specialization of MoE with the expressive power of SwiGLU, creating a highly efficient and scalable FFN.

Architecture Overview

A transformer layer typically consists of a self-attention mechanism followed by an FFN, with layer normalization at various points. In the integrated MoE-SwiGLU design, the FFN is replaced by an MoE layer with the following structure:

1. **Input**: The output $x \in \mathbb{R}^d$ from the self-attention layer (where d is the model dimension, e.g., 768 in BERT).

2. **Gating Network**: A lightweight network (e.g., a single linear layer) computes scores for each of N experts:

$$g(x) = xW_g + b_g, \quad G(x)_i = \frac{\exp(g_i(x))}{\sum_j \exp(g_j(x))}$$

   where $W_g \in \mathbb{R}^{d \times N}, b_g \in \mathbb{R}^N$. The top-$k$ experts (e.g., $k = 2$ or 8) are selected based on $G(x)$.

3. **Experts**: Each expert is a small MLP with SwiGLU activation. For expert i https://www.codecogs.com/eqnedit.php?latex=i#0:

$$z_1 = xW_{i1} + b_{i1}, \quad z_2 = xW_{i2} + b_{i2}$$
$$E_i(x) = \text{Swish}(z_1) \otimes z_2$$
$$E_i(x) = (z_1 \cdot \sigma(z_1)) \otimes z_2$$

   Often, a projection layer is added:

$$E_i(x) = W_{i3} \cdot (\text{Swish}(z_1) \otimes z_2) + b_{i3}$$

   where $W_{i1}, W_{i2} \in \mathbb{R}^{d \times d_h}$, $W_{i3} \in \mathbb{R}^{d_h \times d}$, $b_{i1}, b_{i2} \in \mathbb{R}^{d_h}$, $b_{i3} \in \mathbb{R}^{d}$, and d_h is the hidden dimension (e.g., $4d$).

4. **Output Combination**: The outputs of the top-k experts are weighted by their gating scores:

$$y = \sum_{i \in \text{top-k}} G(x)_i \cdot E_i(x)$$

5. **Layer Normalization**: The output y is normalized before feeding into the next transformer layer.

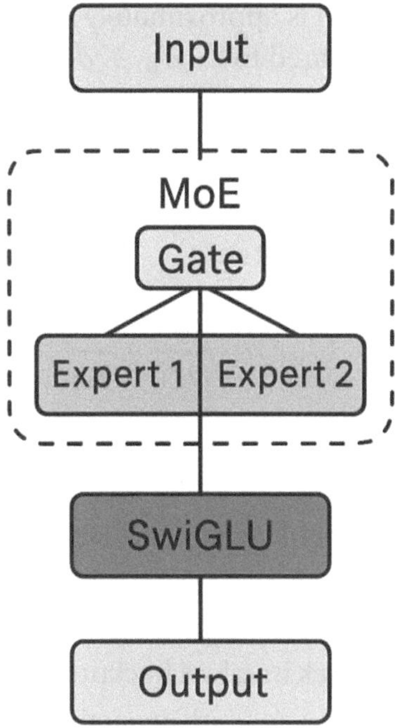

Figure 8-4. *Mixture of Experts Component*

This architecture is typically repeated across multiple transformer layers, with MoE-SwiGLU blocks interspersed (e.g., every other layer in Switch Transformer).

Mathematical Foundations

The integrated model combines MoE's sparse combination with SwiGLU's gated nonlinearity. The output is:

$$y = \sum_{i \in \text{top-k}} G(x)i \cdot [\text{Swish}(xWi1 + b_{i1}) \otimes (xW_{i2} + b_{i2})] \cdot W_{i3} + b_{i3}$$

where

- $G(x)_i = \text{softmax}(xW_g + b_g)_i$.
- $\text{Swish}(z) = z \cdot \sigma(z)$.
- $\sigma(z) = \dfrac{1}{1 + e^{-z}}$.

The parameter count per expert is approximately $2d \cdot d_h + d_h \cdot d$ (for W_{i1}, W_{i2}, W_{i3}) plus biases, compared to $d \cdot d_h + d_h \cdot d$ for a standard MLP. With N experts, the total parameters are large, but only k experts are active, keeping compute cost proportional to k rather than N.

Gradients are computed only for active experts and the gating network, using the chain rule. For example, for W_{i1} in expert i:

$$\frac{\partial L}{\partial W_{i1}} = G(x)i \cdot \frac{\partial L}{\partial E_i} \cdot \frac{\partial E_i}{\partial zi1} \cdot \frac{\partial z_{i1}}{\partial W_{i1}}$$

where $\frac{\partial E_i}{\partial z_{i1}} = [\sigma(z_{i1}) + z_{i1} \cdot \sigma(z_{i1})(1 - \sigma(z_{i1}))] \otimes (xW_{i2} + b_{i2}) \cdot W_{i3}$.

Training Mechanisms

Training an MoE-SwiGLU MLP block involves backpropagation with sparse updates, as only the top-k experts and the gating network are active. Key aspects include:

Forward and Backward Pass

1. **Forward Pass**: Compute gating scores, select top-k https://www.codecogs.com/eqnedit.php?latex=k#0 experts, evaluate their SwiGLU-based MLPs, and combine outputs.
2. **Loss Calculation**: Use task-specific loss (e.g., cross-entropy for language modeling).
3. **Backward Pass**: Compute gradients for active experts' parameters ($W_{i1}, W_{i2}, W_{i3}, b_{i1}, b_{i2}, b_{i3}$) and gating parameters (W_g, b_g).
4. **Parameter Update**: Use optimizers like AdamW, with learning rates around 0.0001–0.001.

Load Balancing

To prevent the gating network from favoring a few experts, a load balancing loss is added:

$$L_{\text{balance}} = \alpha \sum_{i=1}^{N} \left(\frac{1}{m} \sum_{j=1}^{m} G(x_j)_i - \frac{1}{N} \right)^2$$

where m is the batch size, and α is a hyperparameter (e.g., 0.01). This encourages uniform expert utilization, ensuring all experts are trained effectively.

Optimization Enhancements

- **Sparse Optimizers**: SparseAdam or custom sparse optimizers reduce memory usage for inactive experts.
- **Gradient Clipping**: Caps gradients (e.g., at 1.0) to prevent instability.
- **Layer Normalization**: Applied before or after the MoE layer to stabilize activations.
- **Dropout**: Used within experts (e.g., 0.1–0.2) to prevent overfitting.
- **Mixed Precision**: Reduces memory and speeds up training, critical for large MoE models.

Challenges

1. **Routing Collapse**: The gating network may over-rely on a few experts, mitigated by balancing loss or adding noise to gating scores.
2. **Memory Usage**: Despite sparsity, storing N experts requires significant memory, necessitating distributed training.
3. **SwiGLU Overhead**: SwiGLU's sigmoid computation is slightly more expensive than ReLU, though optimized implementations (e.g., CUDA kernels) mitigate this.

Why They Work Together: Rationale

The integration of MLP blocks with MoE and SwiGLU is synergistic due to their complementary strengths:

Scalability and Efficiency

MoE enables massive scaling by activating only k out of N experts, keeping computational cost constant regardless of model size. For example, a model with 1 trillion parameters but $k = 2$ experts per layer uses compute equivalent to a much smaller dense model. SwiGLU enhances this by making each expert more efficient, as its expressive gating allows comparable performance with smaller hidden dimensions or fewer layers compared to ReLU or GELU-based experts.

Specialization and Expressiveness

MoE's experts specialize in different aspects of the input space (e.g., syntax vs. semantics in NLP), allowing the model to handle diverse patterns. Within each expert, SwiGLU's gating mechanism provides fine-grained control over feature transformations, enabling nuanced modeling of complex relationships. This two-level gating—global (MoE) and local (SwiGLU)—creates a hierarchical specialization, where MoE routes inputs to the right experts, and SwiGLU optimizes processing within them.

Gradient Flow and Training Stability

SwiGLU's Swish gate ensures smooth gradients, mitigating vanishing gradient issues that plague deep networks or sigmoid-based GLUs. This is critical in MoE, where experts are smaller but still require effective training. The combination of sparse updates (from MoE) and robust gradients (from SwiGLU) accelerates convergence and stabilizes training, even for models with billions of parameters.

Empirical Performance

Models like Mixtral 8x7B (8 experts, 7B parameters each) outperform dense models like LLaMA 70B on benchmarks like MMLU, thanks to MoE's sparsity and SwiGLU's enhanced expressiveness. The combination allows models to achieve high accuracy with lower inference latency, as only a fraction of parameters is used per input.

Flexibility Across Domains

The architecture is adaptable to various tasks, as MoE can incorporate domain-specific experts (e.g., for medical or legal NLP), and SwiGLU's general-purpose gating suits diverse data distributions. This makes the integrated design ideal for large-scale, multitask, or multimodal applications.

Advantages

The integration offers several benefits:

High Performance

The combination of MoE's specialization and SwiGLU's expressiveness consistently outperforms dense MLPs or ReLU-based FFNs, as seen in models like Mixtral achieving superior results on NLP benchmarks.

Computational Efficiency

Sparse activation ensures constant compute cost, enabling trillion-parameter models with practical inference times. SwiGLU's efficiency reduces the need for oversized experts, further optimizing resources.

Scalability

The architecture scales to massive models without linear increases in compute, making it feasible for large-scale applications in cloud or distributed environments.

Robust Training

SwiGLU's gradient flow and MoE's load balancing ensure stable training, even for deep networks with many experts.

Limitations

Despite its strengths, the integration has challenges:

Training Complexity

Training MoE requires careful load balancing to prevent routing collapse, where the gating network favors a few experts. SwiGLU's additional parameters increase memory demands, complicating optimization.

Hardware Dependency

Sparse operations in MoE benefit from optimized hardware (e.g., GPUs with sparse tensor support), which may not be universally available. SwiGLU's sigmoid computation adds slight overhead compared to ReLU.

Interpretability

The combination of MoE's dynamic routing and SwiGLU's gating makes it difficult to interpret which features or experts drive predictions, limiting applicability in domains requiring explainability (e.g., healthcare).

Memory Requirements

Storing N experts, even with sparsity, requires significant memory, especially for large N . Distributed training across multiple devices is often necessary.

Practical Implementation Considerations

Implementing MoE-SwiGLU MLP blocks requires careful design and tuning:

Architecture Design

Choose $N = 32\#\#28$ experts and $k = 2\#\#$ based on the task and hardware. Use SwiGLU with $\beta = 1$ (fixed) initially, testing learned β for complex datasets. Set hidden dimension $d_h = 4d$ for experts, adjusting based on validation performance.

Framework Support

Use libraries like DeepSpeed, Megatron-LM, or Hugging Face Transformers, which offer MoE and SwiGLU implementations optimized for distributed training and sparse computation. For example, DeepSpeed's MoE module supports top-k routing and load balancing.

Training Workflow

1. Preprocess data (e.g., normalize embeddings, tokenize text).
2. Initialize weights with He initialization for SwiGLU compatibility.
3. Train with AdamW, mixed precision, and gradient clipping.
4. Monitor expert utilization and validation loss to adjust balancing loss weight α.
5. Use early stopping to prevent overfitting.

Hardware Optimization

Leverage GPUs/TPUs with sparse tensor support. Mixed precision training (e.g., FP16) reduces memory usage, critical for large N.

Evaluation

Use task-specific metrics (e.g., BLEU for translation, accuracy for classification) and monitor expert utilization to ensure balanced training. Compare against dense MLP or ReLU/GELU-based MoE to quantify gains.

Future Directions

Research continues to refine MoE-SwiGLU integration:

- **Adaptive Routing**: Dynamic k or learned gating functions.
- **Optimized Sparsity**: Hardware-specific sparse kernels.
- **Interpretability**: Techniques to analyze expert contributions.
- **Small-Scale Models**: Adapting the approach for resource-constrained settings.

The integration of MLP blocks with MoE and SwiGLU creates a highly efficient, scalable, and expressive architecture that powers state-of-the-art models. MoE's sparsity enables massive parameter counts with constant compute, while SwiGLU's gating enhances expert performance. Their synergy drives performance gains in NLP, vision, and multimodal tasks, making this design a cornerstone of modern deep learning.

Despite challenges like training complexity and hardware needs, careful implementation and optimization make MoE-SwiGLU MLPs a transformative approach for large-scale AI systems.

Multilayer Perceptron Block with Mixture of Experts (MoE) and SwiGLU for Our Large Language Model

```
import torch
import torch.nn.functional as F
import math
import torch.distributed as dist

class MLPBlock(torch.nn.Module):
   def __init__(self, config: ModelConfig, device: torch.device | None
   = None):
       super().__init__()
       self.config = config
       self.hidden_size = config.hidden_size                # H
       self.intermediate_size = config.intermediate_size    # I
       self.num_experts = config.num_experts
       self.experts_per_token = config.experts_per_token
       self.swiglu_limit = config.swiglu_limit

       self.world_size = dist.get_world_size() if dist.is_
       initialized() else 1
       if self.intermediate_size % self.world_size != 0:
           raise ValueError("intermediate_size must be divisible by world_
           size for sharding.")
       I_local = self.intermediate_size // self.world_size  # shard of I if
       using tensor-parallel

       self.norm = RMSNorm(config.hidden_size, device=device)

       # Gating network (selects top-k experts per token)
       self.gate = torch.nn.Linear(self.hidden_size, self.num_experts,
       device=device)
```

```
        # Expert params
        # First projection: H -> 2*I  (for SwiGLU: split into [I, I])
        self.mlp1_weight = torch.nn.Parameter(torch.empty(
            self.num_experts, 2 * I_local, self.hidden_size, device=device
        ))
        self.mlp1_bias = torch.nn.Parameter(torch.empty(
            self.num_experts, 2 * I_local, device=device
        ))

        # Second projection: I -> H  (**fixed orientation**)
        self.mlp2_weight = torch.nn.Parameter(torch.empty(
            self.num_experts, I_local, self.hidden_size, device=device
        ))
        self.mlp2_bias = torch.nn.Parameter(torch.empty(
            self.num_experts, self.hidden_size, device=device
        ))

        # Init
        torch.nn.init.kaiming_uniform_(self.mlp1_weight, a=math.sqrt(5))
        torch.nn.init.zeros_(self.mlp1_bias)
        torch.nn.init.kaiming_uniform_(self.mlp2_weight, a=math.sqrt(5))
        torch.nn.init.zeros_(self.mlp2_bias)

    def forward(self, x: torch.Tensor) -> torch.Tensor:
        """
        x: [B, T, H]
        returns: [B, T, H]
        """
        B, T, H = x.shape
        residual = x

        x = self.norm(x)  # [B, T, H]

        # ----- Top-k gating -----
        logits = self.gate(x)                              # [B, T, E]
        topk = torch.topk(logits, self.experts_per_token, dim=-1)
        indices = topk.indices                             # [B, T, k]
        weights = torch.softmax(topk.values, dim=-1)       # [B, T, k]
```

```
# Gather expert params for the chosen experts
# Shapes after gather:
#  w1: [B, T, k, 2*I_local, H]
#  b1: [B, T, k, 2*I_local]
#  w2: [B, T, k, I_local, H]
#  b2: [B, T, k, H]
w1 = self.mlp1_weight[indices]
b1 = self.mlp1_bias[indices]
w2 = self.mlp2_weight[indices]
b2 = self.mlp2_bias[indices]

# Expand input with expert axis for batched matmuls
# x_in: [B, T, 1, H]  (p=1 will be summed over in einsum)
x_in = x.unsqueeze(2)

# ----- First projection: H -> 2I -----
# [B,T,1,H] x [B,T,k,2I,H] -> [B,T,k,2I]
x1 = torch.einsum("btph,btkfh->btkf", x_in, w1) + b1

# ----- SwiGLU (produces shape [B,T,k,I]) -----
x1_glu, x1_lin = x1.chunk(2, dim=-1)  # each [B,T,k,I_local]
# Clamp to keep AMP stable, affine-sigmoid for swish approx
swish = torch.sigmoid(1.702 * x1_glu)
x1_act = x1_glu.clamp(max=self.swiglu_limit) * swish
x1_lin = x1_lin.clamp(min=-self.swiglu_limit) + 1.0
x1 = x1_act * x1_lin                              # [B,T,k,I_local]

# ----- Second projection: I -> H  (FIXED) -----
# [B,T,k,I] x [B,T,k,I,H] -> [B,T,k,H]
x2 = torch.einsum("btkf,btkfh->btkh", x1, w2) + b2

# If tensor-parallel across ranks, sum partials
if dist.is_initialized() and self.world_size > 1:
    dist.all_reduce(x2, op=dist.ReduceOp.SUM)

# Weighted sum across the k experts -> [B,T,H]
out = torch.einsum("btkh,btk->bth", x2, weights)

return residual + out
```

In-Depth Analysis of the MLPBlock with Mixture of Experts (MoE) and SwiGLU

This PyTorch MLPBlock class implements a mixture of experts (MoE) architecture with SwiGLU activation, designed as a feed-forward network (FFN) replacement in transformer models like LLMs. Key features include:

- **MoE sparsity** for scalability at massive scale
- **SwiGLU gating** for enhanced expressiveness
- **Tensor parallelism** for distributed training of billion+ parameter models

The code explanation covers architecture, mathematical operations, initialization, forward pass, training dynamics, practical implications, and edge cases, assuming intermediate familiarity with PyTorch, transformers, MoE, and SwiGLU.

Overview of the MLPBlock

This PyTorch module processes transformer layer inputs with three integrated components:

1. **Mixture of Experts (MoE)**: Multiple expert MLPs with dynamic gating that selects top-k experts per token, enabling trillion-parameter models through computational sparsity
2. **SwiGLU Activation**: Swish-gated activation providing superior nonlinearities and gradient flow vs. ReLU/GELU
3. **Tensor Parallelism**: Distributes intermediate dimensions across devices for large-scale training

Architecture Details

- **Input/Output Shape**: [B, T, H] (batch, sequence length, hidden size)
- Includes residual connections and RMS normalization for training stability

Initialization and Hyperparameters (ModelConfig)

- `hidden_size` (H): Model dimension (e.g., 768, 4096)
- `intermediate_size` (I): MLP projection dimension, typically 4×H
- `num_experts` (E): Total expert MLPs (e.g., 32-128)
- `experts_per_token` (k): Activated experts per token (e.g., 2-8)
- `swiglu_limit`: Numerical stability clamp for AMP training

Initialization Components

The `__init__` method sets up the module's components, including normalization, the gating network, and expert parameters.

1. **World Size and Tensor Parallelism:**
    - `self.world_size = dist.get_world_size() if dist.is_initialized() else 1`: Queries the number of devices (e.g., GPUs) using PyTorch's `torch.distributed` module. If distributed training is not initialized, `world_size` is 1 (single device).
    - `if self.intermediate_size % self.world_size != 0: raise ValueError(...)`: Ensures `intermediate_size` (I) is divisible by `world_size` to enable tensor parallelism. The intermediate dimension is split into shards of size `I_local = I // world_size`. For example, if `I = 4096` and `world_size = 4`, then `I_local = 1024`.
    - This sharding distributes the computational load and memory across devices, critical for large models where a single device cannot store all parameters.
2. **RMS Normalization:**
    - `self.norm = RMSNorm(config.hidden_size, device=device)`: Initializes an RMSNorm layer, which normalizes the input across the H dimension. RMSNorm is defined as:

$$\text{RMSNorm}(x) = \frac{x}{\sqrt{\frac{1}{H}\sum_{i=1}^{H} x_i^2 + \epsilon}} \cdot \gamma$$

and epsilon (typically 1e-5) prevents division by zero. RMSNorm is computationally efficient compared to LayerNorm, as it avoids computing the mean, and is widely used in modern transformers (e.g., LLaMA).

- The layer normalizes the input tensor `[B, T, H]`, stabilizing activations and gradients, especially in deep networks.

3. **Gating Network:**

   - `self.gate = torch.nn.Linear(self.hidden_size, self.num_experts, device=device)`: A linear layer mapping inputs from `H` to `num_experts` (`E`), producing logits `[B, T, E]`. These logits determine which experts are selected for each token. The layer has parameters:
     - **Weight**: Shape `[E, H]`.
     - **Bias**: Shape `[E]`.
   - The gating network is lightweight, with only `H * E` parameters, ensuring minimal overhead compared to the experts.

4. **Expert Parameters:**

   - **First Projection** (`mlp1`): Projects the input from `H` to `2 * I_local` to support SwiGLU's dual components (gated and linear).
     - `self.mlp1_weight`: Shape `[E, 2 * I_local, H]`, a parameter tensor for all experts. The factor of 2 accounts for splitting into two `I_local`-sized parts for SwiGLU.
     - `self.mlp1_bias`: Shape `[E, 2 * I_local]`.
   - **Second Projection** (`mlp2`): Projects from `I_local` back to `H`, restoring the output dimension.
     - `self.mlp2_weight`: Shape `[E, I_local, H]`.
     - `self.mlp2_bias`: Shape `[E, H]`.
     - Each expert has its own MLP with `2 * I_local * H + I_local * H` weights (plus biases), but only k experts are active per token, ensuring sparsity.

5. **Weight Initialization:**

   - `torch.nn.init.kaiming_uniform_(self.mlp1_weight, a=math.sqrt(5))`: Initializes `mlp1_weight` using Kaiming uniform initialization, designed for nonlinear activations (e.g., Swish). The parameter `a=sqrt(5)` sets the bound for uniform sampling, scaled by the fan-in (H), ensuring stable initial gradients:

$$\text{bound} = \sqrt{\frac{6}{(1+a^2)\cdot \text{fan}_\text{i}\text{n}}}$$

   - `torch.nn.init.kaiming_uniform_(self.mlp2_weight, a=math.sqrt(5))`: Similarly initializes `mlp2_weight`, with fan-in `I_local`.
   - `torch.nn.init.zeros_(self.mlp1_bias, self.mlp2_bias)`: Initializes biases to zero, a standard practice to avoid introducing bias at the start of training.
   - Kaiming initialization is critical for SwiGLU, as its Swish component requires careful scaling to prevent vanishing or exploding gradients.

Parameter Sharding and Memory Considerations

In tensor parallelism, the `intermediate_size` (I) is split across `world_size` devices, so each device handles `I_local = I // world_size` dimensions. For example, with `I = 8192` and `world_size = 8`, `I_local = 1024`. This sharding applies to `mlp1_weight`, `mlp1_bias`, and `mlp2_weight`, reducing per-device memory usage. The total parameter count for the experts is:

- `mlp1_weight`: `E * 2 * I * H` (sharded to `E * 2 * I_local * H` per device).
- `mlp1_bias`: `E * 2 * I` (sharded to `E * 2 * I_local`).
- `mlp2_weight`: `E * I * H` (sharded to `E * I_local * H`).
- `mlp2_bias`: `E * H` (not sharded, as `H` is not split).

The gating network adds `H * E + E` parameters, which are typically small compared to the experts. For a model with `H = 4096`, `I = 16384`, `E = 32`, and `world_size = 4`:

- `I_local = 16384 / 4 = 4096`.
- Per-device parameters for `mlp1_weight`: `32 * 2 * 4096 * 4096` ≈ `1.07B`.
- Total parameters remain large, but sparsity (only `k` experts active) and sharding make training feasible.

Forward Pass: Step-by-Step Execution

The `forward` method processes an input tensor `x` of shape `[B, T, H]` and returns an output of the same shape. Below, we break down each operation, including shapes, mathematical details, and implementation nuances.

Step 1: Input and Residual Connection

- **Input**: `x` has shape `[B, T, H]`, where `B` is the batch size, `T` is the sequence length, and `H` is the hidden size. For example, in an LLM, `x` might represent token embeddings after self-attention, with `B=32`, `T=512`, `H=4096`.
- **Residual Connection**: `residual = x` stores the input for a residual connection, added to the output later:

$$\text{output} = x + f(x)$$

  Residual connections, inspired by ResNet, mitigate vanishing gradients and improve training stability in deep transformers.

Step 2: RMS Normalization

- `x = self.norm(x)`: Applies RMSNorm to normalize x across the H dimension, producing `[B, T, H]`. For each token `[b, t, :]`, the operation is:

$$x'b, t = \frac{xb, t}{\sqrt{\frac{1}{H}\sum_{i=1}^{H} x_{b,t,i}^2 + \epsilon}} \cdot \gamma$$

  This normalizes the variance to approximately 1, scaling by learnable (\gamma). RMSNorm is computationally cheaper than LayerNorm (no mean computation) and is common in modern LLMs (e.g., LLaMA, Mixtral).

Step 3: Gating Network and Top-k Selection

- **Gating**: `logits = self.gate(x)` computes logits via a linear transformation:

$$\text{logits} = xW_g + b_g$$

  - **Input**: x is `[B, T, H]`.
  - W_g: `[H, E]`, b_g: `[E]`.
  - **Output**: `logits` is `[B, T, E]`, where each element `[b, t, e]` is the score for expert e on token t in batch b.
- **Top-k Selection**: `topk = torch.topk(logits, self.experts_per_token, dim=-1)` selects the top `k = experts_per_token` experts per token:
  - `indices`: Shape `[B, T, k]`, containing the indices of the top-k experts (e.g., `[3, 7]` for k=2).
  - `values`: Shape `[B, T, k]`, containing the corresponding logits.

- **Gating Weights**: `weights = torch.softmax(topk.values, dim=-1)` applies softmax to the top-k logits:

$$\text{weights}b, t, i = \frac{\exp(\text{values}b, t, i)}{\sum_{j=1}^{k} \exp(\text{values}_{b,t,j})}$$

Output: `weights` is `[B, T, k]`, with each row summing to 1, representing the contribution of each selected expert.

Step 4: Gathering Expert Parameters

- Parameters for the top-k experts are gathered using the `indices` tensor:
    - `w1 = self.mlp1_weight[indices]`: Shape `[B, T, k, 2 * I_local, H]`, selects weights for the first projection.
    - `b1 = self.mlp1_bias[indices]`: Shape `[B, T, k, 2 * I_local]`.
    - `w2 = self.mlp2_weight[indices]`: Shape `[B, T, k, I_local, H]`.
    - `b2 = self.mlp2_bias[indices]`: Shape `[B, T, k, H]`.
- This operation uses advanced indexing to extract parameters for the `k` selected experts for each of the `B * T` tokens, enabling batched processing.

Step 5: First Projection (H → 2 * I_local)

- **Input Expansion**: `x_in = x.unsqueeze(2)` reshapes `x` from `[B, T, H]` to `[B, T, 1, H]` to align with the expert weights for batched matrix multiplication.

- **Matrix Multiplication:**
    - `x1 = torch.einsum("btph,btkfh->btkf", x_in, w1) + b1` computes the first projection:
        - `x_in`: `[B, T, 1, H]` (the `p=1` dimension is summed over).
        - `w1`: `[B, T, k, 2 * I_local, H]`.
        - Einsum operation: For each batch `b`, token `t`, and expert `i`, compute:

$$x1_{b,t,i,f} = \sum_h x_{b,t,1,h} \cdot w1_{b,t,i,f,h} + b1_{b,t,i,f}$$

        - Output: `x1` is `[B, T, k, 2 * I_local]`, representing the projection to the intermediate space for each selected expert.
    - The bias `b1` is broadcast across `B, T, k`.

Step 6: SwiGLU Activation

- **Splitting**: `x1_glu, x1_lin = x1.chunk(2, dim=-1)` splits `x1` along the last dimension into two tensors, each `[B, T, k, I_local]`:
    - `x1_glu`: The component for the Swish gate.
    - `x1_lin`: The linear component.
- **Swish Approximation:**
    - `swish = torch.sigmoid(1.702 * x1_glu)` approximates the Swish function (`z * sigmoid(z)`) using a scaled sigmoid with coefficient 1.702. This is an optimization for AMP, as:

$$\text{Swish}(z) \approx z \cdot \sigma(1.702 \cdot z)$$

    The coefficient 1.702 adjusts the sigmoid to closely match Swish's behavior, improving numerical stability.

- `x1_act = x1_glu.clamp(max=self.swiglu_limit) * swish`: Clamps `x1_glu` to `swiglu_limit` (e.g., 30.0) to prevent overflow in FP16 training, then applies the Swish gate, producing `[B, T, k, I_local]`.

- **Linear Adjustment**: `x1_lin = x1_lin.clamp(min=-self.swiglu_limit) + 1.0` clamps the linear component to prevent extreme negative values and adds 1.0. This adjustment may:
  - Shift the linear component's range to ensure positive outputs, aligning with SwiGLU's design.
  - Enhance stability by avoiding large negative values in the subsequent multiplication.

- **Gating**: `x1 = x1_act * x1_lin` performs element-wise multiplication:

$$x1_{b,t,i,f} = x1_{\text{act},b,t,i,f} \cdot x1_{\text{lin},b,t,i,f}$$

Output: `x1` is `[B, T, k, I_local]`, representing the SwiGLU-activated output for each expert.

Step 7: Second Projection (I_local → H)

- **Matrix Multiplication:**

- `x2 = torch.einsum("btkf,btkfh->btkh", x1, w2) + b2` projects the SwiGLU output back to the hidden size:
  - `x1`: `[B, T, k, I_local]`.
  - `w2`: `[B, T, k, I_local, H]`.
  - Einsum operation:

$$x2_{b,t,i,h} = \sum_{f} x1_{b,t,i,f} \cdot w2_{b,t,i,f,h} + b2_{b,t,i,h}$$

  - Output: `x2` is `[B, T, k, H]`.

- The bias `b2` is broadcast across `B, T, k`.

Step 8: Tensor Parallelism

- `if dist.is_initialized() and self.world_size > 1: dist.all_reduce(x2, op=dist.ReduceOp.SUM)`: In tensor-parallel mode, each device computes a shard of the intermediate dimension (`I_local`). The `all_reduce` operation sums the partial outputs across devices:
    - **Input**: `x2` is `[B, T, k, H]` on each device, representing the contribution of `I_local` dimensions.
    - **Output**: After summation, `x2` includes contributions from all I dimensions, maintaining `[B, T, k, H]`.
- This ensures the output is equivalent to a nonsharded computation, as if each expert processed the full I dimensions.

Step 9: Weighted Sum Across Experts

- `out = torch.einsum("btkh,btk->bth", x2, weights)` combines the outputs of the top-k experts:
    - `x2`: `[B, T, k, H]`, expert outputs.
    - `weights`: `[B, T, k]`, gating weights.
- **Einsum Operation**:

$$\text{out}b, t, h = \sum_{i} \text{weights}b, t, i \cdot x2_{b,t,i,h}$$

- **Output**: `out` is `[B, T, H]`, the weighted sum of expert outputs for each token.

Step 10: Residual Connection and Output

- `return residual + out`: Adds the original input `x` to the MoE output, producing the final output `[B, T, H]`. This residual connection ensures:

- Gradient flow through the identity path, mitigating vanishing gradients.
- Compatibility with transformer layer stacking, as the output shape matches the input.

Training Dynamics

Backpropagation

Training involves computing gradients for the active experts and gating network, leveraging MoE's sparsity:

- **Experts**: Gradients are computed only for the top-k experts' parameters (`mlp1_weight`, `mlp1_bias`, `mlp2_weight`, `mlp2_bias`), weighted by `G(x')_i`. For `W_{i1}`:

  $$\frac{\partial L}{\partial W_{i1}} = G(x')i \cdot \frac{\partial L}{\partial x_i} \cdot \frac{\partial x_i}{\partial z_i} \cdot \frac{\partial z_i}{\partial zi1} \cdot \frac{\partial z_{i1}}{\partial W_{i1}}$$

  The Swish derivative is:

  $$\frac{\partial \text{Swish}(z)}{\partial z} = \sigma(1.702 \cdot z) + 1.702 \cdot z \cdot \sigma(1.702 \cdot z)(1 - \sigma(1.702 \cdot z))$$

  This ensures non-zero gradients, aiding training stability.

- **Gating Network**: Gradients for `gate.weight` and `gate.bias` account for the softmax and top-k selection, which can be complex due to the nondifferentiable nature of top-k but is approximated via straight-through estimation or softmax gradients.

Load Balancing

To prevent routing collapse (where the gating network favors a few experts), a load balancing loss is typically added (though not explicit in the code, its standard in MoE):

$$L_{\text{balance}} = \alpha \sum_{i=1}^{E} \left(\frac{1}{B \cdot T} \sum_{b,t} G(x'_{b,t})_i - \frac{1}{E} \right)^2$$

This encourages uniform expert utilization, ensuring all experts are trained effectively.

Optimization

- **Optimizer**: AdamW is commonly used, with learning rates around 0.0001-0.001, as it adapts to sparse updates.
- **Mixed Precision**: Clamping (`swiglu_limit`) and the 1.702 coefficient in SwiGLU support AMP with FP16, reducing memory usage and speeding up training.
- **Distributed Training**: Tensor parallelism requires synchronization via `all_reduce`, managed by `torch.distributed`. Each device updates its shard of parameters, with gradients aggregated across devices.

Numerical Stability

- **Clamping**: The `swiglu_limit` prevents overflow in FP16 training, critical for large models.
- **Swish Approximation**: The 1.702 coefficient optimizes Swish for AMP, balancing accuracy and stability.
- **RMSNorm**: Stabilizes activations, reducing the risk of exploding gradients.

Edge Cases and Robustness

1. **Small Batch Sizes**: With small B or T, expert utilization may be uneven, requiring stronger balancing loss.
2. **High k Values**: If k approaches E, sparsity benefits diminish, increasing compute cost.
3. **Numerical Issues**: Without clamping, FP16 training may overflow, especially with large I.
4. **Initialization Sensitivity**: Poor initialization (e.g., without Kaiming) can destabilize training, particularly for SwiGLU.

Summary

This chapter provides a comprehensive examination of the multilayer perceptron (MLP) block, emphasizing its evolution and integration with advanced techniques like mixture of experts (MoE) and SwiGLU activation. The chapter begins with a historical overview of MLPs, tracing their roots from Frank Rosenblatt's single-layer perceptron in 1958 to the resurgence in the 1980s with backpropagation, which enabled multilayer networks to model nonlinear problems.

The discussion then delves into the basics of perceptrons and neurons, detailing the single-layer perceptron and modern activation functions like sigmoid, ReLU, tanh, and softmax, including their mathematical formulations, advantages, and challenges (e.g., vanishing gradients). The architecture of MLPs is explored in depth, covering layers' roles, mathematical representations, and the Universal Approximation Theorem, which underscores MLPs' theoretical capability to approximate any continuous function.

Training mechanisms are thoroughly explained, including loss functions (MSE, binary/cross-entropy), backpropagation steps, optimization enhancements (Adam, dropout, batch normalization), and challenges like vanishing/exploding gradients. Practical workflows and advanced techniques, such as residual connections and mixed precision training, are also covered.

Applications of MLPs are discussed across classification, regression, function approximation, and feature learning, with limitations like computational complexity, overfitting, and lack of structure awareness noted. The chapter concludes with an integration of MLPs with MoE and SwiGLU, detailing their synergy for scalability, efficiency, and performance in large models like Grok and Mixtral.

The chapter includes a code implementation of an MLPBlock using MoE and SwiGLU, with tensor parallelism for distributed training, and an in-depth analysis of its forward pass, initialization, and training dynamics. This provides a practical bridge between theory and implementation, emphasizing how MoE's sparsity and SwiGLU's expressiveness enable efficient large-scale AI.

In the next chapter, we will build upon the foundational MLP block explored here by diving into the TransformerBlock, the core building block of transformer architectures. We will examine how self-attention mechanisms integrate with FFN layers (like the MoE-SwiGLU MLP) to create powerful sequence models. From there, we will construct a full transformer model, discussing layer stacking, positional encodings, and overall architecture for tasks like language modeling.

Additionally, we will cover inference utilities, including tokenization, generation strategies (e.g., greedy decoding, beam search), and optimization techniques for efficient deployment, such as quantization and caching. This chapter will provide the tools to assemble and deploy complete transformer-based systems, paving the way for practical applications in natural language processing and beyond.

CHAPTER 9

Transformer Block and Full Transformer Model—It's Time to Put the Puzzle Together

Imagine you've spent the previous sections collecting the scattered pieces of a grand puzzle—fragments of attention mechanisms, glimpses of feed-forward networks, hints of residual connections, and whispers of layer normalization. Each piece, on its own, hints at innovation: attention as the glue that binds distant elements, feed-forwards as the sharpeners of local detail, and residuals as the bridges preventing collapse.

But now it's time to put the puzzle together. Welcome to the **transformer block chapter**, the elegant assembly where these pieces interlock to form a resilient, powerful structure that drives the attention revolution in artificial intelligence.

As we fit these pieces side by side, you'll see how they create something greater than the sum: a parallelizable powerhouse for natural language, vision, and beyond. No more sequential drudgery of RNNs; instead, a harmonious whole that captures the full picture of your data. Let's roll up our sleeves and start connecting the edges.

The Role of the Transformer Block in Sequence Processing

At its core, the transformer block is designed to process sequences of data—think sentences in a language, pixels in an image patch, or even amino acid chains in proteins—by transforming raw input representations into richer, more contextually informed outputs.

D. Grigorov, *Building Large Language Models from Scratch*, https://doi.org/10.1007/979-8-8688-2297-1_9

The transformer block flips this paradigm on its head. It enables **parallel processing** of the entire sequence, allowing all elements to be computed simultaneously. This is achieved through a mechanism that dynamically weighs the relevance of different parts of the input to each other, a process rooted in the concept of attention.

By stacking multiple transformer blocks—typically 6 to 12 in early models and up to hundreds in modern giants like PaLM or GPT-4—the architecture progressively refines representations, layer by layer, turning simple embeddings into sophisticated encodings that capture syntactic, semantic, and even pragmatic nuances.

In practical terms, imagine feeding a sentence like "The cat sat on the mat" into a transformer-based model. Each word starts as a dense vector (an embedding). As it passes through the first block, the model begins to understand basic relationships, such as "cat" relating to "sat."

By the final block, it has woven a tapestry of meaning, potentially inferring that this is a scene of domestic tranquility. This layered refinement is what makes transformers so versatile, applicable not just to text but to any data with sequential or relational structure.

The Architecture of the Transformers Block

The architecture of a transformer block embodies the distilled wisdom of the attention mechanism, serving as the atomic yet extensible unit that underpins the entire transformer family's prowess in handling sequential and relational data. This block's design philosophy—prioritizing modularity, parallelism, and gradient-friendly shortcuts—allows it to be stacked vertically into deep hierarchies (from a modest 6 layers in the original transformer to 96+ in behemoths like GPT-4 or PaLM), while horizontally scaling via multihead parallelism.

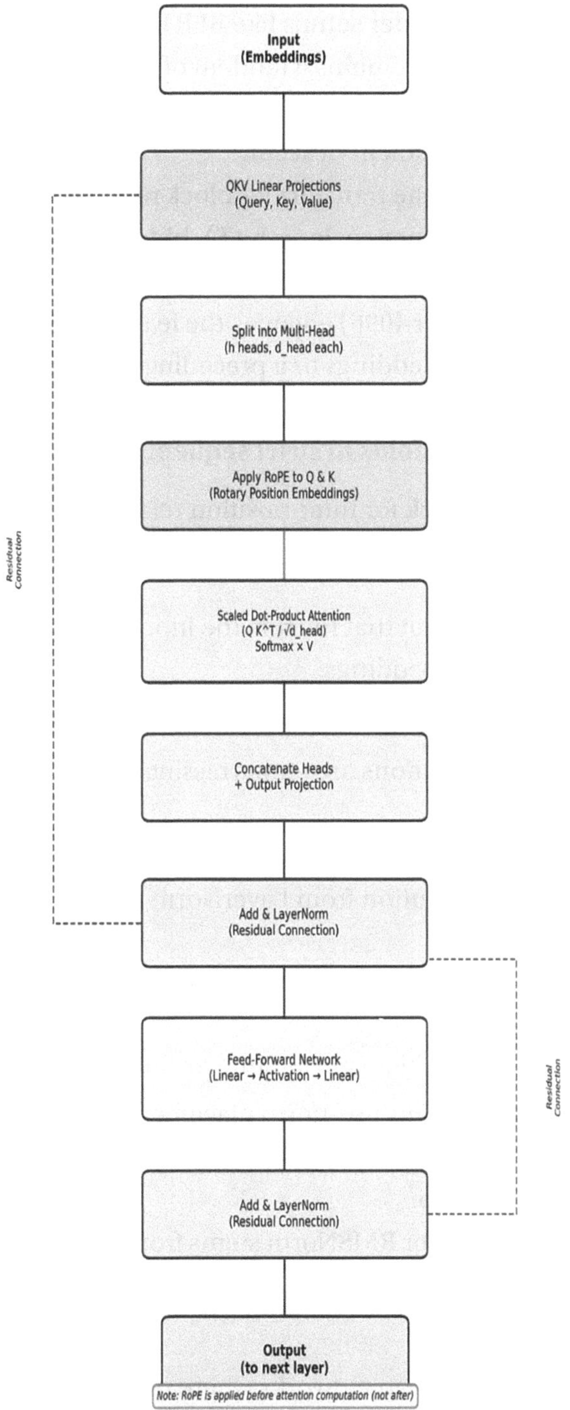

Figure 9-1. *Transformer Layer Architecture*

In decoder-only configurations prevalent in generative models, the block processes autoregressive sequences; in encoder setups like BERT, it enables bidirectional context. Regardless of variant, its skeleton remains a tandem of self-attention and feed-forward components, laced with residuals, dropouts, and normalizations to foster robust optimization under stochastic gradient descent.

Commencing at the ingress, the transformer block receives a tensor x x x of dimensions **[batch_size (B), sequence_length (T), hidden_size (d_model)]**, where B batches parallelize training, T encapsulates the temporal or positional span (e.g., 1024 tokens), and d_model (e.g., 768 or 4096) encodes the feature richness per position. This input might stem from prior embeddings or a preceding block, already infused with token-level semantics.

The block's forward pass unfolds in strict sequence:

- **first the AttentionBlock** for inter-position relational modeling
- **then the MLPBlock** for intra-position feature transformation
- culminating in an output that mirrors the input shape but harbors deeper contextual embeddings.

Absent from this core is any recurrent loop, eschewing RNN pitfalls like sequential bottlenecks; instead, all computations are embarrassingly parallel, leveraging matrix operations on accelerators.

Zooming into the AttentionBlock, the pipeline opens with a pre-normalization step via RMSNorm, a lightweight evolution from LayerNorm that eschews mean subtraction for pure variance scaling: for each position, compute $\sigma = \sqrt{\frac{1}{d_{\text{model}}} \sum_{i=1}^{d_{\text{model}}} x_i^2 + \epsilon}$, then $\hat{x} = x/\sigma \odot g$, where g is a learnable scale parameter (initialized to ones) and $\epsilon \approx 10^{-6}$ guards against division by zero. This pre-norm placement—unlike the original post-norm—accelerates convergence by stabilizing early-layer activations, a tweak popularized in models like Llama.

The theoretical justification for RMSNorm stems from the observation that the recentering operation in LayerNorm—computing $\mu = \frac{1}{d_{\text{model}}} \sum_{i=1}^{d_{\text{model}}} x_i$ and subtracting it—provides minimal benefit in deep networks where residual connections already center the distribution.

Empirically, RMSNorm achieves comparable perplexity while reducing computational overhead by approximately 7–10% per layer, a non-trivial saving when scaling to $N > 80$ layers. The learnable scale $g \in \mathbb{R}^{d_{\text{model}}}$ grants each dimension adaptive expressivity, often converging to values near $[0.8, 1.2]$ during training, suggesting the initial unit scale is near-optimal.

Query-Key-Value Projections and Multihead Decomposition

The normalized $\hat{x}$ then bifurcates into query, key, and value pathways through bias-free linear projections: $Q = \hat{x}W_Q$, $K = \hat{x}W_K$, $V = \hat{x}W_V$, with each $W \in \mathbb{R}^{d_{\text{model}} \times d_{\text{model}}}$. The deliberate omission of bias terms—contrasting with classical feed-forward networks—reflects the finding that biases contribute negligibly to representational capacity in post-norm architectures while inflating parameter count by $3d_{\text{model}}$ per block.

To harness multihead attention (h heads, e.g., 12-128), these are reshaped to $[B, h, T, d_{\text{head}}]$ where $d_{\text{head}} = d_{\text{model}}/h$ (ensuring integer divisibility, like 64 for $d_{\text{model}} = 768, h = 12$), and transposed to $[B, h, T, d_{\text{head}}]$ for batched head-wise ops. The multihead decomposition can be understood through a representational diversity lens: each head $i \in [1, h]$ learns a distinct attention pattern, with empirical studies revealing specialization—some heads attend locally (syntactic dependencies), others globally (semantic coherence), and yet others exhibit task-specific patterns like coreference resolution. Mathematically, the parallel heads compute:

$$Q^{(i)} = \hat{x}W_Q^{(i)}, \quad K^{(i)} = \hat{x}W_K^{(i)}, \quad V^{(i)} = \hat{x}W_V^{(i)}$$

where $W_Q^{(i)}, W_K^{(i)}, W_V^{(i)} \in \mathbb{R}^{d_{\text{model}} \times d_{\text{head}}}$. In practice, these are implemented as single large weight matrices partitioned along the output dimension, enabling efficient batched matrix multiplication on modern accelerators.

Rotary Position Embeddings: Geometric Foundations

Positional awareness, vital to break attention's permutation invariance, is elegantly woven via rotary position embeddings (RoPE), a parameter-efficient alternative to absolute sinusoidal additions. RoPE treats the head dimension as a 2D plane, applying a rotation matrix $R(\theta_m) = \begin{pmatrix}\cos m\theta & -\sin m\theta \ \sin m\theta & \cos m\theta\end{pmatrix}$ per pair of dimensions, where $\theta_i = 10^{-2i/d_{\text{head}}}$ (base 10000) and m indexes position.

The geometric intuition underlying RoPE derives from complex number multiplication: representing each dimension pair as $z = x_1 + ix_2$, a rotation by angle $m\theta$ corresponds to multiplication by $e^{im\theta}$. Expanding this yields:

$$z' = z \cdot e^{im\theta} = (x_1 + ix_2)(\cos m\theta + i \sin m\theta) = (x_1 \cos m\theta - x_2 \sin m\theta) + i(x_1 \sin m\theta + x_2 \cos m\theta)$$

This formulation ensures that the inner product between queries and keys at positions m and n depends only on their relative distance $m - n$:

$$\langle R_m q, R_n k \rangle = \langle R_{m-n} q, k \rangle$$

where R_m denotes the rotation at position m. In practice, precompute cos and sin buffers up to T_{max} (e.g., 2048, extensible via caching), then for queries/keys: split into first/last halves (x_1, x_2) yielding:

$$q_{\text{embed}} = [x_1 \odot \cos(m\theta) - x_2 \odot \sin(m\theta), , x_1 \odot \sin(m\theta) + x_2 \odot \cos(m\theta)]$$

where $\odot$ denotes element-wise multiplication. This relative encoding excels in length generalization, as rotations compound additively for distances, outperforming absolute PE in long-context models like RWKV hybrids. The base frequency $\theta_0 = 10^{-4}$ (corresponding to base 10000) was chosen to ensure the longest wavelength $2\pi/\theta_{\text{min}} \approx 10000$ tokens, though recent work explores adaptive bases (10^5 for 32k contexts) and interpolation techniques.

Scaled Dot-Product Attention: The Core Mechanism

The computational fulcrum is scaled dot-product attention: raw scores $S = (QK^T)/\sqrt{d_{\text{head}}} \in \mathbb{R}^{[B,h,T,T]}$, softened to probabilities via row-wise softmax (with causal masking: upper triangle set to $-\infty$ for autoregression, implemented as a lower-triangular boolean mask).

The scaling factor $1/\sqrt{d_{\text{head}}}$ is theoretically motivated by variance stabilization. Assuming Q and K have unit variance entries, their dot product $q^{\top}k = \sum_{i=1}^{d_{\text{head}}} q_i k_i$ has variance $\text{Var}(q^{\top}k) = d_{\text{head}}$ (sum of d_{head} independent unit-variance terms). Without scaling, as d_{head} increases, the logits grow large, pushing the softmax into saturation regions where gradients vanish. The scaling normalizes this to:

$$\text{Var}\left(\frac{q^{\top}k}{\sqrt{d_{\text{head}}}}\right) = \frac{d_{\text{head}}}{d_{\text{head}}} = 1$$

ensuring gradient flow remains stable across varying model dimensions.

The causal mask $M \in 0, -\infty^{T\times T}$ enforces autoregressive constraints:

$$M\,ij{=}\{0 \text{ if } i{\geq}j \ -\infty \text{ if } i{<}j$$

applied before the softmax: $S_{\text{masked}} = S + M$. The $-\infty$ entries ensure $\exp(-\infty) = 0$ in the subsequent softmax, preventing position i from attending to future positions $j > i$. In encoder-only architectures (BERT, ViT), the mask is omitted for bidirectional attention.

The attention weights are then computed as:

$$A = \text{softmax}\left(\frac{QK^{T}}{\sqrt{d_{\text{head}}}} + M\right) = \frac{\exp\left(\frac{q_i k_j^{T}}{\sqrt{d_{\text{head}}}} + M_{ij}\right)}{\sum_{k=1}^{T} \exp\left(\frac{q_i k_k^{T}}{\sqrt{d_{\text{head}}}} + M_{ik}\right)}$$

where A_{ij} represents the attention weight from position i to position j. These weights are then applied to the values:

The attended output per head is $A_h = \text{softmax}(S)V$, concatenated across heads to $[B, T, d_{\text{model}}]$, and linearly mixed via $W_O \in \mathbb{R}^{d_{\text{model}}\times d_{\text{model}}}$ to yield the sub-layer result. The output projection can be expressed as:

$$\text{MultiHead}(Q, K, V) = \text{Concat}(A_1, A_2, \ldots, A_h)W_O$$

where each $A_i \in \mathbb{R}^{B\times T\times d_{\text{head}}}$ and the concatenation yields $\mathbb{R}^{B\times T\times d_{\text{model}}}$.

Training-time dropout ($p_{\text{attn}} \approx 0.1$) sparsifies the softmax matrix, while a post-projection residual dropout (p_{resid}) randomizes the additive skip:

$$\text{attn}_{\text{out}} = x + \text{Dropout}(W_O \cdot \text{MultiHead}(Q, K, V))$$

This sandwich—norm around the attention core, residual bridging—empowers the block to learn subtle perturbations atop identity, crucial for $N > 100$ layer depths without degradation. The residual connection provides a gradient highway: during backpropagation, gradients can bypass the attention sublayer entirely via the identity path, with effective gradient flow satisfying:

$$\frac{\partial \mathcal{L}}{\partial x_\ell} = \frac{\partial \mathcal{L}}{\partial x_{\ell+1}} \left(I + \frac{\partial f_\ell}{\partial x_\ell} \right)$$

where f_ℓ represents the attention transformation at layer ℓ The additive identity I ensures $|\partial \mathcal{L}/\partial x_\ell| \geq |\partial \mathcal{L}/\partial x_{\ell+1}|$, preventing gradient vanishing.

The Feed-Forward Block: Position-Wise Transformations

Cascading to the MLPBlock, the attnout *feeds another RMSNorm, mirroring the attention's stabilizer. The feed-forward sublayers then unleash capacity: a "SwiGLU" or vanilla expansion via* $\text{fc1} : [d\text{model} \to d\text{ff}]$ https://www.codecogs.com/eqnedit.php?latex=%5Ctext%7Bfc%7D1%3A%20%5Bd%7B%5Ctext%7Bmodel%7D%7D%20%5Cto%20d%7B%5Ctext%7Bff%7D%7D%5D#0 where $d_{\text{ff}} = \alpha \cdot d_{\text{model}}$ https://www.codecogs.com/eqnedit.php?latex=d_%7B%5Ctext%7Bff%7D%7D%20%3D%20%5Calpha%20%5Ccdot%20d_%7B%5Ctext%7Bmodel%7D%7D#0 with $\alpha \in [4, 8]$ https://www.codecogs.com/eqnedit.php?latex=%5Calpha%20%5Cin%20%5B4%2C%208%5D#0 (modern ratios like 8× in Mistral vs. classic 4× in GPT-2), but sticking to GELU classics:

$$\text{intermediate} = \text{GELU}(\text{fc}_1(\text{norm}(x)))$$

where $\text{GELU}(x) \approx 0.5x \left(1 + \tanh \left(\sqrt{\frac{2}{\pi}} \left(x + 0.044715x^3 \right) \right) \right)$ https://www.codecogs.com/eqnedit.php?latex=%5Ctext%7BGELU%7D(x)%20%5Capprox%200.5x%20%5Cleft(1%20%2B%20%5Ctanh%5Cleft(%5Csqrt%7B%5Cfrac%7B2%7D%7B%5Cpi%7D%7D%5Cleft(x%20%2B%200.044715x%5E3%5Cright)%5Cright)%5Cright)#0 approximates probabilistic neuron firing. This nonlinearity, smoother than ReLU, gates information flow.

The GELU activation function can be derived from stochastic regularization: considering each neuron as randomly dropping out with probability $\Phi(-x)$ where Φ is the standard Gaussian CDF, the expected output is $x\Phi(x)$. The practical approximation above closely matches $x\Phi(x)$ while being computationally efficient. An alternative exact formulation uses the error function:

$$\text{GELU}(x) = x \cdot \Phi(x) = x \cdot \frac{1}{2}\left[1 + \text{erf}\left(\frac{x}{\sqrt{2}}\right)\right]$$

Contraction follows with $\text{fc2} : [d\text{ff} \to d_{\text{model}}]$ https://www.codecogs.com/eqnedit.php?latex=%5Ctext%7Bfc%7D2%3A%20%5Bd%7B%5Ctext%7Bff%7D%7D%20%5Cto%20d_%7B%5Ctext%7Bmodel%7D%7D%5D#0, sans bias for efficiency, trailed by dropout ($p_{\text{ff}} \approx 0.1$ https://www.codecogs.com/eqnedit.php?latex=p_%7B%5Ctext%7Bff%7D%7D%20%5Capprox%200.1#0):

$$\text{FFN}(x) = \text{fc}_2(\text{Dropout}(\text{GELU}(\text{fc}_1(x))))$$

Residual summation closes the loop:

$$\text{mlp}_{\text{out}} = x + \text{Dropout}(\text{FFN}(\text{norm}(x)))$$

where x is the attnout. *The MLP's position-wise nature (broadcast over T) contrasts attention's globalism, jointly modeling local hierarchies—syntax in shallow layers, semantics deeper. This complementarity can be formalized: attention provides a data-dependent routing mechanism (Aij* depends on input content), while the FFN applies a fixed transformation at each position, creating a two-stage processing pipeline akin to biological neural circuits.

Alternative Activations: SwiGLU and Gated Variants

Modern architectures increasingly adopt gated linear units, particularly SwiGLU (Swish-Gated Linear Unit), which computes:

$$\text{SwiGLU}(x, W, V, b, c) = \text{Swish}(xW + b) \odot (xV + c)$$

where $\mathrm{Swish}(x) = x \cdot \sigma(x) = \frac{x}{1+e^{-x}}$ and $\odot$ denotes element-wise multiplication.

This can be rewritten compactly as:

$$\mathrm{FFN}_{\mathrm{SwiGLU}}(x) = (\mathrm{Swish}(xW_1) \odot xW_3)W_2$$

requiring three weight matrices instead of two. The gating mechanism $\odot xW_3$ provides dynamic modulation, allowing the network to selectively amplify or suppress features. Empirical comparisons show SwiGLU improves perplexity by 0.5-1.0 points over GELU in large-scale pretraining, though at 50% increased FFN parameter cost. The expansion ratio is often reduced to $\alpha = \frac{2}{3} \cdot 8 \approx 5.33$ to maintain total parameter parity.

Block Composition and Information Flow

In the holistic block forward pass:

$$x' = x + \mathrm{Attention}(\mathrm{RMSNorm}(x)) \; x'' \quad = x' + \mathrm{FFN}(\mathrm{RMSNorm}(x'))$$

No terminal norm appears within the block itself; the subsequent block or final layer handles it, chaining pre-norms for end-to-end stability. This pre-norm configuration can be contrasted with the original post-norm architecture:

$$\text{Post-norm: } x' = \mathrm{RMSNorm}(x + \mathrm{Attention}(x)) \; x'' \quad = \mathrm{RMSNorm}(x' + \mathrm{FFN}(x'))$$

The pre-norm variant exhibits superior training dynamics, particularly in deep networks ($N > 24$ layers), as the normalization layers protect the residual stream from exploding activations early in training.

Hyperparameters like $\mathrm{layer}_{\mathrm{idx}}$ (for potential depth-wise adaptations, e.g., grouped query attention in later heads) and device placement (GPU offload strategies) are configurable, often via dataclass wrappers or configuration dictionaries. Modern frameworks support layer-wise learning rate scaling:

$$\eta_\ell = \eta_0 \cdot \gamma^{N-\ell}$$

where $\gamma \in [0.95, 0.999]$ applies exponentially decaying learning rates to earlier layers, reflecting the observation that shallow layers learn faster and benefit from lower learning rates to maintain stable feature representations.

Full Model Architecture and Training Dynamics

Encapsulating the model, a transformer orchestrates blocks in a ModuleList: embed tokens $[B, T] \to [B, T, d_{\text{model}}]$ (via $\text{nn.Embedding}(\text{vocabsize}, d\text{model})$), iterate through N layers, apply terminal RMSNorm, and project to logits $[B, T, \text{vocab}_{\text{size}}]$ for cross-entropy loss in language modeling:

$$\mathcal{L} = -\frac{1}{BT}\sum_{b=1}^{B}\sum_{t=1}^{T}\log P(x_t^{(b)} \mid x_{<t}^{(b)})$$

where $P(x_t \mid x_{<t}) = \text{softmax}(\text{logitst})x_t$ and x_t is the target token at position t.

The embedding layer $E \in \mathbb{R}^{\text{vocabsize} \times d\text{model}}$ maps discrete tokens to continuous representations. In tied-embedding architectures, the output projection W_{out} shares weights with E^T, reducing parameters by $d_{\text{model}} \times \text{vocabsize}$ *while enforcing a symmetry: tokens with similar embeddings produce similar output probabilities. The embedding initialization typically follows* $\mathcal{N}(0, \text{dmodel}^{-1/2})$ to maintain unit variance in the initial forward pass.

Building the Transformers Block for Our LLM from Scratch

```
class TransformerBlock(torch.nn.Module):
    def __init__(self, config: ModelConfig, layer_idx: int, device: torch.
    device | None = None):
        super().__init__()
        self.attn = AttentionBlock(config, layer_idx, device)
        self.mlp = MLPBlock(config, device)

    def forward(self, x: torch.Tensor) -> torch.Tensor:
        x = self.attn(x)  # [B, T, H]
        x = self.mlp(x)   # [B, T, H]
        return x
```

```
# assumes RMSNorm, RotaryEmbedding, sdpa, MLPBlock, TransformerBlock
already defined
class Transformer(torch.nn.Module):
    def __init__(self, config: ModelConfig, device: torch.device | None
    = None):
        super().__init__()
        self.config = config
        self.embedding = torch.nn.Embedding(config.vocab_size, config.
        hidden_size, device=device)
        self.layers = torch.nn.ModuleList([TransformerBlock(config, i,
        device=device) for i in range(config.num_hidden_layers)])
        self.norm = RMSNorm(config.hidden_size, device=device)
        self.output_proj = torch.nn.Linear(config.hidden_size, config.
        vocab_size, bias=False, device=device)

    def forward(self, tokens: torch.Tensor) -> torch.Tensor:
        x = self.embedding(tokens)
        for layer in self.layers:
            x = layer(x)
        x = self.norm(x)
        return self.output_proj(x)

    @classmethod
    def from_checkpoint(cls, path: str, device: str | torch.device =
    "cuda", strict: bool = True) -> "Transformer":
        import os, json
        if isinstance(device, str):
            device = torch.device(device)
        with open(os.path.join(path, "config.json"), "r") as f:
            cfg = json.load(f)
        config = ModelConfig(**cfg)
        model = cls(config, device=device).to(device)
        sd = torch.load(os.path.join(path, "model.pt"), map_
        location=device)
        missing, unexpected = model.load_state_dict(sd, strict=False)
        if strict and (missing or unexpected):
```

```
        raise RuntimeError(f"Missing: {missing}, Unexpected:
        {unexpected}")
    if missing or unexpected:
        print(f"[load_warning] missing={missing}
        unexpected={unexpected}")
    model.eval()
    return model
```

The TransformerBlock Class

The `TransformerBlock` class represents a single layer in the transformer architecture. In modern language models, these blocks are stacked on top of each other to create deep neural networks capable of learning complex language patterns and representations. Each block is essentially identical in structure but learns different patterns at different levels of abstraction.

Initialization Method (`__init__`)

```
def __init__(self, config: ModelConfig, layer_idx: int, device: torch.
device | None = None):
    super().__init__()
    self.attn = AttentionBlock(config, layer_idx, device)
    self.mlp = MLPBlock(config, device)
```

The initialization method sets up the two fundamental components that make up a transformer layer:

The `config` parameter is a configuration object (likely a dataclass or similar structure) that contains all the hyperparameters needed to construct the model. This typically includes settings like the hidden dimension size, number of attention heads, vocabulary size, number of layers, dropout rates, and various other architectural choices. By encapsulating all these settings in a single config object, the code follows a clean design pattern that makes it easy to experiment with different model configurations without changing the code structure.

The `layer_idx` parameter is particularly important and represents the index of this specific layer within the full transformer stack (0 for the first layer, 1 for the second, and so on). This parameter is crucial for certain operations that need to be position-aware within the network.

For example, if the model uses rotary position embeddings (RoPE), which is mentioned in the code comments and is a modern alternative to absolute position embeddings used in models like LLaMA and GPT-NeoX, different layers might need to know their position in the stack. Some advanced architectures also use layer-specific learning rates or apply different operations at different depths, making this layer index valuable. Additionally, during debugging and logging, having the layer index helps track which specific layer might be causing issues or exhibiting interesting behavior.

The `device` parameter allows for explicit specification of where the model components should be allocated in memory—either on a CPU or a specific GPU. This is important for efficient training and inference, especially when working with large models that need to be distributed across multiple GPUs or when you want to ensure that all components of a layer reside on the same device to avoid expensive data transfers between devices during computation.

The method creates two sub-modules:

- **`self.attn` (AttentionBlock)**: This is the self-attention mechanism, which is the core innovation of the transformer architecture. The attention mechanism allows each position in the sequence to attend to all other positions, enabling the model to capture long-range dependencies and relationships in the input text. The attention block computes queries, keys, and values from the input, then uses these to determine how much each token should "pay attention to" every other token in the sequence. This is done through a similarity calculation (typically scaled dot-product attention) followed by a weighted sum. The `layer_idx` is passed here because, as mentioned, certain positional encoding schemes like RoPE need to know which layer they're operating in to correctly apply rotational position information.

- **`self.mlp` (MLPBlock)**: This is the feed-forward network, also called the position-wise feed-forward network. While the attention mechanism handles inter-token relationships, the MLP processes each position independently, applying the same transformation to every token. Typically, this consists of two linear transformations with a nonlinear activation function in between (commonly GELU or SiLU in modern models). The MLP usually expands the hidden dimension by a factor of 4 (so if hidden_size is 768, it might expand to 3072 internally), applies the nonlinearity, then projects back down to the original hidden dimension. This expansion and contraction allow the model to learn complex, nonlinear transformations of the representations. Note that `layer_idx` is not passed to the MLP because MLPs typically don't need layer-specific behavior.

Forward Pass Method

```
def forward(self, x: torch.Tensor) -> torch.Tensor:
    x = self.attn(x)  # [B, T, H]
    x = self.mlp(x)   # [B, T, H]
    return x
```

The forward method defines how data flows through a single transformer block. This is where the actual computation happens when the model processes input.

Input shape `[B, T, H]` means:

- **B (Batch Size)**: The number of independent sequences being processed simultaneously. Modern deep learning leverages parallelism by processing multiple examples at once, which is more efficient than processing them one at a time. Batch size is crucial for both training speed and model convergence—larger batches provide more stable gradient estimates but require more memory.
- **T (Sequence Length)**: The number of tokens in each sequence. This could be 512, 1024, 2048, or even longer in modern models. Each token represents a piece of text (could be a word, subword, or character depending on the tokenization scheme used). The sequence length determines the context window—how much text the model can consider at once.
- **H (Hidden Dimension)**: The size of the representation vector for each token. This is the fundamental "width" of the model and is a key hyperparameter that determines model capacity. Common values are 768 (BERT-base), 1024, 2048, 4096, or even larger for massive models. Larger hidden dimensions allow the model to capture more nuanced information but require more computation and memory.

The data flow is straightforward but important to understand:

First, the input `x` is passed through the attention block. The attention mechanism allows each token to gather information from all other tokens in the sequence based on learned attention patterns. This is where the model learns which parts of the input are relevant to which other parts—for example, learning that pronouns should attend to their antecedents or that verbs should attend to their subjects and objects.

The output maintains the same shape `[B, T, H]`—we still have the same number of tokens and the same hidden dimension, but now each token's representation has been enriched with information from the entire sequence context.

Then, the output from attention is passed through the MLP block. This feed-forward network applies a nonlinear transformation to each token position independently. While attention handles the "horizontal" relationships between tokens, the MLP handles the "vertical" transformation of each token's representation into a more useful form. Again, the shape remains `[B, T, H]`.

Important architectural note: This implementation shows a simplified version of the transformer block. In most real-world implementations, including the original "Attention Is All You Need" paper and modern models like GPT and LLaMA, there are additional components that are crucial for training stability and performance:

1. **Residual Connections (Skip Connections)**: The typical pattern is `x = x + self.attn(x)` rather than just `x = self.attn(x)`. These residual connections allow gradients to flow more easily during backpropagation, enabling training of much deeper networks. Without residual connections, very deep networks suffer from vanishing gradients and are difficult to train.

2. **Layer Normalization**: Usually applied before each sub-layer (pre-norm) or after (post-norm). The code mentions RMSNorm, which is a modern variant of layer normalization used in models like LLaMA. Layer normalization stabilizes training by normalizing the activations, preventing them from growing too large or too small, which could cause numerical instability or slow learning.

The fact that these components aren't explicitly shown in the `TransformerBlock.forward()` method suggests they're likely implemented inside the `AttentionBlock` and `MLPBlock` classes themselves. Modern implementations often encapsulate the full sub-layer logic (including pre-norm and residual connections) within each block for cleaner code organization.

Transformer Class—The Complete Language Model

The `Transformer` class represents the full model architecture that combines all the pieces into a complete language model capable of processing text and making predictions.

Initialization Method—Building the Full Architecture

```
def __init__(self, config: ModelConfig, device: torch.device | None
= None):
    super().__init__()
    self.config = config
    self.embedding = torch.nn.Embedding(config.vocab_size, config.hidden_
    size, device=device)
    self.layers = torch.nn.ModuleList([TransformerBlock(config, i,
    device=device)
                                       for i in range(config.num_hidden_
                                       layers)])
    self.norm = RMSNorm(config.hidden_size, device=device)
    self.output_proj = torch.nn.Linear(config.hidden_size, config.vocab_
    size, bias=False, device=device)
```

This initialization method constructs the complete model architecture by creating and organizing all the necessary components. Let's examine each component in detail:

- **`self.config = config`**: Storing the configuration object as an instance variable is good practice. It allows other methods to access configuration parameters without having them passed as arguments, and it makes it easy to inspect the model's configuration after it's been created. This is particularly useful when loading models from checkpoints or when debugging—you can always check what hyperparameters the model was built with.
- **`self.embedding`**: This is the token embedding layer, which is the entry point for all text data into the model. In language models, text is first tokenized into discrete units (tokens), which are represented as integer IDs. The embedding layer is essentially a lookup table that maps each token ID to a learned dense vector representation.

For example, if the vocabulary size is 50,000 and the hidden dimension is 768, this creates a matrix of shape `[50000, 768]`. When you pass in a token ID like 1523, it retrieves the 1523rd row of this matrix, giving you a 768-dimensional vector. These embeddings are learned during training—initially they might be random or pre-initialized with some method, but as the model trains, similar words end up with similar embedding vectors because they're used in similar contexts.

The `vocab_size` determines how many unique tokens the model can represent. This is determined by the tokenization algorithm used (could be word-level, subword-level like BPE or WordPiece, or character-level). Modern models typically use subword tokenization with vocabularies ranging from 30,000 to 100,000 tokens or more.

`self.layers`: This is where the actual depth of the model comes from. It creates a `ModuleList` (PyTorch's way of storing a list of modules that should be registered as part of the model) containing `num_hidden_layers` transformer blocks. Each block is a complete `TransformerBlock` instance with its own attention and MLP components.

The list comprehension `[TransformerBlock(config, i, device=device) for i in range(config.num_hidden_layers)]` creates these blocks, passing each one its position index `i`. Modern language models can have anywhere from 12 layers (BERT-base) to 24 (BERT-large), 32, 80 (LLaMA-2 70B), or even more than 100 layers in the largest models. Each layer learns to process information at a different level of abstraction—lower layers tend to learn syntactic patterns and simple features, while higher layers learn more semantic and contextual relationships.

Using `ModuleList` is important because it ensures PyTorch properly registers all these sub-modules as part of the parent model. This means their parameters will be included when you call `model.parameters()` for optimization, and they'll be properly moved to the correct device when you call `model.to(device)`, and they'll be included in state dictionaries when saving/loading the model.

`self.norm`: This is the final layer normalization applied after all transformer blocks but before the output projection. The code specifically uses RMSNorm (Root Mean Square Layer Normalization), which is a simpler and more efficient variant of the standard LayerNorm used in the original transformer.

RMSNorm simplifies the standard layer normalization by removing the mean centering step and only performing RMS (root mean square) normalization. The formula is roughly: `output = input / RMS(input) * scale`, where RMS is the root mean square of the input and scale is a learnable parameter. This has been shown to work just as well as full LayerNorm in many cases while being computationally cheaper. Models like LLaMA, Mistral, and other modern open-source models use RMSNorm.

This final normalization is important for stabilizing the final hidden states before they're projected to logits. Without it, the pre-output representations might have high variance, which could lead to numerical issues or make the output projection harder to learn.

`self.output_proj`: This is the language modeling head—the final linear layer that transforms the hidden representations into predictions over the vocabulary. It takes the hidden dimension `H` and projects it to `vocab_size` dimensions, producing a score (logit) for every possible token in the vocabulary.

For example, with hidden_size=768 and vocab_size=50000, this is a weight matrix of shape `[768, 50000]`. For each token position, it multiplies the 768-dimensional hidden state by this matrix to get 50000 scores, one for each possible next token. These raw scores (logits) are then typically passed through a softmax function to convert them into probabilities, though during training we often skip the softmax and use the logits directly with cross-entropy loss for numerical stability.

The parameter `bias=False` is significant—this layer doesn't include a bias term. This has become standard practice in modern language models for a few reasons: (1) it reduces the number of parameters slightly (though bias is tiny compared to the weight matrix), (2) it can improve training stability, and (3) empirically, models seem to work just as well or better without bias in the output projection. The lack of bias means that the logit for each token depends purely on how similar the hidden state is to that token's "direction" in the weight matrix, without any constant offset.

Forward Pass—From Tokens to Predictions

```
def forward(self, tokens: torch.Tensor) -> torch.Tensor:
    x = self.embedding(tokens)
    for layer in self.layers:
        x = layer(x)
    x = self.norm(x)
    return self.output_proj(x)
```

The forward method defines the complete forward pass of the model—how raw token IDs are transformed into predicted probability distributions over the vocabulary. This is the heart of the model's inference process.

Step 1: Embedding Lookup

```
x = self.embedding(tokens)
```

The input `tokens` has shape `[B, T]`—a batch of B sequences, each containing T token IDs (integers). The embedding layer transforms this into `[B, T, H]` by replacing each token ID with its corresponding embedding vector. This is a very fast operation—it's just array indexing, no matrix multiplication involved. After this step, each token is represented by a dense, continuous vector that the model can process.

Step 2: Processing Through Transformer Layers

```
for layer in self.layers:
    x = layer(x)
```

This loop is where the magic happens. The input sequentially passes through all transformer layers, with each layer refining and enriching the representations. The shape remains `[B, T, H]` throughout—we're not changing dimensionality, just transforming the representations to be more useful.

In the first few layers, the model typically learns to capture local syntactic patterns—things like phrase structure, part-of-speech patterns, and simple grammatical relationships. As you go deeper into the network, the representations become more semantic and contextual—the model learns about entities, relationships, long-range dependencies, and abstract concepts.

Each layer's attention mechanism allows information to flow between tokens, enabling the model to build contextualized representations where each token's vector incorporates information from relevant parts of the sequence. Each layer's MLP then performs nonlinear transformations on these contextualized representations.

The sequential nature (as opposed to having all layers process in parallel) is important because it allows for hierarchical feature learning. Each layer builds upon the representations created by previous layers, similar to how convolutional neural networks in computer vision learn hierarchical visual features (edges ➤ textures ➤ parts ➤ objects).

Step 3: Final Normalization

```
x = self.norm(x)
```

After all layers have processed the input, we apply final layer normalization. This ensures that the hidden states going into the output projection have stable statistics (normalized variance), which helps with the quality and stability of the final predictions. Without this normalization, different positions or different examples in the batch might have very different scales, which could cause the output projection to behave poorly.

Step 4: Output Projection

```
return self.output_proj(x)
```

Finally, we project the hidden states to vocabulary-sized logits. The output shape is `[B, T, vocab_size]`—for every position in every sequence in the batch, we now have a score for each possible token in the vocabulary.

These logits represent the model's raw predictions. For language modeling, we typically care about the logits at the last position (predicting the next token), but having logits for all positions is useful for training, where we can compute loss at every position simultaneously. During generation, you would take the logits at the last position, apply softmax to convert to probabilities, then sample or take the argmax to select the next token.

It's worth noting that this forward pass doesn't include any position embeddings explicitly. This is because the code comments mention "RotaryEmbedding", which suggests the model uses rotary position embeddings (RoPE). RoPE is applied within the attention mechanism itself (inside `AttentionBlock`) rather than being added to the token embeddings at the start. This is different from the original transformer, which added sinusoidal position embeddings or learned position embeddings to the token embeddings at the input.

from_checkpoint Class Method—Loading Pretrained Models

```
@classmethod
def from_checkpoint(cls, path: str, device: str | torch.device = "cuda",
strict: bool = True) -> "Transformer":
```

This is a class method (note the `@classmethod` decorator) that provides a convenient way to load a pretrained model from disk. Class methods are called on the class itself rather than on an instance, and they receive the class as their first argument (`cls`) rather than an instance (`self`). This makes them perfect for factory methods that create and return new instances.

The method signature tells us a lot about its intended use:

- **`path`**: A string specifying the directory containing the saved model checkpoint
- **`device`**: Where to load the model—defaults to "cuda" (GPU), but can be "cpu" or a specific torch.device
- **`strict`**: Whether to enforce strict loading (all parameters must match exactly)
- **Returns**: A fully initialized and loaded `Transformer` instance ready for inference

Step-by-Step Loading Process

Import Necessary Modules

```
import os, json
```

These imports are done inside the method rather than at the module level. This is sometimes done to avoid circular imports or to keep imports local to where they're needed, though it's more common to import at the top of the file.

Device Handling

```
if isinstance(device, str):
    device = torch.device(device)
```

This converts string device specifications like "cuda", "cpu", "cuda:0", and "cuda:1" into proper `torch.device` objects. This normalization ensures that the rest of the method can assume it's working with a device object, which has useful methods and properties. It's a nice defensive programming practice that makes the API more flexible—users can pass either strings or device objects, and both will work correctly.

Load Configuration from JSON

```
with open(os.path.join(path, "config.json"), "r") as f:
    cfg = json.load(f)
config = ModelConfig(**cfg)
```

This assumes the checkpoint directory contains a `config.json` file with all the model hyperparameters saved as a JSON object. Using `os.path.join` is the correct, platform-independent way to construct file paths (works on Windows, Linux, Mac).

The `json.load(f)` reads the JSON file and converts it to a Python dictionary. Then `ModelConfig(**cfg)` uses dictionary unpacking to pass the configuration parameters as keyword arguments to the ModelConfig constructor. This pattern assumes ModelConfig is either a dataclass or has an `__init__` method that accepts all the configuration parameters as keyword arguments.

Saving and loading configuration separately from weights is good practice because

1. The configuration is human-readable and can be inspected or modified easily.
2. It's much smaller than the weight file.
3. It's necessary to construct the model architecture before loading weights.
4. It makes the checkpoint self-documenting—you can see exactly how the model was configured.

Initialize Model Architecture

```
model = cls(config, device=device).to(device)
```

Here, `cls` refers to the `Transformer` class itself (since this is a class method). This line creates a new transformer instance with the loaded configuration, specifying the target device. The `.to(device)` call then moves all model parameters and buffers to the specified device.

There's actually a subtle redundancy here—the `__init__` method already accepts a device parameter and creates modules on that device, so the explicit `.to(device)` call might be redundant in most cases. However, it's defensive programming that ensures the model is definitely on the correct device, which doesn't hurt (moving to the same device is a no-op).

Load Model Weights

```
sd = torch.load(os.path.join(path, "model.pt"), map_location=device)
```

This loads the state dictionary (saved model weights) from the `model.pt` file. The `map_location=device` parameter is crucial—it tells PyTorch where to load the tensors. If you trained on GPU and are loading on CPU, or vice versa, this handles the device mapping. Without `map_location`, PyTorch would try to load tensors to the same device they were saved from, which would fail if that device isn't available.

The state dictionary is a Python dict that maps parameter names (strings) to their corresponding tensors. For example:

```
{
    'embedding.weight': tensor(...),
    'layers.0.attn.q_proj.weight': tensor(...),
    'layers.0.attn.k_proj.weight': tensor(...),
    ...
    'output_proj.weight': tensor(...)
}
```

Load State Dictionary into Model

```
missing, unexpected = model.load_state_dict(sd, strict=False)
```

This is where the actual parameter loading happens. The `load_state_dict` method takes each tensor from the state dictionary and copies it to the corresponding parameter in the model.

Setting `strict=False` allows for some flexibility:

- **Missing Keys**: Parameters in the model that aren't in the state dict will remain at their initialized values (usually random).
- **Unexpected Keys**: Parameters in the state dict that don't correspond to any model parameter are ignored.

The method returns two lists:

- **`missing`**: List of parameter names that exist in the model but weren't found in the state dict
- **`unexpected`**: List of parameter names that were in the state dict but don't exist in the model

Error Handling and Validation

```
if strict and (missing or unexpected):
    raise RuntimeError(f"Missing: {missing}, Unexpected: {unexpected}")
if missing or unexpected:
    print(f"[load_warning] missing={missing} unexpected={unexpected}")
```

This implements a two-tier validation system:

1. If `strict=True` (the default), any mismatch between the model and saved weights causes an error. This is the safe default—you probably don't want to silently initialize some parameters randomly when you thought you were loading a fully trained model.
2. If `strict=False`, mismatches are allowed but warned about. This is useful for transfer learning scenarios, like when you're loading a pretrained model but adding new layers on top, or when you're loading weights from a slightly different architecture.

The warning message shows exactly which parameters are problematic, which is very helpful for debugging. Common scenarios include:

- Adding or removing layers
- Changing the vocabulary size
- Loading from a different but compatible architecture
- Loading a base model into a model with task-specific heads

Set to Evaluation Mode

```
model.eval()
return model
```

Calling `eval()` is crucial when loading a model for inference. This method:

- Disables dropout layers (you don't want random neuron dropping during inference)
- Puts batch normalization layers in inference mode (uses running statistics rather than batch statistics)
- Sets various flags that other layers might check to behave differently during training vs. inference

Finally, the fully loaded, properly configured, and evaluation-ready model is returned.

Architecture Design Choices and Modern Practices

This implementation reflects several modern best practices and design choices that are worth understanding:

1. **RMSNorm Instead of LayerNorm**: This is a modern optimization that simplifies layer normalization by removing the mean-centering step. It's used in LLaMA, Mistral, and other state-of-the-art models because it's faster and seems to work just as well in practice.

2. **No Bias in Output Projection**: Modern large language models typically omit the bias term in the final linear layer. This is one of many small optimizations that collectively make training more stable and efficient at scale.

3. **Rotary Position Embeddings (Implied)**: The code comments mention RoPE, which is a modern alternative to the sinusoidal or learned position embeddings used in the original transformer. RoPE encodes position information directly into the attention mechanism through rotation matrices, and it generalizes better to longer sequences than the model was trained on.

4. **Configuration-Driven Design**: Using a config object to pass hyperparameters is clean, extensible, and makes experimentation easy. You can save/load configurations easily and create different model variants by just changing the config.

5. **Device-Aware Initialization**: Supporting explicit device specification during initialization is important for efficient training and inference, especially when working with large models that need careful device management.

This code represents a clean, modern implementation of a transformer language model that incorporates current best practices while remaining readable and maintainable.

Advantages, Challenges, and Broader Impact

Why has the transformer block become ubiquitous across machine learning domains? Its advantages are manifold, stemming from fundamental architectural properties that address long-standing limitations of prior sequence models while introducing novel capabilities. However, this dominance comes with significant challenges—computational, environmental, and societal—that shape the trajectory of modern AI research.

Parallelism and Scalability

Scalability: The transformer's most transformative property is its inherent parallelism, enabled by replacing recurrent connections with attention mechanisms. Unlike RNNs and LSTMs, which process sequences sequentially through hidden states $h_t = f(h_{t-1}, x_t)$, transformers compute all positions simultaneously. For a sequence of length T, an RNN requires $O(T)$ sequential steps, each dependent on the previous state, fundamentally limiting parallelization. In contrast, the transformer's self-attention computes:

$$\text{Attention}(Q, K, V) = \text{softmax}\left(\frac{QK^T}{\sqrt{d_{\text{head}}}}\right)V$$

as a single batched matrix multiplication, achieving $O(1)$ sequential depth. This architectural shift translates to dramatic speedups: training a 12-layer transformer on modern GPUs can be 50-100× faster than an equivalently sized LSTM, despite the $O(T^2)$ attention complexity.

This parallelism enables training on massive datasets that would be intractable for recurrent models. Models like GPT-3 with 175 billion parameters (96 layers, $d_{\text{model}} = 12288, h = 96$) were trained on 300 billion tokens using thousands of GPUs. The training throughput scales nearly linearly with hardware: doubling GPUs approximately halves training time, whereas RNN scaling saturates due to sequential dependencies. Empirically, the Chinchilla scaling laws suggest optimal compute allocation follows:

$$N_{\text{optimal}} \propto D^{0.5}, \quad \text{FLOPs} \propto D^2$$

where N is model size and D is dataset size, demonstrating transformers' ability to efficiently leverage both dimensions.

Modern distributed training strategies exploit this parallelism through multiple axes:

- **Data Parallelism**: Batch sharded across n devices, each processing B/n examples.
- **Tensor Parallelism**: Weight matrices partitioned column-wise or row-wise across devices.
- **Pipeline Parallelism**: Layers distributed across stages, with micro-batching for efficiency.
- **Sequence Parallelism**: Long sequences partitioned across devices, particularly for activations.

These techniques enable models exceeding one trillion parameters (e.g., Switch Transformer with 1.6T parameters using 2048-way MoE), a scale unimaginable for recurrent architectures.

Transfer Learning and Few-Shot Generalization

Transfer Learning: The transformer block's learned representations exhibit remarkable transferability across tasks and domains. Pre-trained blocks (e.g., via self-supervised objectives like masked language modeling or next-token prediction) encode rich linguistic, visual, or temporal structures that generalize to downstream tasks with minimal fine-tuning. This paradigm—pre-train then fine-tune—has democratized AI for low-resource tasks.

The theoretical foundation lies in the lottery ticket hypothesis and task-invariant feature learning: early layers learn universal features (edges in vision, syntax in language), while later layers specialize. Empirically, frozen BERT embeddings achieve 85-90% of fully fine-tuned performance on GLUE benchmarks, and parameter-efficient fine-tuning methods like LoRA (Low-Rank Adaptation) modify only $0.1-1\%$ of parameters:

$$W' = W + \Delta W = W + BA$$

where $B \in \mathbb{R}^{d \times r}, A \in \mathbb{R}^{r \times k}$ with rank $r \ll \min(d, k)$. This reduces trainable parameters from $\sim 10^9$ to $\sim 10^6$ while maintaining performance within 1-2% of full fine-tuning.

Moreover, emergent few-shot learning capabilities arise at scale: models like GPT-3 achieve non-trivial performance on tasks with zero to few examples in context, without gradient updates. The in-context learning mechanism can be formalized as implicit Bayesian inference, where the model treats the context as evidence and computes:

$$P(y \mid x, \text{context}) \propto P(x, y \mid \text{context}) = \int P(x, y \mid \theta) P(\theta \mid \text{context}) d\theta$$

This meta-learning behavior emerges around 10^{10} parameters and 10^{11} training tokens, suggesting transformers learn task-agnostic reasoning algorithms.

Interpretability and Mechanistic Understanding

Interpretability: Attention weights offer interpretable glimpses into decision-making, unlike the opaque hidden states of RNNs. The attention matrix $A \in \mathbb{R}^{T \times T}$ can be visualized to reveal which tokens the model deems relevant for each position. Empirical studies have identified specialized attention patterns:

- **Positional Heads**: Attend primarily to absolute positions (e.g., first/last token)
- **Syntactic Heads**: Track grammatical dependencies (subject-verb agreement, coreference)

- **Semantic Heads**: Aggregate information broadly for meaning composition
- **Delimiter Heads**: Focus on special tokens (punctuation, separators)

For instance, in BERT, specific heads in layer 7 consistently attend from pronouns to their antecedents with > 0.8 https://www.codecogs.com/eqnedit.php?latex=%3E0.8#0 attention weight, enabling coreference resolution. This interpretability has spawned the field of mechanistic interpretability, which reverse-engineers neural networks to understand algorithms encoded in weights. Recent work has discovered

- **Induction Circuits**: Multilayer compositions detecting and completing repeated patterns
- **Feature Superposition**: Individual neurons encode multiple features through sparse activation
- **Modular Circuits**: Task-specific sub-networks that can be ablated or enhanced

Beyond attention visualization, probing classifiers $f_{\text{probe}} : h_\ell \to y$ applied to intermediate representations h_ℓ reveal what information is encoded. Studies show:

- **Layers 0-4 (Shallow)**: Part-of-speech, syntax trees
- **Layers 5-8 (Middle)**: Named entities, semantic roles
- **Layers 9-12 (Deep)**: Sentiment, discourse relations

This hierarchical encoding mirrors linguistic theory, validating transformers as cognitively plausible models.

Long-Range Dependency Modeling

Unlike RNNs, which suffer from vanishing gradients over long sequences (effective horizon $\sim 20 - 50$ tokens), transformers theoretically access all prior context via direct attention paths. The gradient flow from position t to t' is:

$$\frac{\partial \mathcal{L}t}{\partial xt'} = \frac{\partial \mathcal{L}t}{\partial A_t}\frac{\partial A_t}{\partial xt'} \propto A_{t,t'}\frac{\partial \mathcal{L}t}{\partial Vt'}$$

which does not decay exponentially with distance $|t - t'|$ as in RNNs. This enables modeling dependencies spanning thousands of tokens, crucial for tasks like document-level translation, long-form question answering, and coherent story generation.

Fundamental Challenges

Quadratic Complexity and Context Length Limitations

The transformer's Achilles' heel is the $O(T^2)$ memory and compute complexity of self-attention. For a sequence of length T, storing the attention matrix requires $O(BhT^2)$ memory, where B is the batch size and h is the number of heads. This quadratic scaling creates hard limits:

- **GPT-3**: $T_{\max} = 2048$ tokens (cost to extend to 4096: $4\times$ memory)
- **Claude**: $T_{\max} = 100k$ tokens via attention approximations
- **Gemini 1.5**: $T_{\max} = 1M$ tokens using sparse attention

The memory wall is hit around $T \approx 4096$ on consumer GPUs (24GB VRAM), and the compute wall around $T \approx 32768$ where attention dominates total FLOPs. Addressing this has spawned numerous approximations:

Sparse Attention: Only attend to $k \ll T$ positions. Variants include:

- **Local Windows**: Attend within radius w, reducing to $O(Tw)$
- **Strided Patterns**: Attend every s -th position
- **Fixed Patterns**: Combinations of local + global (Longformer, BigBird)
- **Learned Sparsity**: Dynamic routing (Routing Transformer)

Linear Attention: Replace $\mathrm{softmax}(QK^T)V$ with kernelized features:

$$\mathrm{Attention}(Q, K, V) = \phi(Q)(\phi(K)^T V)$$

where $\phi : \mathbb{R}^d \to \mathbb{R}^{d'}$ is a feature map (e.g., $\phi(x) = \mathrm{elu}(x) + 1$). Reordering associatively yields $O(Td^2)$ complexity but empirically degrades performance 5-10% on perplexity.

Memory and Computational Costs

Training large transformers requires immense resources. GPT-3 (175B parameters) consumed an estimated 3.14×10^{23} FLOPs, costing $4.6M in compute at 2020 rates. Carbon footprint estimates suggest:

$$\mathrm{CO_2\ emissions} \approx 552\ \mathrm{metric\ tons\ for\ GPT\text{-}3\ training}$$

equivalent to 123 gasoline cars driven for one year. Inference costs compound: serving GPT-3 at scale requires thousands of GPUs, with estimated costs of $100-$700 per million tokens.

The parameter count scaling:

$$\text{Parameters}(N \text{ layers}, d_{\text{model}}, d_{\text{ff}}) \approx 12Nd_{\text{model}}^2 + 4Nd_{\text{model}}d_{\text{ff}}$$

grows quadratically in d_{model}, leading to models like GPT-4 (rumored 1.76T parameters) and PaLM-2 (340B). Memory requirements include:

- **Weights**: $N_{\text{params}} \times \text{precision}$ (2 bytes for FP16, 1 for INT8)
- **Activations**: $O(BTd_{\text{model}}N)$ for gradients in backprop
- **Optimizer States**: $2N_{\text{params}}$ for Adam (momentum + variance)

Total training memory for GPT-3: ~ 12 TB across 1024 GPUs. Techniques like gradient checkpointing trade compute for memory (recompute activations during backward pass), reducing memory to $O(\sqrt{N}BTd_{\text{model}})$ at 33% compute overhead.

Data Efficiency and Sample Complexity

Despite their power, transformers are data-hungry. GPT-3 trained on 300B tokens, Chinchilla on 1.4T tokens, and LLaMA-2 on 2T tokens. For comparison, a human child is exposed to $\sim 10^8$ words by age 10—four orders of magnitude less. This sample inefficiency stems from

1. **Lack of Inductive Biases**: CNNs encode translation equivariance; transformers learn it from data.
2. **Overparameterization**: Models with $N_{\text{params}} \gg N_{\text{data}}$ rely on implicit regularization.
3. **Weak Supervision**: Self-supervised objectives (next-token prediction) are less efficient than supervised learning.

Data scaling laws quantify this:

$$\text{Loss}(D) = L_{\infty} + \alpha D^{-\beta}$$

where $\beta \approx 0.095$ for language models, meaning a 10× data increase yields only $\sim 1.25\times$ loss reduction. This diminishing returns curve motivates high-quality data curation (e.g., filtering CommonCrawl from 400TB to 5TB for GPT-3).

Catastrophic Forgetting and Continual Learning

Fine-tuning transformers on new tasks often causes catastrophic forgetting of pretraining knowledge. When adapting to domain-specific data, perplexity on general corpora can increase by 20-50%. This stems from high plasticity: gradient updates in Adam move parameters significantly (effective learning rate $\sim 10^{-4}$), overwriting previous representations.

Mitigation strategies include:

- **Regularization**: Elastic Weight Consolidation (EWC) penalizes changes to important parameters
- **Replay Buffers**: Mix new data with old samples
- **Progressive Training**: Freeze early layers, fine-tune later ones
- **Adapters**: Insert small trainable modules ($d_{\text{adapter}} = 64$) while freezing backbone

However, no solution fully resolves the plasticity-stability trade-off, limiting lifelong learning applications.

Alignment and Control

Transformers optimize for next-token prediction, which doesn't inherently align with human preferences. Models can generate:

- **Hallucinations**: Plausible-sounding but factually incorrect text (15-20% of GPT-3.5 outputs contain factual errors)
- **Toxicity**: Offensive or harmful content (6.7% of outputs flagged on RealToxicityPrompts)
- **Bias**: Gender, racial, and cultural stereotypes amplified from training data

Reinforcement learning from human feedback (RLHF) mitigates this by fine-tuning with a reward model $R_\phi(x, y)$ trained on preference comparisons:

$$\mathcal{L}\text{RLHF} = \mathbb{E}x \sim \mathcal{D}, y \sim \pi_\theta[R_\phi(x, y)] - \beta\mathbb{D}\text{KL}[\pi\theta|\pi_{\text{ref}}]$$

where π_θ is the policy (language model) and π_{ref} is the reference model. This improves safety and helpfulness but introduces challenges:

- **Reward Hacking**: Models exploit reward model failures
- **Brittleness**: Performance degrades on distribution shift
- **Value Alignment**: Human preferences are inconsistent and culturally dependent

Robustness and Adversarial Vulnerabilities

Transformers exhibit fragility to adversarial examples. Small perturbations invisible to humans can flip predictions:

- **Textual Adversaries**: Synonym substitution can change sentiment classification from positive to negative
- **Prompt Injection**: Malicious instructions embedded in inputs can hijack model behavior
- **Jailbreaks**: Carefully crafted prompts bypass safety guardrails

Certified robustness is difficult: smoothed classifiers and interval bound propagation provide worst-case guarantees but reduce accuracy by 10-20%. This vulnerability limits deployment in high-stakes domains (medical diagnosis, autonomous vehicles).

Future Trajectories

Architectural Innovations on the Horizon

The transformer architecture, despite its current dominance, continues evolving through incremental refinements and radical reimaginings. Research groups worldwide are pushing beyond vanilla attention mechanisms to address fundamental limitations and unlock new capabilities.

Adaptive computation represents one promising direction, recognizing that not all tokens require equal processing depth. Early exiting mechanisms allow shallow layers to handle simple patterns while reserving deep processing for complex inputs. A token like "the" might exit after layer 3, having formed sufficient representation for grammatical function, while "antidisestablishmentarianism" might require full depth to properly encode semantic content.

This approach can reduce average computation by 30-40% while maintaining accuracy. Mixture-of-depths extends this concept further, dynamically allocating different amounts of computation per token within each layer—some positions might skip the FFN sublayer entirely, others might use expanded capacity. These methods effectively implement conditional computation, adapting processing intensity to input difficulty.

Neural memory systems augment transformers with external storage, addressing the $O(T^2)$ context length bottleneck. Rather than attending over the entire sequence, models attend to a compressed memory bank storing historical context. Memorizing transformers maintain a key-value store of past activations, using approximate nearest-neighbor search to retrieve relevant memories in $O(\log M)$ time, where M is memory size. This enables unbounded context length ($T \to \infty$ in principle) while maintaining tractable computation.

The system learns what to store, when to retrieve, and how to integrate retrieved information—essentially implementing differentiable database operations. Applications include book-length document understanding, multisession conversations maintaining coherent personas across months, and personalized agents that accumulate user-specific knowledge.

Multimodal fusion moves beyond single-modality processing toward unified architectures handling text, images, audio, and video simultaneously. Models like GPT-4V, Gemini, and Claude 3 implement cross-modal attention, where image patches attend to text tokens and vice versa, learning joint representations that capture relationships between visual concepts and linguistic descriptions.

The tokenization challenge varies by modality: text uses subword vocabularies (32k-100k tokens), images use patch embeddings or VQ-VAE codes (8k-16k visual tokens per image), and audio uses mel-spectrogram frames or neural codec tokens (50-75 Hz).

Aligning these heterogeneous tokenization schemes into coherent sequences enables models to perform tasks like image captioning, visual question answering, audio transcription, and video understanding within a single architecture. The shared transformer backbone learns cross-modal semantics: that "dog" tokens relate to canine visual features, that upward melodic contours correlate with questioning intonation patterns, and that motion vectors in video correspond to action verbs in text.

Hardware co-design recognizes that architectural choices and hardware capabilities coevolve. Custom accelerators optimize for specific operations: Google's TPU v5 features specialized matrix multiplication units for attention and FFN layers, achieving 2× throughput versus previous generations.

Cerebras's Wafer-Scale Engine places 850,000 cores on a single silicon wafer, eliminating off-chip memory access for models up to 120B parameters—the entire model fits in on-chip SRAM, enabling 1000× faster weight access versus GPU DRAM. Graphcore's Intelligence Processing Unit uses fine-grained parallelism across 1,400 independent processors for efficient batch-1 inference.

These architectural specializations yield 5-10× efficiency improvements over general-purpose GPUs for transformer workloads. Future co-design might include analog computing for matrix multiplication (exploiting Ohm's law for constant-time multiply-accumulate), photonic interconnects reducing communication latency, or neuromorphic chips implementing spiking attention mechanisms.

Beyond incremental improvements, emerging paradigms challenge transformer dominance by addressing fundamental limitations. Liquid neural networks, inspired by biological neurons, implement continuous-time dynamics governed by ordinary differential equations:

$$\frac{dx}{dt} = -\lambda x + f(Wx + b, t)$$

where λ controls time constant and f is a nonlinearity. These systems adapt processing speed to input characteristics—slowing down for complex patterns, accelerating for simple ones—and naturally handle irregular sampling rates common in robotics and control applications. Unlike transformers' discrete token steps, liquid networks provide smooth trajectories suitable for continuous control and real-time adaptation.

Kolmogorov-Arnold Networks (KANs) replace fixed activation functions with learnable univariate functions on edges. Instead of $y = \sigma(Wx + b)$, KANs compute $y = \sum_i \phi_i(x_i)$ where each ϕ_i is represented as a spline or Fourier series. This design, inspired by the Kolmogorov-Arnold representation theorem (any multivariate continuous function can be decomposed into univariate functions), promises improved interpretability—each learned function ϕ_i has a clear input-output relationship visualizable as a curve. Early results show KANs match MLP performance with 100× fewer parameters on certain tasks, though scaling to transformer-size models remains unexplored.

Analogical reasoning networks incorporate explicit relational structure inspired by cognitive science. These models represent entities and relationships separately, performing operations like relation composition, analogy-making, and systematic generalization.

For example, learning "cat:kitten :: dog:?" should enable inferring "puppy" even without seeing dog-puppy pairs during training. This contrasts with transformers' statistical pattern matching, which struggles with systematic compositional generalization. Hybrid architectures might combine transformer perception (learning representations from data) with structured reasoning modules (performing relational operations), leveraging the strengths of both approaches.

Toward Artificial General Intelligence

Transformers represent significant progress toward artificial general intelligence—systems matching human cognitive breadth and flexibility. Current state-of-the-art models exhibit impressive capabilities across diverse domains, yet clear gaps remain before achieving human-level general intelligence.

In natural language processing, transformers achieve near-human performance on reading comprehension (SQuAD, RACE), generation (producing coherent long-form text), and translation (matching professional translators on common language pairs). They pass many theory-of-mind tests, generate plausible analogies, and engage in nuanced conversation. However, they struggle with tasks requiring grounded semantics: understanding novel verb usage, reasoning about physical constraints, or tracking dynamic world state across narratives. The representations remain primarily statistical associations rather than compositional semantic structures.

Computer vision capabilities are similarly mixed. On ImageNet classification, models achieve 90%+ accuracy, surpassing average human performance. Object detection, segmentation, and visual question answering reach practical utility.

Yet 3D reasoning—inferring object structure from partial views, predicting physical dynamics, understanding occlusion and transparency—remains challenging. Adversarial robustness is weak: imperceptible perturbations fool models, and out-of-distribution images (novel viewpoints, lighting, and artistic styles) degrade performance significantly.

Mathematical reasoning shows particularly striking capability discontinuities. GPT-4 solves approximately 50% of problems on the MATH dataset (competition-level high school mathematics), including multistep algebraic manipulation, geometric proofs, and word problems.

Chain-of-thought prompting further improves performance by eliciting step-by-step reasoning. Yet models fail on trivial variations: changing problem phrasing, introducing irrelevant information, or requiring novel problem-solving strategies exposes brittleness. The systems pattern-match against training examples rather than forming robust problem-solving procedures.

The remaining gaps toward AGI are substantial and multifaceted. Common-sense reasoning—understanding that unsupported objects fall, that solid objects cannot pass through each other, that people's beliefs may differ from reality—requires grounded physical and psychological models absent in text-only training. Current approaches like vision-language pretraining and robotics integration provide richer grounding, but comprehensive world models remain elusive.

Continual learning poses another fundamental challenge. Humans accumulate knowledge across lifetimes, integrating new information while retaining prior learning. Transformers exhibit catastrophic forgetting: fine-tuning on new tasks degrades prior capabilities. This stems from distributed representation—each weight influences many behaviors, so updates for new tasks overwrite existing knowledge.

Biological neural networks employ complementary learning systems (fast hippocampal encoding, slow cortical consolidation) and synaptic homeostasis, maintaining stable long-term memories while enabling rapid adaptation. Replicating these mechanisms in artificial systems remains an open problem. Promising directions include memory-augmented networks storing episodic experiences, progressive neural networks adding capacity for new tasks, and meta-learning algorithms that learn how to learn without forgetting.

Sample efficiency represents a glaring disparity between human and artificial intelligence. Children learn language from approximately 10^8 words by age 10, develop visual object recognition from limited exposure, and acquire physical intuitions from direct interaction.

Transformers require $10^{12} - 10^{13}$ tokens for comparable language proficiency—four to five orders of magnitude more data. This inefficiency stems from weak inductive biases and a lack of active learning. Humans leverage strong priors (objects persist, causes precede effects, others have minds) and learn through experimentation and causal intervention, not passive observation. Incorporating these principles—through hybrid neuro-symbolic architectures, causal representation learning, and interactive embodied training—could dramatically improve sample efficiency.

Embodied intelligence integration remains perhaps the deepest challenge. Current transformers process disembodied text and images, lacking grounded sensorimotor experience connecting perceptions to actions and world states. Embodied cognition theories suggest intelligence emerges from body-environment coupling: representations encode action affordances, not abstract symbols.

A fully general intelligence must integrate perception (multimodal sensing), action (motor control, manipulation), and world models (predicting environmental dynamics). Robotics provides embodied testbeds, but current systems remain narrow: manipulation succeeds in structured environments but fails with novel objects; navigation works in mapped spaces but struggles with dynamic obstacle avoidance; task planning succeeds for short horizons but collapses for complex long-term goals.

The path from current transformers to AGI likely requires architectural innovations beyond scaling. While capability improvements from increasing parameters and data continue following power laws, certain cognitive functions seem to require qualitatively different mechanisms.

Working memory maintaining active goals, episodic memory retrieving specific experiences, causal reasoning distinguishing correlation from causation, and hierarchical planning decomposing complex objectives into subgoals—these might demand explicit computational structures rather than emerging from scaled-up attention and feedforward layers.

Multitask learning across diverse domains with shared representations could accelerate progress. Rather than training separate models for language, vision, robotics, and reasoning, unified architectures learning all modalities simultaneously develop more abstract, general-purpose representations.

Gato, a generalist agent from DeepMind, demonstrates this approach: a single transformer processes text, images, proprioceptive state, and action commands, performing 604 distinct tasks, including chatbot conversation, image captioning, Atari gameplay, and robotic manipulation. While performance on individual tasks trails specialized models, the shared backbone learns transferable structure benefiting all domains.

Curriculum design and training paradigms may prove as important as architecture. Human development follows staged progression: infants first learn sensorimotor coordination, then object permanence, later language, and finally abstract reasoning. Analogous curricula for AI—starting with simple physical prediction, progressing to multiobject reasoning, then grounded language, and finally abstract problem-solving—might enable more efficient learning of complex capabilities. Self-supervised objectives capturing increasingly sophisticated world structure (spatial relationships, temporal dynamics, agent intentionality, causal mechanisms) could provide scaffolding for general intelligence.

Timeline estimates for achieving AGI vary enormously among experts and reflect deep uncertainty about both technical requirements and pace of progress. Optimistic forecasts, often from practitioners observing rapid capability scaling, suggest 5-15 years, reasoning that continued exponential compute growth plus targeted algorithmic innovations could reach human-level performance by 2030–2040.

Conservative estimates, typically from researchers emphasizing remaining gaps, project 50+ years or question whether current architectures can ever achieve general intelligence at any scale. Median expert surveys suggest 2040-2060 timelines with high variance.

These predictions are contingent on multiple factors: availability of computational resources (continued exponential growth in specialized AI chips versus saturation of semiconductor scaling), algorithmic breakthroughs (whether qualitatively new architectures are required), data availability (whether useful training signal saturates before AGI capability emerges), and societal choices (investment levels, regulatory constraints, safety prioritization).

The transformative impact of AGI—systems matching or exceeding human cognitive capabilities across all domains—motivates intense focus on both acceleration (for economic and scientific benefits) and caution (for safety and alignment). Navigating this tension will shape the trajectory of intelligent systems and, potentially, the future of human civilization.

The transformer block's ubiquity reflects a rare convergence: theoretical elegance, empirical success, and practical scalability. Yet this dominance comes with responsibilities—environmental stewardship, equitable access, safety assurance, and alignment with human values. As models grow more capable, navigating these challenges will determine whether transformers become a tool for collective flourishing or a source of systemic risk. Architecture's future lies not just in technical refinement but in the sociotechnical ecosystems we construct around it.

Summary

This chapter brings together all the previously discussed components to build the complete transformer architecture. The chapter thoroughly explains how attention mechanisms, feed-forward networks, residual connections, and layer normalization combine into a powerful, production-ready language model.

The **TransformerBlock** consists of two main components:

- **AttentionBlock**: Self-attention with RMSNorm, multihead attention (12-128 heads), rotary position embeddings (RoPE), and scaled dot-product attention with causal masking
- **MLPBlock**: Position-wise feed-forward network with expansion (4-8×), GELU/SwiGLU activation, and contraction

The complete **Transformer** model orchestrates these blocks through:

- Token embedding layer mapping IDs to dense vectors
- Stack of TransformerBlocks (12 to 100+ layers)
- Final RMSNorm for stability
- Output projection to vocabulary logits

Advantages and Challenges

- **Strengths**: Parallelism (50-100× faster than RNNs), transfer learning capability, interpretability through attention weights, and versatility across domains (vision, audio, proteins).
- **Limitations**: Quadratic complexity limiting context length, massive resource requirements ($4.6M for GPT-3), data inefficiency (10,000× more data than humans), and catastrophic forgetting during fine-tuning.

The chapter addresses societal implications, including misinformation risks, power concentration among tech giants, governance challenges, economic transformation, and ethical questions about machine consciousness and existential risk.

In the next chapter, we will move from architecture to practical application. **Chapter 10** will cover **training the transformer model**, where we'll build the **TokenGenerator** for autoregressive text generation (with temperature, top-k, top-p sampling techniques), prepare **training datasets** with proper tokenization and batching, and implement the complete **training loop** with optimization strategies and progress monitoring. You'll transform this architectural knowledge into a fully functional language model you can train and use for text generation.

CHAPTER 10

Dataset Preparation, Model Training, Token Generator for Inference and Prompting—The BIG Moment

Dataset Preparation for LLM Training

Dataset preparation for LLM training is a complex, multifaceted process that extends far beyond simply collecting large amounts of text from the internet. It involves careful curation, cleaning, filtering, deduplication, and balancing of data from diverse sources. The process requires both sophisticated computational techniques and thoughtful human judgment to ensure that the resulting dataset will produce a model that is capable, safe, and aligned with intended use cases.

The Importance of Dataset Quality

Before diving into specific techniques, it's essential to understand why dataset preparation matters so profoundly. The training dataset serves as the foundation upon which an LLM builds its understanding of language, knowledge, and reasoning patterns. Poor dataset quality manifests in multiple ways in the final model, including factual inaccuracies, toxic outputs, reasoning failures, and limited generalization capabilities.

D. Grigorov, *Building Large Language Models from Scratch*, https://doi.org/10.1007/979-8-8688-2297-1_10

High-quality datasets enable models to learn robust representations of language that generalize well to diverse tasks and domains. They help models distinguish between reliable and unreliable information, understand nuanced contexts, and generate responses that are both informative and safe. Conversely, datasets contaminated with misinformation, biased content, or low-quality text will produce models that perpetuate and amplify these issues.

The scale of modern LLM training introduces unique challenges. Training datasets often contain hundreds of billions or even trillions of tokens, making manual review impossible. This necessitates automated quality control systems that can operate at scale while maintaining high standards. The investment in dataset preparation pays dividends throughout the model's lifecycle, reducing the need for extensive post-training corrections and improving overall performance.

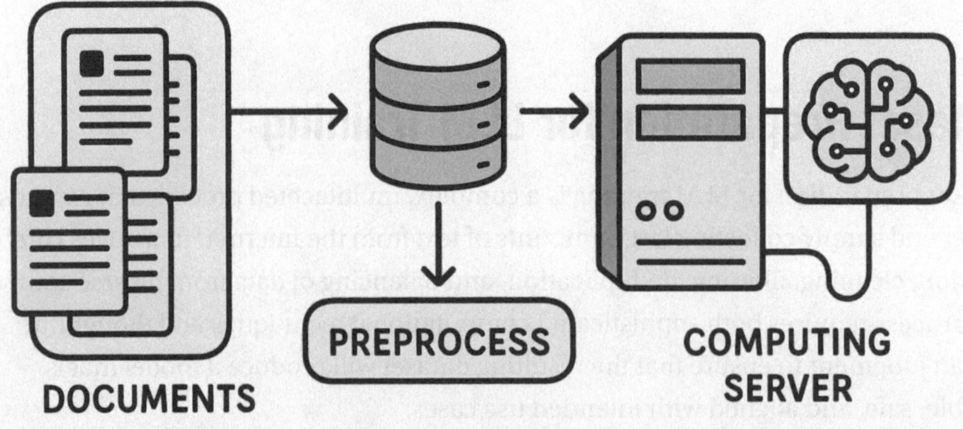

Figure 10-1. *Dataset Preparation for LLM Training*

Data Collection and Sourcing

LLM training datasets combine diverse sources—web pages, books, academic papers, code, and conversational data—each contributing distinct capabilities.

Web Crawl Data provides massive scale and topic diversity (e.g., Common Crawl) but requires filtering for spam, duplicates, and quality issues using domain reputation scoring and content-type filtering.

Books and Long-Form Content teach narrative structure and deep reasoning through coherent, in-depth coverage. Copyright restrictions limit collection to public domain works, licensed materials, or synthetic alternatives.

Academic and Technical Literature (arXiv, PubMed, patents) offers specialized knowledge, formal reasoning, and factual grounding, helping models understand scientific methodology and technical argumentation despite specialized jargon.

Code Repositories (GitHub) provide structured, logical sequences teaching formal systems and programming patterns. Selection considers license compatibility, quality indicators (stars, documentation, tests), and language distribution.

Conversational and Instructional Data from forums, transcripts, or synthetic creation teaches interactive communication patterns. This carefully curated data directly shapes model behavior and requires higher quality standards.

Data Cleaning and Filtering

Raw collected data undergoes cleaning and filtering to remove harmful content, low-quality text, and artifacts.

Text Normalization standardizes character encodings (UTF-8), normalizes whitespace, removes HTML artifacts, and eliminates boilerplate (navigation menus, ads, headers) that adds noise without value.

Language Identification and Filtering ensure datasets maintain the intended language composition. Multilingual models require deliberate oversampling of lower-resource languages to avoid English dominance while balancing data quality.

Quality Filtering employs multiple approaches:

- **Heuristic filters** examine word count, sentence length, punctuation, and capitalization patterns
- **Perplexity-based filtering** scores text using language models to identify incoherent patterns
- **Classifier-based filtering** uses trained models to distinguish quality writing from spam or machine-generated text

Content Safety Filtering removes explicit adult content, violence, hate speech, and harmful material using keyword blocking, classifier detection, and manual review, balancing thoroughness with avoiding over-censorship.

Personal Information and Privacy protection identifies and removes/redacts names, addresses, emails, and sensitive identifiers using named entity recognition and pattern matching, with difficulty achieving perfect removal given varied formats.

Deduplication

Deduplication removes redundant content, preventing overfitting and overrepresentation, addressing extensive web duplication from mirrored sites, quoted content, and republished material.

Document-Level Deduplication identifies perfect copies using hash-based methods (MD5, SHA-256) for efficient exact matching across large datasets.

Near-Deduplication detects substantially similar documents using MinHash and locality-sensitive hashing (LSH), enabling efficient detection without exhaustive comparisons. Similarity thresholds balance removing redundancy with preserving valuable variations. Strategies for handling near-duplicate clusters include keeping the highest-quality version or proportionally downweighting all members.

Substring Deduplication addresses repeated passages within and across documents using suffix arrays and string-matching algorithms. The challenge is distinguishing problematic repetition from legitimate recurring phrases—preserving common expressions and standard formulations while removing verbatim copying of substantial passages.

Deduplication Across Splits prevents contamination between training and evaluation sets, ensuring test data doesn't appear in training. This is critical for benchmark datasets that might have been crawled with web data. Aggressive deduplication generally improves performance by reducing overfitting, though extreme approaches risk removing legitimately repeated content.

Data Formatting and Tokenization

Preparing text for models requires formatting and tokenization decisions affecting training efficiency and capabilities.

Document Formatting involves decisions on document boundaries, metadata inclusion, and context windows. Long documents may be split into chunks; related documents concatenated. Special tokens mark boundaries, transitions, or metadata. Preserving document structure helps models learn coherence, while random shuffling destroys discourse structure.

Tokenization Strategy breaks text into processable units using subword methods (Byte-Pair Encoding, WordPiece, SentencePiece). Key considerations include:

- Vocabulary size (balancing model size, training efficiency, and representation granularity)
- Handling rare words, new terms, and multilingual coverage
- Treatment of code versus natural language
- Special cases: numbers, URLs, code, mixed-language text

Poor tokenization severely impairs domain-specific performance—inefficient number tokenization harms mathematical reasoning; poor code tokenization affects programming capabilities.

Sequence Length and Packing addresses fixed-length sequence processing. Short documents can be packed together for efficiency but require careful handling to prevent spurious associations between unrelated concatenated documents. Longer context windows enable processing lengthy documents and capturing long-range dependencies but increase computational cost significantly. Training sequence length balances these factors against resources and intended use cases.

Dataset Composition and Mixing

Proportions of different data sources significantly impact model characteristics, requiring careful consideration.

Domain Balancing varies by intended application—web data provides broad coverage, books enable long-form understanding, code teaches logical reasoning, academic papers provide technical knowledge, and conversational data shapes interactions. Oversampling valuable but scarce data (like reasoning examples) improves specific capabilities without proportional data amounts.

Temporal Considerations balance recent data (current knowledge, conventions) with historical data (breadth, perspective). Training cutoff dates define model knowledge of current events and should be documented for users.

Curriculum Learning structures training from simpler to complex content but faces challenges at the LLM scale. More commonly, strategies emphasize certain data types in different training stages (e.g., general language understanding before specialized technical content).

Multilingual Balance requires deliberate oversampling of lower-resource languages beyond natural web proportions (which favor English/Chinese). Excessive oversampling risks memorization; optimal balance requires empirical experimentation.

Bias and Safety Considerations address inevitable societal biases in training data. Mitigation strategies include balanced demographic representation, removing explicitly biased content, and evaluating subtle associations. Ensuring fair representation requires examining implicit associations and omissions, deliberately including diverse perspectives while avoiding tokenization.

Evaluation Set Creation requires even more careful curation:

- **Representative sampling** balances typical content with edge cases, revealing limitations.
- **Contamination prevention** through rigorous deduplication, temporal separation, and tracking data provenance; benchmark datasets pose particular risks.
- **Diversity and coverage** across linguistic phenomena, reasoning types, domains, user groups, and cultural contexts to avoid missing significant weaknesses.

Best Practices and Recommendations

Drawing from the extensive discussion above, several best practices emerge for effective dataset preparation:

- **Start with clear objectives.** Understanding the intended use cases, target capabilities, and deployment context guides dataset preparation choices. Different applications require different data characteristics and balances.
- **Invest in infrastructure early.** Robust data processing infrastructure pays dividends throughout the project lifecycle. Building scalable, reproducible pipelines saves time and reduces errors.

- **Automate with oversight.** Automated systems enable scaling, but human judgment remains crucial for establishing policies, reviewing edge cases, and validating outcomes. Combine automation with thoughtful human oversight.
- **Document everything.** Comprehensive documentation of data sources, processing steps, filtering decisions, and dataset characteristics enables reproducibility, helps diagnose issues, and provides transparency to stakeholders and users.
- **Iterate and evaluate.** Dataset preparation benefits from iteration. Initial choices may prove suboptimal, and evaluation reveals areas for improvement. Building flexibility for refinement into the process enables continuous improvement.
- **Prioritize quality over quantity.** While scale matters, quality often matters more. Smaller datasets of high-quality, relevant data frequently outperform larger datasets of mixed quality. Focus on curation alongside collection.
- **Consider downstream effects.** Dataset decisions affect not just immediate model performance but also longer-term properties like robustness, fairness, and safety. Think holistically about how preparation choices manifest in model behavior.
- **Stay current with research.** Dataset preparation techniques continue to evolve rapidly. Staying engaged with research communities provides insights into emerging best practices and potential pitfalls.

Dataset preparation for LLM training is a sophisticated discipline that combines elements of data engineering, natural language processing, machine learning, and human judgment. The quality, diversity, and careful curation of training data fundamentally determine the capabilities and characteristics of the resulting models.

For practitioners embarking on LLM training, investing time and resources in dataset preparation is not optional but essential. The effort pays dividends in model performance, safety, and reliability. For researchers, dataset preparation remains a rich area for innovation with significant practical impact. For policymakers and stakeholders, understanding dataset preparation provides crucial context for evaluating LLM capabilities and limitations.

Ultimately, dataset preparation embodies a fundamental truth of machine learning: models learn what we teach them, and teaching requires careful preparation of lessons. As we continue building increasingly powerful language models, our diligence in preparing their training data will shape the future of AI and its impact on society.

Preparing Our Dataset

The text file under analysis represents a complete short story from early 20th-century literary fiction, exhibiting characteristics typical of what is often called "literary realism" or "psychological fiction." This document provides a detailed examination of how such a text would be prepared for inclusion in a large language model training dataset, and more importantly, why texts of this nature are considered valuable training material despite constituting a relatively small portion of most training corpora.

Here is the full text:

I HAD always thought Jack Gisburn rather a cheap genius—though a good fellow enough—so it was no great surprise to me to hear that, in the height of his glory, he had dropped his painting, married a rich widow, and established himself in a villa on the Riviera. (Though I rather thought it would have been Rome or Florence.)

"The height of his glory"—that was what the women called it. I can hear Mrs. Gideon Thwing—his last Chicago sitter—deploring his unaccountable abdication. "Of course it's going to send the value of my picture 'way up; but I don't think of that, Mr. Rickham—the loss to Arrt is all I think of." The word, on Mrs. Thwing's lips, multiplied its _rs_ as though they were reflected in an endless vista of mirrors. And it was not only the Mrs. Thwings who mourned. Had not the exquisite Hermia Croft, at the last Grafton Gallery show, stopped me before Gisburn's "Moon-dancers" to say, with tears in her eyes: "We shall not look upon its like again"?

Well!—even through the prism of Hermia's tears I felt able to face the fact with equanimity. Poor Jack Gisburn! The women had made him—it was fitting that they should mourn him. Among his own sex fewer regrets were heard, and in his own trade hardly a murmur. Professional jealousy? Perhaps. If it were, the honour of the craft was vindicated by little Claude Nutley, who, in all good faith, brought out in the Burlington a very handsome "obituary" on Jack—one of those showy articles stocked with random technicalities that I have heard (I won't say by whom) compared to Gisburn's painting. And so—his resolve being apparently irrevocable—the discussion gradually died out, and, as Mrs. Thwing had predicted, the price of "Gisburns" went up.

It was not till three years later that, in the course of a few weeks' idling on the Riviera, it suddenly occurred to me to wonder why Gisburn had given up his painting. On reflection, it really was a tempting problem. To accuse his wife would have been too easy—his fair sitters had been denied the solace of saying that Mrs. Gisburn had "dragged him down." For Mrs. Gisburn—as such—had not existed till nearly a year after Jack's resolve had been taken. It might be that he had married her—since he liked his ease—because he didn't want to go on painting; but it would have been hard to prove that he had given up his painting because he had married her.

Of course, if she had not dragged him down, she had equally, as Miss Croft contended, failed to "lift him up"—she had not led him back to the easel. To put the brush into his hand again—what a vocation for a wife! But Mrs. Gisburn appeared to have disdained it—and I felt it might be interesting to find out why.

The desultory life of the Riviera lends itself to such purely academic speculations; and having, on my way to Monte Carlo, caught a glimpse of Jack's balustraded terraces between the pines, I had myself borne thither the next day.

I found the couple at tea beneath their palm-trees; and Mrs. Gisburn's welcome was so genial that, in the ensuing weeks, I claimed it frequently. It was not that my hostess was "interesting": on that point I could have given Miss Croft the fullest reassurance. It was just because she was _not_ interesting—if I may be pardoned the bull—that I found her so. For Jack, all his life, had been surrounded by interesting women: they had fostered his art, it had been reared in the hot-house of their adulation. And it was therefore instructive to note what effect the "deadening atmosphere of mediocrity" (I quote Miss Croft) was having on him.

I have mentioned that Mrs. Gisburn was rich; and it was immediately perceptible that her husband was extracting from this circumstance a delicate but substantial satisfaction. It is, as a rule, the people who scorn money who get most out of it; and Jack's elegant disdain of his wife's big balance enabled him, with an appearance of perfect good-breeding, to transmute it into objects of art and luxury. To the latter, I must add, he remained relatively indifferent; but he was buying Renaissance bronzes and eighteenth-century pictures with a discrimination that bespoke the amplest resources.

"Money's only excuse is to put beauty into circulation," was one of the axioms he laid down across the Sevres and silver of an exquisitely appointed luncheon-table, when, on a later day, I had again run over from Monte Carlo; and Mrs. Gisburn, beaming on him, added for my enlightenment: "Jack is so morbidly sensitive to every form of beauty."

Poor Jack! It had always been his fate to have women say such things of him: the fact should be set down in extenuation. What struck me now was that, for the first time, he resented the tone. I had seen him, so often, basking under similar tributes—was it the conjugal note that robbed them of their savour? No—for, oddly enough, it became apparent that he was fond of Mrs. Gisburn—fond enough not to see her absurdity. It was his own absurdity he seemed to be wincing under—his own attitude as an object for garlands and incense.

"My dear, since I've chucked painting people don't say that stuff about me—they say it about Victor Grindle," was his only protest, as he rose from the table and strolled out onto the sunlit terrace.

I glanced after him, struck by his last word. Victor Grindle was, in fact, becoming the man of the moment—as Jack himself, one might put it, had been the man of the hour. The younger artist was said to have formed himself at my friend's feet, and I wondered if a tinge of jealousy underlay the latter's mysterious abdication. But no—for it was not till after that event that the _rose Dubarry_ drawing-rooms had begun to display their "Grindles."

I turned to Mrs. Gisburn, who had lingered to give a lump of sugar to her spaniel in the dining-room.

"Why _has_ he chucked painting?" I asked abruptly.

She raised her eyebrows with a hint of good-humoured surprise.

"Oh, he doesn't _have_ to now, you know; and I want him to enjoy himself," she said quite simply.

I looked about the spacious white-panelled room, with its _famille-verte_ vases repeating the tones of the pale damask curtains, and its eighteenth-century pastels in delicate faded frames.

"Has he chucked his pictures too? I haven't seen a single one in the house."

A slight shade of constraint crossed Mrs. Gisburn's open countenance. "It's his ridiculous modesty, you know. He says they're not fit to have about; he's sent them all away except one—my portrait—and that I have to keep upstairs."

His ridiculous modesty—Jack's modesty about his pictures? My curiosity was growing like the bean-stalk. I said persuasively to my hostess: "I must really see your portrait, you know."

She glanced out almost timorously at the terrace where her husband, lounging in a hooded chair, had lit a cigar and drawn the Russian deerhound's head between his knees.

"Well, come while he's not looking," she said, with a laugh that tried to hide her nervousness; and I followed her between the marble Emperors of the hall, and up the wide stairs with terra-cotta nymphs poised among flowers at each landing.

In the dimmest corner of her boudoir, amid a profusion of delicate and distinguished objects, hung one of the familiar oval canvases, in the inevitable garlanded frame. The mere outline of the frame called up all Gisburn's past!

Mrs. Gisburn drew back the window-curtains, moved aside a _jardiniere_ full of pink azaleas, pushed an arm-chair away, and said: "If you stand here you can just manage to see it. I had it over the mantel-piece, but he wouldn't let it stay."

Yes—I could just manage to see it—the first portrait of Jack's I had ever had to strain my eyes over! Usually they had the place of honour—say the central panel in a pale yellow or _rose Dubarry_ drawing-room, or a monumental easel placed so that it took the light through curtains of old Venetian point. The more modest place became the picture better; yet, as my eyes grew accustomed to the half-light, all the characteristic qualities came out—all the hesitations disguised as audacities, the tricks of prestidigitation by which, with such consummate skill, he managed to divert attention from the real business of the picture to some pretty irrelevance of detail. Mrs. Gisburn, presenting a neutral surface to work on—forming, as it were, so inevitably the background of her own picture—had lent herself in an unusual degree to the display of this false virtuosity. The picture was one of Jack's "strongest," as his admirers would have put it—it represented, on his part, a swelling of muscles, a congesting of veins, a balancing, straddling and straining, that reminded one of the circus-clown's ironic efforts to lift a feather. It met, in short, at every point the demand of lovely woman to be painted "strongly" because she was tired of being painted "sweetly"—and yet not to lose an atom of the sweetness.

"It's the last he painted, you know," Mrs. Gisburn said with pardonable pride. "The last but one," she corrected herself—"but the other doesn't count, because he destroyed it."

"Destroyed it?" I was about to follow up this clue when I heard a footstep and saw Jack himself on the threshold.

As he stood there, his hands in the pockets of his velveteen coat, the thin brown waves of hair pushed back from his white forehead, his lean sunburnt cheeks furrowed by a smile that lifted the tips of a self-confident moustache, I felt to what a degree he had the same quality as his pictures—the quality of looking cleverer than he was.

His wife glanced at him deprecatingly, but his eyes travelled past her to the portrait.

"Mr. Rickham wanted to see it," she began, as if excusing herself. He shrugged his shoulders, still smiling.

"Oh, Rickham found me out long ago," he said lightly; then, passing his arm through mine: "Come and see the rest of the house."

He showed it to me with a kind of naive suburban pride: the bath-rooms, the speaking-tubes, the dress-closets, the trouser-presses—all the complex simplifications of the millionaire's domestic economy. And whenever my wonder paid the expected tribute he said, throwing out his chest a little: "Yes, I really don't see how people manage to live without that."

Well—it was just the end one might have foreseen for him. Only he was, through it all and in spite of it all—as he had been through, and in spite of, his pictures—so handsome, so charming, so disarming, that one longed to cry out: "Be dissatisfied with your leisure!" as once one had longed to say: "Be dissatisfied with your work!"

But, with the cry on my lips, my diagnosis suffered an unexpected check.

"This is my own lair," he said, leading me into a dark plain room at the end of the florid vista. It was square and brown and leathery: no "effects"; no bric-a-brac, none of the air of posing for reproduction in a picture weekly—above all, no least sign of ever having been used as a studio.

The fact brought home to me the absolute finality of Jack's break with his old life.

"Don't you ever dabble with paint any more?" I asked, still looking about for a trace of such activity.

"Never," he said briefly.

"Or water-colour—or etching?"

His confident eyes grew dim, and his cheeks paled a little under their handsome sunburn.

"Never think of it, my dear fellow—any more than if I'd never touched a brush."

And his tone told me in a flash that he never thought of anything else.

I moved away, instinctively embarrassed by my unexpected discovery; and as I turned, my eye fell on a small picture above the mantel-piece—the only object breaking the plain oak panelling of the room.

"Oh, by Jove!" I said.

It was a sketch of a donkey—an old tired donkey, standing in the rain under a wall.

"By Jove—a Stroud!" I cried.

He was silent; but I felt him close behind me, breathing a little quickly.

"What a wonder! Made with a dozen lines—but on everlasting foundations. You lucky chap, where did you get it?"

He answered slowly: "Mrs. Stroud gave it to me."

"Ah—I didn't know you even knew the Strouds. He was such an inflexible hermit."

"I didn't—till after. . . . She sent for me to paint him when he was dead."

"When he was dead? You?"

I must have let a little too much amazement escape through my surprise, for he answered with a deprecating laugh: "Yes—she's an awful simpleton, you know, Mrs. Stroud. Her only idea was to have him done by a fashionable painter—ah, poor Stroud! She thought it the surest way of proclaiming his greatness—of forcing it on a purblind public. And at the moment I was _the_ fashionable painter."

"Ah, poor Stroud—as you say. Was _that_ his history?"

"That was his history. She believed in him, gloried in him—or thought she did. But she couldn't bear not to have all the drawing-rooms with her. She couldn't bear the fact that, on varnishing days, one could always get near enough to see his pictures. Poor woman! She's just a fragment groping for other fragments. Stroud is the only whole I ever knew."

"You ever knew? But you just said—"

Gisburn had a curious smile in his eyes.

"Oh, I knew him, and he knew me—only it happened after he was dead."

I dropped my voice instinctively. "When she sent for you?"

"Yes—quite insensible to the irony. She wanted him vindicated—and by me!"

He laughed again, and threw back his head to look up at the sketch of the donkey. "There were days when I couldn't look at that thing—couldn't face it. But I forced myself to put it here; and now it's cured me—cured me. That's the reason why I don't dabble any more, my dear Rickham; or rather Stroud himself is the reason."

For the first time my idle curiosity about my companion turned into a serious desire to understand him better.

"I wish you'd tell me how it happened," I said.

He stood looking up at the sketch, and twirling between his fingers a cigarette he had forgotten to light. Suddenly he turned toward me.

"I'd rather like to tell you—because I've always suspected you of loathing my work."

I made a deprecating gesture, which he negatived with a good-humoured shrug.

"Oh, I didn't care a straw when I believed in myself—and now it's an added tie between us!"

He laughed slightly, without bitterness, and pushed one of the deep arm-chairs forward. "There: make yourself comfortable—and here are the cigars you like."

He placed them at my elbow and continued to wander up and down the room, stopping now and then beneath the picture.

"How it happened? I can tell you in five minutes—and it didn't take much longer to happen. . . . I can remember now how surprised and pleased I was when I got Mrs. Stroud's note. Of course, deep down, I had always _felt_ there was no one like him—only I had gone with the stream, echoed the usual platitudes about him, till I half got to think he was a failure, one of the kind that are left behind. By Jove, and he _was_ left behind—because he had come to stay! The rest of us had to let ourselves be swept along or go under, but he was high above the current—on everlasting foundations, as you say.

"Well, I went off to the house in my most egregious mood—rather moved, Lord forgive me, at the pathos of poor Stroud's career of failure being crowned by the glory of my painting him! Of course I meant to do the picture for nothing—I told Mrs. Stroud so when she began to stammer something about her poverty. I remember getting off a prodigious phrase about the honour being _mine_—oh, I was princely, my dear Rickham! I was posing to myself like one of my own sitters.

"Then I was taken up and left alone with him. I had sent all my traps in advance, and I had only to set up the easel and get to work. He had been dead only twenty-four hours, and he died suddenly, of heart disease, so that there had been no preliminary work of destruction—his face was clear and untouched. I had met him once or twice, years before, and thought him insignificant and dingy. Now I saw that he was superb.

"I was glad at first, with a merely aesthetic satisfaction: glad to have my hand on such a 'subject.' Then his strange life-likeness began to affect me queerly—as I blocked the head in I felt as if he were watching me do it. The sensation was followed by the thought: if he _were_ watching me, what would he say to my way of working? My strokes began to go a little wild—I felt nervous and uncertain.

"Once, when I looked up, I seemed to see a smile behind his close grayish beard—as if he had the secret, and were amusing himself by holding it back from me. That exasperated me still more. The secret? Why, I had a secret worth twenty of his! I dashed at the canvas furiously, and tried some of my bravura tricks. But they failed me, they crumbled. I saw that he wasn't watching the showy bits—I couldn't distract his attention; he just kept his eyes on the hard passages between. Those were the ones I had always shirked, or covered up with some lying paint. And how he saw through my lies!

"I looked up again, and caught sight of that sketch of the donkey hanging on the wall near his bed. His wife told me afterward it was the last thing he had done—just a note taken with a shaking hand, when he was down in Devonshire recovering from a previous heart attack. Just a note! But it tells his whole history. There are years of patient scornful persistence in every line. A man who had swum with the current could never have learned that mighty up-stream stroke. . . .

"I turned back to my work, and went on groping and muddling; then I looked at the donkey again. I saw that, when Stroud laid in the first stroke, he knew just what the end would be. He had possessed his subject, absorbed it, recreated it. When had I done that with any of my things? They hadn't been born of me—I had just adopted them. . . .

"Hang it, Rickham, with that face watching me I couldn't do another stroke. The plain truth was, I didn't know where to put it—_I had never known_. Only, with my sitters and my public, a showy splash of colour covered up the fact—I just threw paint into their faces. . . . Well, paint was the one medium those dead eyes could see through—see straight to the tottering foundations underneath. Don't you know how, in talking a foreign language, even fluently, one says half the time not what one wants to but what one can? Well—that was the way I painted; and as he lay there and watched me, the thing they called my 'technique' collapsed like a house of cards. He didn't sneer, you understand, poor Stroud—he just lay there quietly watching, and on his lips, through the gray beard, I seemed to hear the question: 'Are you sure you know where you're coming out?'

"If I could have painted that face, with that question on it, I should have done a great thing. The next greatest thing was to see that I couldn't—and that grace was given me. But, oh, at that minute, Rickham, was there anything on earth I wouldn't have given to have Stroud alive before me, and to hear him say: 'It's not too late—I'll show you how'?

"It _was_ too late—it would have been, even if he'd been alive. I packed up my traps, and went down and told Mrs. Stroud. Of course I didn't tell her _that_—it would have been Greek to her. I simply said I couldn't paint him, that I was too moved. She rather liked the idea—she's so romantic! It was that that made her give me the donkey. But she was terribly upset at not getting the portrait—she did so want him 'done' by some one showy! At first I was afraid she wouldn't let me off—and at my wits' end I suggested Grindle. Yes, it was I who started Grindle: I told Mrs. Stroud he was the 'coming' man, and she told somebody else, and so it got to be true. . . . And he painted Stroud without wincing; and she hung the picture among her husband's things. . . ."

He flung himself down in the arm-chair near mine, laid back his head, and clasping his arms beneath it, looked up at the picture above the chimney-piece.

"I like to fancy that Stroud himself would have given it to me, if he'd been able to say what he thought that day."

And, in answer to a question I put half-mechanically—"Begin again?" he flashed out. "When the one thing that brings me anywhere near him is that I knew enough to leave off?"

He stood up and laid his hand on my shoulder with a laugh. "Only the irony of it is that I _am_ still painting—since Grindle's doing it for me! The Strouds stand alone, and happen once—but there's no exterminating our kind of art."

Text Characteristics and Classification of the Dataset

Genre and Style Identification

This text is a literary short story, specifically a psychological narrative exploring themes of artistic integrity, self-awareness, and the conflict between commercial success and genuine artistic achievement. The narrative voice is first-person, employing sophisticated vocabulary and complex sentence structures characteristic of early modernist fiction.

Key Characteristics:

- **Literary Quality**: Elevated prose with careful attention to linguistic craft
- **Narrative Complexity**: Multilayered storytelling with retrospective framing
- **Psychological Depth**: Exploration of internal conflict and self-realization
- **Temporal Setting**: References to specific artistic movements and social contexts (the Riviera, drawing-rooms, Chicago sitters) suggest the early 1900s
- **Cultural References**: Art world terminology, European locations, social hierarchies of the period

Data Preparation Pipeline for This Text

Stage 1: Initial Collection and Extraction. Likely sources: Public domain archives (pre-1928), Project Gutenberg, academic databases, or web scraping. Text shows clean extraction with complete sentences, intact punctuation, no OCR errors, and proper character encoding.

Stage 2: Text Cleaning and Normalization. Minimal cleaning required due to high quality. UTF-8 encoding correct, special characters (em dashes, quotation marks) properly represented, Italian phrases correctly marked. Consistent paragraph breaks and spacing. Unlike web content, literary texts need minimal artifact removal (checking for headers, page numbers, copyright notices).

Stage 3: Quality Assessment. Scores exceptionally high: rich vocabulary, varied syntax, strong narrative coherence, and professional grammar. Contains narrative completeness, thematic depth, psychological complexity, and vivid descriptions. Moderate perplexity scores reflect sophisticated but clear prose.

Stage 4: Content Classification and Metadata. Domain: Literary Fiction, Psychological Drama, Early 20th Century. Metadata tags: ~1900s timeframe, French Riviera/Chicago/Rome/Florence settings, themes of artistic integrity and authenticity, first-person narrative, American English, ~4,200 words.

Stage 5: Safety and Appropriateness Filtering. Passes standard filters: no graphic violence, explicit content, hate speech, personal information, or harmful instructions. Period-appropriate social attitudes depicted critically. Demonstrates sophisticated moral reasoning about integrity over commercial success.

Stage 6: Deduplication Analysis. Full text hashed and checked for duplicates. Short quoted passages might appear in criticism, but substantial unique content remains after substring deduplication.

Stage 7: Tokenization Considerations. Challenges include italicized foreign phrases, period-specific hyphenation, complex dialogue punctuation, em dashes, and proper names. Modern subword tokenizers (BPE, WordPiece, SentencePiece) handle well, breaking rare words into learnable subword units.

Dataset Balancing Considerations

Representativeness vs. Oversampling: Literary fiction typically comprises 1-5% of training corpora. Overrepresentation risks overly formal/archaic language, inappropriate literary style, and reduced contemporary informal language ability. Balance early 20th-century texts with contemporary literature, modern web content, current conversational patterns, and up-to-date references. Balance high literary style with technical writing, casual conversation, journalism, academic prose, and code.

Complementary Data Types: Works synergistically with contemporary fiction, literary criticism, art history, conversational data, and instructional content.

Specific Training Benefits: Models learn: narrative competence, psychological realism, social awareness, aesthetic judgment, ethical reasoning, historical context, sophisticated expression, irony/subtext, character development, and descriptive precision.

Potential Issues and Mitigation

- **Dated Language/References**: Balance with contemporary content; models learn temporal context through multiperiod exposure.
- **Social Attitudes**: Historical perspectives teach cultural change; balance with diverse contemporary voices.
- **Specialized Domain Knowledge**: Expands domain coverage; context provides implicit definitions.
- **Rarified Settings**: Diverse training data provides counterbalance; teaches transferable social dynamics.

Quality Assurance Validation: Manual review verifies complete narrative, no extraction errors, appropriate content, high linguistic quality, educational value, no copyright restrictions, and accurate metadata. Automated metrics confirm moderate perplexity, high vocabulary diversity, varied sentence length, excellent grammar, strong coherence, and clean toxicity screening.

Value of Literary Training Data: Despite small corpus percentage, provides disproportionate value: linguistic sophistication, narrative complexity, psychological depth, cultural knowledge, ethical reasoning, stylistic range, and symbolic thinking. Quality training data—not just quantity—determines sophisticated language understanding.

Training Large Language Models

Training LLMs transforms prepared datasets and architectures into models capable of human-like text understanding and generation through billions of parameters learning language patterns, knowledge, and reasoning. Modern scale creates qualitative differences: parameter counts reaching trillions, datasets with trillions of tokens, and months-long training consuming thousands of accelerators and millions of dollars.

Model Architecture

Critical decisions: depth (20–100+ layers), width (2,048–12,288 hidden dimensions), attention heads (16–128), context length (2,048–128,000+ tokens). Recent innovations include flash attention, mixture of experts, rotary position embeddings, and grouped query attention.

Training Objectives

Causal language modeling dominates—predicting next tokens given previous tokens with cross-entropy loss. Self-supervised nature creates unlimited training examples without manual annotation, enabling models to learn representations capturing statistical regularities, semantic relationships, and pragmatic patterns.

Optimization

Adam optimizer variants (especially AdamW) adapt learning rates per parameter. Learning rate schedules use warmup then decay (cosine/linear). Batch sizes reach millions of tokens through gradient accumulation. Gradient clipping prevents catastrophic updates. Mixed precision training (FP16/BF16) reduces memory and increases throughput while maintaining numerical stability.

Distributed Training

Parallelization strategies: data parallelism (replicate model across devices), tensor parallelism (partition layers), pipeline parallelism (divide into sequential stages), and Fully Sharded Data Parallelism (FSDP/ZeRO sharding optimizer states, gradients, parameters). The largest runs combine strategies in three-dimensional configurations optimized for hardware topology.

Infrastructure

Specialized accelerators (NVIDIA A100/H100, Google TPUs), high-performance interconnects (NVLink, InfiniBand), frameworks (PyTorch, Megatron-LM, DeepSpeed), and comprehensive monitoring. Regular checkpointing enables fault tolerance for weeks-long runs.

Training Stages

1. **Pretraining**: Initial phase on vast unlabeled corpora consuming the majority of compute (trillions of tokens, thousands of GPU-days).
2. **Instruction Fine-Tuning**: Smaller curated datasets teaching task-following and interaction style.
3. **RLHF**: Uses human preferences to refine behavior through reward models and reinforcement learning (PPO). Alternatives include direct preference optimization and constitutional AI.

Evaluation Training/validation loss and perplexity track progress. Standardized benchmarks test language understanding (GLUE, SuperGLUE), question answering, reasoning, mathematics, code generation, and multitask knowledge (MMLU). Limitations include shortcuts, saturation, and data contamination. Human evaluation assesses fluency, coherence, accuracy, and safety.

Challenges

Loss spikes from numerical issues or bad data, divergence from excessive learning rates, hardware failures, communication bottlenecks, and data quality problems. Exact reproduction is notoriously difficult despite documentation.

Hyperparameter Selection

Scaling laws (Chinchilla) predict optimal settings, suggesting model size and data size should increase in tandem for compute budgets. Key hyperparameters: learning rate (1e-4 to 3e-4), batch size (millions of tokens), weight decay (~0.1), and weight initialization (Xavier/He).

Compute and Efficiency

Frontier models consume 10,000–100,000+ petaflop-days, costing millions to tens of millions of dollars. Well-optimized training reaches 40-60% model FLOPS utilization. Optimization strategies: mixed precision (2–3x throughput), activation checkpointing,

Flash Attention, and careful hardware selection. Environmental impact: hundreds of tons of CO_2 per large model; mitigation through cleaner energy grids, efficiency improvements, and model sharing.

Advanced Techniques

Mixture of Experts increases capacity while managing costs. Extended context lengths through sparse/sliding window attention. Multimodal training processes text, images, audio, and video. Continual learning updates models without complete retraining while preventing catastrophic forgetting.

Post-training

Model compression: knowledge distillation (90–95% performance retention at a fraction of the size), quantization (8-bit or lower), and pruning (sparse models). Safety testing: red teaming, automated evaluations for toxicity/bias/privacy. Deployment requires inference optimization and monitoring.

Ethical Considerations

Training data reflects societal biases manifesting as stereotypes and differential performance. Misuse potential for misinformation, harassment, and dangerous instructions. Resource concentration among well-funded organizations raises equity concerns. Field must balance technical excellence with safety, bias mitigation, environmental responsibility, and appropriate governance as systems grow more capable and influential.

Source Code for Training Our LLM

```
import os
import json
import math
import torch
import pickle
from tqdm import tqdm
```

```
from torch.utils.data import Dataset, DataLoader
from torch.amp import autocast
from torch.cuda.amp import GradScaler
from dataclasses import asdict

# --- assumes BPETokenizer, ModelConfig, Transformer, and save_checkpoint()
from earlier parts are available ---

# Mixed Precision Settings

USE_AMP = torch.cuda.is_available()
DEVICE = torch.device("cuda" if torch.cuda.is_available() else "cpu")
scaler = GradScaler(enabled=USE_AMP)

# Config
VOCAB_SIZE = 2000
BATCH_SIZE = 4
BLOCK_SIZE = 64
EPOCHS = 20
LR = 3e-4
ENCODED_PATH = "book_tokens.pkl"
TOKENIZER_PREFIX = "bpe_tokenizer"   # keep consistent with saved files
RAW_TEXT_PATH = "book.txt"

# Dataset Definition
class TextDataset(Dataset):
    def __init__(self, token_ids, block_size):
        self.block_size = block_size
        # windows of length block_size+1 for next-token prediction
        N = len(token_ids)
        if N < block_size + 1:
            raise ValueError(f"Not enough tokens ({N}) for block_
            size={block_size}.")
        self.data = [token_ids[i:i + block_size + 1] for i in range(N -
        block_size - 1)]

    def __len__(self):
        return len(self.data)
```

```
    def __getitem__(self, idx):
        chunk = self.data[idx]
        x = torch.tensor(chunk[:-1], dtype=torch.long)
        y = torch.tensor(chunk[1:], dtype=torch.long)
        return x, y

# Tokenizer: Train or Load
tokenizer = BPETokenizer()

if not (os.path.exists(f"{TOKENIZER_PREFIX}_vocab.json") and os.path.
exists(f"{TOKENIZER_PREFIX}_merges.json")):
    print("[*] Training tokenizer...")
    with open(RAW_TEXT_PATH, "r", encoding="utf-8") as f:
        lines = [ln.strip() for ln in f if ln.strip()]
    tokenizer.train(lines, vocab_size=VOCAB_SIZE)
    tokenizer.save(TOKENIZER_PREFIX)
else:
    print("[*] Loading tokenizer...")
    tokenizer.load(TOKENIZER_PREFIX)

# Encode Dataset and Cache
if os.path.exists(ENCODED_PATH):
    print("[*] Loading cached tokenized dataset...")
    with open(ENCODED_PATH, "rb") as f:
        token_ids = pickle.load(f)
else:
    print("[*] Encoding and caching dataset...")
    with open(RAW_TEXT_PATH, "r", encoding="utf-8") as f:
        raw_text = f.read()
    # add BOS/EOS once around the whole corpus (your BPETokenizer from Part
    6 supports this)
    token_ids = tokenizer.encode(raw_text, add_bos_eos=True)
    with open(ENCODED_PATH, "wb") as f:
        pickle.dump(token_ids, f)

# Dataset & DataLoader
dataset = TextDataset(token_ids, BLOCK_SIZE)
```

```
num_workers = 4 if torch.cuda.is_available() else 0
pin_mem = torch.cuda.is_available()
dataloader = DataLoader(
    dataset,
    batch_size=BATCH_SIZE,
    shuffle=True,
    num_workers=num_workers,
    pin_memory=pin_mem,
    drop_last=True,
    persistent_workers=pin_mem and num_workers > 0
)

# Model Setup
config = ModelConfig(
    # capacity (keep tiny)
    num_hidden_layers=2,           # ↓ from 4
    hidden_size=128,               # ↓ from 256
    intermediate_size=256,         # 2× hidden is fine

    # attention (keep simple & valid)
    head_dim=32,                   # must be even (RoPE)
    num_attention_heads=4,         # total heads
    num_key_value_heads=2,         # GQA: must divide num_attention_heads

    # MoE (disable for tiny data)
    num_experts=2,                 # 1 = dense MLP (no MoE)
    experts_per_token=1,

    # tokenizer
    vocab_size=len(tokenizer.token_to_id),  # after retraining BPE at ~300

    # context & rope (defaults are fine)
    sliding_window=0,
    initial_context_length=1024,  # smaller is ok; you trained with
    short blocks
    rope_theta=10000.0,
    rope_scaling_factor=1.0,
```

```
    rope_ntk_alpha=1.0,
    rope_ntk_beta=32.0,

    # activation clamp (stable)
    swiglu_limit=7.0
)

model = Transformer(config).to(DEVICE)
model.output_proj.weight = model.embedding.weight
# AdamW tuned for transformers
optimizer = torch.optim.AdamW(
    model.parameters(),
    lr=LR,
    betas=(0.9, 0.95),
    eps=1e-8,
    weight_decay=0.0,
)
loss_fn = torch.nn.CrossEntropyLoss()

# Cosine decay with warmup
from torch.optim.lr_scheduler import LambdaLR
total_steps = max(1, EPOCHS * len(dataloader))
warmup_steps = max(1, int(0.06 * total_steps))

def lr_lambda(step):
    if step < warmup_steps:
        return float(step) / float(max(1, warmup_steps))
    progress = float(step - warmup_steps) / float(max(1, total_steps -
    warmup_steps))
    return 0.5 * (1.0 + math.cos(math.pi * progress))

scheduler = LambdaLR(optimizer, lr_lambda)

# Optional compile (comment out if PyTorch version/hardware complains)
try:
    model = torch.compile(model)
except Exception:
    pass
```

```
# Training Loop
print("[*] Training model...")
model.train()
max_grad_norm = 1.0
global_step = 0

for epoch in range(EPOCHS):
    total_loss = 0.0
    progress_bar = tqdm(dataloader, desc=f"Epoch {epoch+1}", leave=False)

    for x, y in progress_bar:
        x = x.to(DEVICE, non_blocking=True)
        y = y.to(DEVICE, non_blocking=True)

        optimizer.zero_grad(set_to_none=True)

        with autocast(device_type="cuda", enabled=USE_AMP):
            logits = model(x)  # [B, T, vocab_size]
            loss = loss_fn(logits.view(-1, logits.size(-1)), y.view(-1))

        scaler.scale(loss).backward()

        # clip after unscale
        scaler.unscale_(optimizer)
        torch.nn.utils.clip_grad_norm_(model.parameters(), max_grad_norm)

        scaler.step(optimizer)
        scaler.update()
        scheduler.step()

        total_loss += float(loss.item())
        perplexity = math.exp(loss.item())
        progress_bar.set_postfix(loss=float(loss.item()), ppl=perplexity,
        lr=scheduler.get_last_lr()[0])
        global_step += 1

    avg_loss = total_loss / len(dataloader)
    avg_perplexity = math.exp(avg_loss)
    print(f"[+] Epoch {epoch+1} complete. Avg Loss: {avg_loss:.4f} |
    Perplexity: {avg_perplexity:.2f} | LR: {scheduler.get_last_lr()[0]:.3e}")
```

```
# Save Model and Config (folder)
def save_checkpoint(path: str, model: torch.nn.Module, config:
ModelConfig):
    os.makedirs(path, exist_ok=True)
    torch.save(model.state_dict(), os.path.join(path, "model.pt"))
    with open(os.path.join(path, "config.json"), "w") as f:
        json.dump(asdict(config), f)

save_checkpoint("chkpt", model, config)
print("[*] Saved to ./chkpt")
```

```
Output:
[+] Epoch 1 complete. Avg Loss: 24.2249 | Perplexity: 3.29e+10 | LR:
2.500e-04
[+] Epoch 2 complete. Avg Loss: 3.6564 | Perplexity: 38.71 | LR: 2.987e-04
[+] Epoch 3 complete. Avg Loss: 2.6101 | Perplexity: 13.60 | LR: 2.933e-04
[+] Epoch 4 complete. Avg Loss: 1.7325 | Perplexity: 5.65 | LR: 2.839e-04
[+] Epoch 5 complete. Avg Loss: 1.0923 | Perplexity: 2.98 | LR: 2.708e-04
[+] Epoch 6 complete. Avg Loss: 0.7433 | Perplexity: 2.10 | LR: 2.543e-04
[+] Epoch 7 complete. Avg Loss: 0.5481 | Perplexity: 1.73 | LR: 2.349e-04
[+] Epoch 8 complete. Avg Loss: 0.4319 | Perplexity: 1.54 | LR: 2.131e-04
[+] Epoch 9 complete. Avg Loss: 0.3502 | Perplexity: 1.42 | LR: 1.896e-04
[+] Epoch 10 complete. Avg Loss: 0.2931 | Perplexity: 1.34 | LR: 1.650e-04
[+] Epoch 11 complete. Avg Loss: 0.2469 | Perplexity: 1.28 | LR: 1.400e-04
[+] Epoch 12 complete. Avg Loss: 0.2142 | Perplexity: 1.24 | LR: 1.152e-04
[+] Epoch 13 complete. Avg Loss: 0.1881 | Perplexity: 1.21 | LR: 9.144e-05
[+] Epoch 14 complete. Avg Loss: 0.1681 | Perplexity: 1.18 | LR: 6.929e-05
[+] Epoch 15 complete. Avg Loss: 0.1532 | Perplexity: 1.17 | LR: 4.938e-05
[+] Epoch 16 complete. Avg Loss: 0.1408 | Perplexity: 1.15 | LR: 3.228e-05
[+] Epoch 17 complete. Avg Loss: 0.1316 | Perplexity: 1.14 | LR: 1.846e-05
[+] Epoch 18 complete. Avg Loss: 0.1249 | Perplexity: 1.13 | LR: 8.300e-06
[+] Epoch 19 complete. Avg Loss: 0.1208 | Perplexity: 1.13 | LR: 2.089e-06
[+] Epoch 20 complete. Avg Loss: 0.1183 | Perplexity: 1.13 | LR: 0.000e+00
[*] Saved to ./chkpt
```

Understanding the Code Step-by-Step

While the model is intentionally kept small for educational purposes and limited computational resources, it demonstrates all the key concepts used in training large-scale language models like GPT or Claude. The code trains a transformer-based model on text data, implementing modern best practices including mixed precision training, gradient clipping, learning rate scheduling, and efficient data loading.

The training process follows these major steps:

- **Preparation Phase**: We configure the training environment, set up mixed precision training if a GPU is available, and define all hyperparameters that control the training process.
- **Tokenization**: The raw text must be converted into numerical tokens that the model can process. We either train a new Byte-Pair Encoding (BPE) tokenizer on our text or load a previously trained one.
- **Dataset Creation**: We organize the tokenized text into training examples of fixed length, where each example contains an input sequence and the target next tokens.
- **Model Initialization**: We create a transformer model with specified architecture parameters and move it to the GPU if available.
- **Training Loop**: We repeatedly feed batches of data through the model, compute loss measuring prediction quality, calculate gradients, and update model parameters to improve predictions.
- **Checkpoint Saving**: Finally, we save the trained model and its configuration so it can be loaded and used later for text generation or further training.

Let's examine each component in detail.

Part 1: Environment Setup and Configuration

```
import os
import json
import math
import torch
```

```
import pickle
from tqdm import tqdm
from torch.utils.data import Dataset, DataLoader
from torch.amp import autocast
from torch.cuda.amp import GradScaler
from dataclasses import asdict
```

The imports bring in essential libraries. PyTorch provides the deep learning framework, while `tqdm` gives us nice progress bars during training. The `autocast` and `GradScaler` enable mixed precision training, a crucial optimization for modern GPU training.

```
USE_AMP = torch.cuda.is_available()
DEVICE = torch.device("cuda" if torch.cuda.is_available() else "cpu")
scaler = GradScaler(enabled=USE_AMP)
```

Mixed precision training uses 16-bit floating point (FP16) for most operations instead of 32-bit (FP32), reducing memory usage and increasing speed by 2-3x with minimal impact on model quality. The `GradScaler` handles the numerical challenges of training in lower precision by scaling the loss to prevent gradient underflow. We only enable this if a CUDA-capable GPU is available, as it doesn't benefit CPU training.

```
VOCAB_SIZE = 2000
BATCH_SIZE = 4
BLOCK_SIZE = 64
EPOCHS = 20
LR = 3e-4
```

These hyperparameters control the training process:

- **VOCAB_SIZE = 2000**: The tokenizer will learn 2000 different tokens. This is quite small compared to production models (which use 32,000-100,000+ tokens) but sufficient for our small dataset. A smaller vocabulary means the model has fewer output options to choose from but may need to break words into more pieces.

- **BATCH_SIZE = 4**: We process 4 sequences simultaneously in each training step. Larger batches provide more stable gradients and better GPU utilization but require more memory. For a small model and limited resources, 4 is reasonable. Production training uses batch sizes in the millions of tokens by accumulating gradients across many steps.
- **BLOCK_SIZE = 64**: Each training example contains 64 tokens. The model learns to predict the next token given up to 64 previous tokens of context. This is the sequence length or context window for training. Modern LLMs use much longer contexts (2048-128000 tokens) but require proportionally more memory and computation.
- **EPOCHS = 20**: We'll pass through the entire dataset 20 times. Each epoch represents one complete iteration over all training data. More epochs allow the model to learn better but risk overfitting if training data is limited.
- **LR = 3e-4**: The learning rate of 0.0003 controls how much we update model parameters at each step. This value (3e-4) is a standard starting point for transformer models discovered through extensive experimentation. Too high causes instability; too low wastes time with slow learning.

Part 2: Dataset Preparation

```
class TextDataset(Dataset):
    def __init__(self, token_ids, block_size):
        self.block_size = block_size
        N = len(token_ids)
        if N < block_size + 1:
            raise ValueError(f"Not enough tokens ({N}) for block_
            size={block_size}.")
        self.data = [token_ids[i:i + block_size + 1] for i in range(N -
        block_size - 1)]
```

The TextDataset class organizes our tokenized text into training examples. It creates overlapping windows of block_size + 1 tokens sliding across the entire token sequence. Why the extra +1 token? Because for next-token prediction, we need both input tokens and target tokens. The first block_size tokens serve as input, and tokens 1 through block_size + 1 serve as targets (shifted by one position).

For example, if block_size = 64 and we have the tokens [10, 20, 30, 40, 50, ...], we'd create:

- Example 1: Input [10, 20, ..., 73] (64 tokens), Target [20, 30, ..., 74] (the next 64 tokens)
- Example 2: Input [20, 30, ..., 74], Target [30, 40, ..., 75]
- And so on...

The sliding window approach means adjacent training examples overlap significantly, which is efficient for learning from limited data.

```
def __getitem__(self, idx):
    chunk = self.data[idx]
    x = torch.tensor(chunk[:-1], dtype=torch.long)
    y = torch.tensor(chunk[1:], dtype=torch.long)
    return x, y
```

When the training loop requests a specific example, this method returns the input sequence x (all but the last token) and target sequence y (all but the first token). This creates the shifted relationship needed for next-token prediction: given tokens at positions 0 through 63, predict tokens at positions 1 through 64.

Part 3: Tokenizer Training or Loading

```
tokenizer = BPETokenizer()

if not (os.path.exists(f"{TOKENIZER_PREFIX}_vocab.json") and
        os.path.exists(f"{TOKENIZER_PREFIX}_merges.json")):
    print("[*] Training tokenizer...")
    with open(RAW_TEXT_PATH, "r", encoding="utf-8") as f:
        lines = [ln.strip() for ln in f if ln.strip()]
    tokenizer.train(lines, vocab_size=VOCAB_SIZE)
    tokenizer.save(TOKENIZER_PREFIX)
```

```
else:
    print("[*] Loading tokenizer...")
    tokenizer.load(TOKENIZER_PREFIX)
```

Before training the model, we need a tokenizer to convert text into numbers. Byte-Pair Encoding (BPE) is a popular tokenization algorithm that learns common subword units from the training data. It starts with individual characters and iteratively merges the most frequent pairs, building a vocabulary of 2000 tokens in our case.

The code checks if a tokenizer already exists (by looking for saved vocabulary and merge files). If found, it loads the existing tokenizer to ensure consistency—using different tokenizers would make previous training incompatible. If not found, it trains a new tokenizer from scratch on the raw text file.

Training the tokenizer is separate from training the model. The tokenizer learns statistical patterns about which character sequences appear together frequently, creating efficient representations. Common words might become single tokens, while rare words get broken into multiple subword pieces.

Part 4: Dataset Encoding and Caching

```
if os.path.exists(ENCODED_PATH):
    print("[*] Loading cached tokenized dataset...")
    with open(ENCODED_PATH, "rb") as f:
        token_ids = pickle.load(f)
else:
    print("[*] Encoding and caching dataset...")
    with open(RAW_TEXT_PATH, "r", encoding="utf-8") as f:
        raw_text = f.read()
    token_ids = tokenizer.encode(raw_text, add_bos_eos=True)
    with open(ENCODED_PATH, "wb") as f:
        pickle.dump(token_ids, f)
```

Tokenizing text takes time, especially for large datasets. To avoid re-tokenizing every time we run training, this code caches the tokenized result. On the first run, it reads the raw text file, applies the tokenizer to convert text into token IDs (integers), and saves the result in a pickle file. On subsequent runs, it simply loads the pretokenized data.

The add_bos_eos=True parameter adds special tokens marking the beginning (BOS) and end (EOS) of the sequence. These special tokens help the model learn where text begins and ends, which is particularly important for generation.

Part 5: DataLoader Setup

```
dataset = TextDataset(token_ids, BLOCK_SIZE)
num_workers = 4 if torch.cuda.is_available() else 0
pin_mem = torch.cuda.is_available()
dataloader = DataLoader(
    dataset,
    batch_size=BATCH_SIZE,
    shuffle=True,
    num_workers=num_workers,
    pin_memory=pin_mem,
    drop_last=True,
    persistent_workers=pin_mem and num_workers > 0
)
```

The DataLoader handles the mechanics of feeding data to the model during training. Let's examine each parameter:

- **shuffle=True**: Randomly reorders training examples each epoch. This prevents the model from learning spurious patterns based on the order of examples and generally improves generalization. The randomness acts as a form of regularization.
- **num_workers=4**: Uses 4 parallel processes to load and prepare data while the GPU trains on the previous batch. This overlap hides data loading latency. We only use multiple workers on GPU systems where this parallelism provides benefit.
- **pin_memory=True**: For GPU training, this allocates data in page-locked RAM, enabling faster transfers to GPU memory. This is a standard optimization for GPU training.

- **drop_last=True**: Discards the last batch if it's smaller than BATCH_SIZE. This ensures all batches have a consistent size, simplifying the training code and avoiding edge cases with different batch sizes.
- **persistent_workers**: Keeps the worker processes alive between epochs rather than recreating them, reducing overhead when using multiple workers.

Part 6: Model Configuration and Initialization

```
config = ModelConfig(
    num_hidden_layers=2,
    hidden_size=128,
    intermediate_size=256,
    head_dim=32,
    num_attention_heads=4,
    num_key_value_heads=2,
    num_experts=2,
    experts_per_token=1,
    vocab_size=len(tokenizer.token_to_id),
    sliding_window=0,
    initial_context_length=1024,
    rope_theta=10000.0,
    rope_scaling_factor=1.0,
    rope_ntk_alpha=1.0,
    rope_ntk_beta=32.0,
    swiglu_limit=7.0
)
```

The model configuration defines the transformer architecture. This configuration creates an intentionally small model suitable for limited data and compute:

- **num_hidden_layers=2**: The model has only 2 transformer layers. Production models use 20-100+ layers. Fewer layers mean less capacity to learn complex patterns but faster training and less risk of overfitting on small datasets.

- **hidden_size=128**: Each token is represented by a 128-dimensional vector internally. This is the model's "width." Production models use dimensions of 2048-12288. Wider models can represent more nuanced information but require more memory and computation.
- **intermediate_size=256**: The feed-forward network in each layer expands to 256 dimensions (2x the hidden size). The standard ratio is 4x, but for such a tiny model, 2x suffices. This controls the model's capacity within each layer.
- **head_dim=32 and num_attention_heads=4**: The attention mechanism uses 4 heads, each with 32 dimensions. Multiple heads let the model attend to different aspects of the input simultaneously. With `hidden_size=128` and 4 heads of 32 dimensions, we have 4 × 32 = 128, perfectly utilizing the hidden dimension.
- **num_key_value_heads=2**: This implements Grouped Query Attention (GQA), where 4 query heads share 2 key-value heads. This reduces memory bandwidth requirements during attention computation. Each key-value head serves 2 query heads.
- **num_experts=2 and experts_per_token=1**: These parameters control Mixture of Experts (MoE). With 2 experts but only 1 active per token, the model can specialize different experts for different types of input. However, for such a tiny model and dataset, MoE may not provide significant benefits.
- **vocab_size**: Set to match the tokenizer's vocabulary size, ensuring the model's output layer can predict all possible tokens.
- **Positional encoding parameters** (rope_theta, scaling factors): These control Rotary Position Embeddings (RoPE), which encode position information into the attention mechanism. The defaults are standard values that work well for most applications.
- **swiglu_limit=7.0**: Clamps activations in the SwiGLU activation function to prevent numerical instabilities. This is a stability mechanism for training.

```
model = Transformer(config).to(DEVICE)
model.output_proj.weight = model.embedding.weight
```

We instantiate the transformer model with our configuration and move it to the GPU if available. The second line implements weight tying, where the output projection layer (converting hidden states to vocabulary logits) shares weights with the input embedding layer (converting tokens to hidden states). This reduces parameters and often improves performance since both layers map between the vocabulary space and the hidden space.

Part 7: Optimizer Configuration

```
optimizer = torch.optim.AdamW(
    model.parameters(),
    lr=LR,
    betas=(0.9, 0.95),
    eps=1e-8,
    weight_decay=0.0,
)
```

AdamW is the standard optimizer for transformer training. It adapts the learning rate for each parameter based on gradient statistics:

- **lr=3e-4**: The base learning rate, which will be modulated by the learning rate scheduler.
- **betas=(0.9, 0.95)**: Control the exponential decay rates for the first and second moment estimates of gradients. These specific values (particularly the second beta being 0.95 instead of the default 0.999) are tuned for transformer training and help with training stability.
- **eps=1e-8**: A small constant preventing division by zero in the optimizer's calculations.
- **weight_decay=0.0**: L2 regularization strength. Set to 0 here, though production training often uses small values like 0.01 or 0.1 to prevent overfitting. For our small model and dataset, we're not applying weight decay.

The "W" in AdamW stands for "weight decay," and AdamW implements it more correctly than the original Adam by decoupling weight decay from the gradient-based updates.

Part 8: Learning Rate Scheduling

```
total_steps = max(1, EPOCHS * len(dataloader))
warmup_steps = max(1, int(0.06 * total_steps))

def lr_lambda(step):
    if step < warmup_steps:
        return float(step) / float(max(1, warmup_steps))
    progress = float(step - warmup_steps) / float(max(1, total_steps -
    warmup_steps))
    return 0.5 * (1.0 + math.cos(math.pi * progress))

scheduler = LambdaLR(optimizer, lr_lambda)
```

This implements a warmup followed by cosine decay, the standard learning rate schedule for transformer training:

Warmup Phase: For the first 6% of training steps (warmup_steps), the learning rate increases linearly from near zero to the full learning rate (3e-4). This gradual ramp-up prevents instabilities when training starts from random initialization. Large updates to random parameters can send optimization into problematic regions, so we start gently.

Cosine Decay: After warmup, the learning rate follows a cosine curve, smoothly decreasing from the peak value toward zero. The formula `0.5 * (1.0 + math.cos(math.pi * progress))` creates a smooth curve from 1.0 to 0.0 as progress goes from 0 to 1. This gradual decrease allows the model to make large exploratory updates early while taking smaller, refinement-oriented steps later.

The cosine schedule is preferred over linear decay because the smooth, continuous curve often leads to better final performance. The gradual nature of the decay prevents sudden changes in optimization dynamics.

Part 9: Model Compilation (Optional)

```
try:
    model = torch.compile(model)
except Exception:
    pass
```

PyTorch 2.0+ includes a compilation feature that optimizes the model's computational graph for faster execution. The `torch.compile` function analyzes the model's forward pass and generates optimized code, potentially improving training speed by 20–50%. The try-except block ensures the code works even if compilation isn't supported (older PyTorch versions or certain hardware configurations).

Compilation is purely an optimization—it doesn't change what the model learns, just how fast it learns it.

Part 10: The Training Loop

```
print("[*] Training model...")
model.train()
max_grad_norm = 1.0
global_step = 0
```

Before training begins, we call `model.train()` to put the model in training mode. This affects certain layers like dropout (though our small model might not have any) and ensures gradients are computed. The `max_grad_norm` sets the threshold for gradient clipping, and `global_step` tracks the total number of optimization steps across all epochs.

```
for epoch in range(EPOCHS):
    total_loss = 0.0
    progress_bar = tqdm(dataloader, desc=f"Epoch {epoch+1}", leave=False)
```

The outer loop iterates over epochs. Each epoch represents one complete pass through all training data. We track cumulative loss to compute the average for the epoch and create a progress bar for visual feedback during training.

```
    for x, y in progress_bar:
        x = x.to(DEVICE, non_blocking=True)
        y = y.to(DEVICE, non_blocking=True)
```

The inner loop iterates over batches from the dataloader. Each iteration retrieves a batch of input sequences (x) and target sequences (y), both of shape [BATCH_SIZE, BLOCK_SIZE]. We transfer them to the GPU using non_blocking=True, which allows the CPU to continue preparing the next batch while data transfers, improving efficiency.

```
optimizer.zero_grad(set_to_none=True)
```

Before computing gradients for the new batch, we must zero out gradients from the previous batch. The set_to_none=True parameter is a performance optimization—instead of filling gradient tensors with zeros, it sets them to None, saving memory and computation. They'll be properly initialized when .backward() runs.

```
with autocast(device_type="cuda", enabled=USE_AMP):
    logits = model(x)  # [B, T, vocab_size]
    loss = loss_fn(logits.view(-1, logits.size(-1)), y.view(-1))
```

This is where the actual forward pass happens. The autocast context manager enables mixed precision training if a GPU is available:

Forward Pass: model(x) feeds the input through all layers of the transformer, producing logits (unnormalized predictions) of shape [BATCH_SIZE, BLOCK_SIZE, vocab_size]. For each position in each sequence, we get a score for every possible next token.

Loss Computation: Cross-entropy loss measures how well the model's predictions match the actual next tokens. We reshape the predictions and targets from 3D to 2D: [BATCH_SIZE * BLOCK_SIZE, vocab_size] and [BATCH_SIZE * BLOCK_SIZE]. This treats each position independently when computing loss. The loss averages across all positions and sequences, giving us a single number indicating prediction quality.

```
scaler.scale(loss).backward()
```

Backpropagation computes gradients of the loss with respect to all model parameters. The `scaler.scale(loss)` multiplies the loss by a large constant before backpropagation. Why? In mixed precision training using FP16, gradients can be very small and underflow to zero. Scaling prevents this underflow, and we'll unscale before the actual parameter update.

```
scaler.unscale_(optimizer)
 torch.nn.utils.clip_grad_norm_(model.parameters(), max_grad_norm)
```

Before updating parameters, we perform two critical operations:

Unscaling: Divides gradients by the scaling factor applied earlier, restoring their true magnitude.

Gradient Clipping: Rescales gradients if their total magnitude exceeds `max_grad_norm` (1.0). This prevents exploding gradients that can occur with deep networks or unusual data batches. If the total gradient norm is, say, 5.0, all gradients get multiplied by 1.0/5.0 = 0.2, bringing the total norm down to 1.0. This stabilizes training significantly.

```
scaler.step(optimizer)
scaler.update()
scheduler.step()
```

Parameter Update: The optimizer uses the clipped gradients to update model parameters. The scaler checks for infinities or NaNs that would indicate numerical problems; if found, it skips the update and adjusts the scaling factor for the next iteration.

Scaler Update: Adjusts the loss scaling factor based on whether this iteration succeeded. If several iterations succeed without overflow, it increases the scaling factor for better precision. If overflow occurs, it decreases the scaling factor.

Learning Rate Step: Advances the learning rate schedule by one step, implementing the warmup and cosine decay we configured earlier.

```
total_loss += float(loss.item())
perplexity = math.exp(loss.item())
progress_bar.set_postfix(loss=float(loss.item()), ppl=perplexity,
lr=scheduler.get_last_lr()[0])
 global_step += 1
```

We accumulate the loss for computing the epoch average and update the progress bar with the current batch's loss and learning rate. This provides real-time feedback on training progress. The `global_step` counter increments, tracking total optimization steps across all epochs.

```
avg_loss = total_loss / len(dataloader)
avg_perplexity = math.exp(avg_loss)
print(f"[+] Epoch {epoch+1} complete. Avg Loss: {avg_loss:.4f} |
Perplexity: {avg_perplexity:.2f} | LR: {scheduler.get_last_lr()[0]:.3e}")
```

At the end of each epoch, we compute the average loss across all batches and print a summary. Watching this average loss decrease over epochs confirms the model is learning. Typical behavior shows a rapid initial decrease as the model learns basic patterns, then slower decrease as it learns more subtle patterns.

Part 11: Saving the Model

```
def save_checkpoint(path: str, model: torch.nn.Module, config:
ModelConfig):
    os.makedirs(path, exist_ok=True)
    torch.save(model.state_dict(), os.path.join(path, "model.pt"))
    with open(os.path.join(path, "config.json"), "w") as f:
        json.dump(asdict(config), f)

save_checkpoint("chkpt", model, config)
```

After training completes, we save two files:

- **model.pt**: Contains all learned parameter values (weights and biases). This is the actual trained model that took hours or days to create. The file size depends on model size—our tiny model might be a few megabytes, while production models can be hundreds of gigabytes.

- **config.json**: Contains the model architecture specification. This is essential for loading the model later because we need to reconstruct the exact same architecture before loading the weights. Without the config, we wouldn't know how many layers, what the hidden size, etc.

Saving both pieces separately is standard practice. The config is human-readable JSON and can be inspected or modified, while the model weights are in an efficient binary format.

Understanding Training Dynamics

As training progresses, several things happen inside the model:

- **Early Training (Epochs 1-5)**: The model learns basic statistics of the data. It discovers which tokens appear frequently, which tokens often follow others, and basic syntactic patterns. Loss decreases rapidly as the model moves from random guessing to somewhat informed prediction.
- **Middle Training (Epochs 6-15)**: The model learns more complex patterns, including multi-word phrases, syntactic structures, and correlations between distant tokens. Loss decrease slows as easy patterns are exhausted and harder patterns require more training.
- **Late Training (Epochs 16-20)**: The model refines its knowledge, learning subtle patterns and edge cases. Loss decrease becomes very gradual. Depending on dataset size, the model might start overfitting—memorizing specific training examples rather than learning general patterns. This is why we'd typically monitor validation loss (which we don't implement in this simple example).

What the Model Learns

Through next-token prediction, the model develops several capabilities:

- **Syntactic Knowledge**: Understanding grammar, sentence structure, and valid word orderings. The model learns that adjectives precede nouns, verbs require subjects, and how to balance parentheses or quotes.

- **Semantic Understanding**: Learning word meanings and relationships. Words appearing in similar contexts get similar representations, allowing the model to understand synonyms, antonyms, and conceptual relationships.
- **Contextual Reasoning**: Using previous tokens to inform predictions about next tokens. The model learns that after "The capital of France is" the next token should be "Paris."
- **Stylistic Patterns**: Mimicking the writing style, vocabulary, and structures present in the training data. If trained on formal text, it generates formally; if trained on casual text, it generates casually.
- **Factual Knowledge**: Implicitly encoding facts encountered during training. The model doesn't have an explicit knowledge base, but patterns in its parameters reflect facts seen in the training data.

This training code demonstrates the complete pipeline from raw text to trained model. While the model is small and the dataset limited, the code implements all the essential techniques used in training frontier LLMs: transformer architecture, causal language modeling objective, AdamW optimization, learning rate scheduling with warmup and cosine decay, mixed precision training, gradient clipping, and efficient data loading.

The same principles scale up to massive models trained on trillions of tokens. The main differences are scale (more layers, wider dimensions, larger vocabularies, longer contexts, bigger batches) and infrastructure (distributed training across thousands of GPUs, sophisticated data preprocessing pipelines, extensive monitoring systems). But fundamentally, training GPT-4 or Claude follows this same pattern: prepare data, initialize model, repeatedly compute predictions and gradients, and update parameters to improve predictions.

Understanding this code provides a foundation for comprehending state-of-the-art LLM training. The concepts transfer directly—when reading about how GPT or Claude were trained, you'll recognize the optimizer choices, learning rate schedules, loss functions, and training dynamics. The path from this educational example to frontier models is continuous, not requiring fundamentally different techniques but rather careful engineering to make these same techniques work at an unprecedented scale.

Training language models remains computationally expensive and requires significant expertise, but the core ideas are accessible. This tutorial bridges the gap between theoretical understanding and practical implementation, providing hands-on experience with the techniques that power modern AI.

TokenGenerator for Inference

What Is TokenGenerator?

TokenGenerator is an inference engine that loads a trained language model and generates text token by token. After spending hours or days training a model, this class brings it to life by producing actual text. It takes a prompt (sequence of tokens) and autoregressively generates new tokens by repeatedly predicting the next most likely token, adding it to the sequence, and using the extended sequence to predict the next token.

```
import torch
import os, json
from collections import defaultdict
import torch.nn.functional as F

class TokenGenerator:
    def __init__(self, checkpoint_path: str, device: str | torch.device =
    "cuda"):
        if isinstance(device, str):
            device = torch.device(device)

        # --- Load config ---
        cfg_path = os.path.join(checkpoint_path, "config.json")
        if not os.path.exists(cfg_path):
            raise FileNotFoundError(f"Config not found: {cfg_path}")
        with open(cfg_path, "r") as f:
            cfg_dict = json.load(f)
        config = ModelConfig(**cfg_dict)

        # --- Build and load model ---
        self.model = Transformer(config, device=device).to(device)
```

```
        weights_path = os.path.join(checkpoint_path, "model.pt")
        if not os.path.exists(weights_path):
            raise FileNotFoundError(f"Weights not found: {weights_path}")
        state = torch.load(weights_path, map_location=device)
        self.model.load_state_dict(state, strict=False)
        self.model.eval()

        self.device = device

    def generate(
        self,
        prompt_tokens: list[int],
        stop_tokens: list[int] = None,
        temperature: float = 0.6,
        top_p: float = 0.85,
        max_tokens: int = None,
        *,
        stop_ids: list[int] = None,
        max_new_tokens: int = None,
        repetition_penalty: float = 2.4,
        frequency_penalty: float = 0.6,
        no_repeat_ngram: int = 12,
        loop_patience: int = 1,
        anti_copy_indices: dict[int, dict] | None = None,
        anti_copy_max_n: int = 16,
        penalty_window: int = 256,
        loop_min_period: int = 8,
        loop_max_period: int = 64,
        self_anticopy_max_L: int = 64,
        self_anticopy_min_L: int = 6,
        # new sampling knobs
        typical_p: float = 0.95,
        top_k: int = 50,
    ):
        """
```

```
Strong repetition guardrails:
  - Windowed repetition & frequency penalties over
    (prompt+generated)
  - Whole-history no-repeat n-gram
  - Self bigram/trigram blocking
  - Dataset anti-copy (k-gram continuation ban)
  - Self anti-copy (ban continuing earlier in-context substrings)
  - Variable-period loop detector (8..64)
  - Typical sampling + top-k + top-p with robust fallbacks
"""
if stop_ids is not None and stop_tokens is None:
    stop_tokens = stop_ids
if max_new_tokens is not None and max_tokens is None:
    max_tokens = max_new_tokens
if stop_tokens is None:
    stop_tokens = []
if max_tokens is None:
    max_tokens = 50

device = self.device
tokens = torch.tensor([prompt_tokens], dtype=torch.long,
device=device)

generated: list[int] = []
loop_hits = 0

def _apply_windowed_penalties(logits_1v: torch.Tensor, hist:
list[int]):
    if not hist:
        return
    window = hist[-penalty_window:] if len(hist) > penalty_window
    else hist
    freq = defaultdict(int)
    for t in window:
        freq[t] += 1
```

```
    rp = float(repetition_penalty) if repetition_penalty else 1.0
    fp = float(frequency_penalty) if frequency_penalty else 0.0
    if rp != 1.0:
        for tid in freq.keys():
            logits_1v[tid] = logits_1v[tid] / rp
    if fp > 0.0:
        for tid, c in freq.items():
            if c > 0:
                logits_1v[tid] = logits_1v[tid] - fp * float(c)

def _ban_ids(logits: torch.Tensor, ids):
    if not ids:
        return
    logits[0, torch.tensor(list(ids), device=logits.device,
    dtype=torch.long)] = float("-inf")

def _no_repeat_ngram_ban(logits, hist, n):
    if n and len(hist) >= n - 1:
        plen = n - 1
        seen = defaultdict(set)
        for i in range(len(hist) - n + 1):
            prefix = tuple(hist[i:i + plen])
            nxt = hist[i + plen]
            seen[prefix].add(nxt)
        cur_prefix = tuple(hist[-plen:])
        _ban_ids(logits, seen.get(cur_prefix, ()))

def _self_ngram_ban(logits, hist):
    # Cheap bigram/trigram self-ban: prevent immediate echoing
    bans = set()
    if len(hist) >= 1:
        bans.add(hist[-1])  # avoid repeating the last token
        directly
    if len(hist) >= 2:
        # Ban the token that would recreate the last bigram
        # i.e., if ... x,y | ban any token z s.t. (y)->z seen after
          (...,y) earlier
```

```
        y = hist[-1]
        x = hist[-2]
        for i in range(len(hist) - 2):
            if hist[i] == x and hist[i + 1] == y and i + 2 <
            len(hist):
                bans.add(hist[i + 2])
    if len(hist) >= 3:
        # Ban the token that would recreate the last trigram
          continuation
        a, b, c = hist[-3], hist[-2], hist[-1]
        for i in range(len(hist) - 3):
            if hist[i] == a and hist[i + 1] == b and hist[i + 2] ==
            c and i + 3 < len(hist):
                bans.add(hist[i + 3])
    _ban_ids(logits, bans)

def _dataset_anticopy_ban(logits, hist):
    if not anti_copy_indices:
        return
    nmax = min(anti_copy_max_n, max(len(hist), 2))
    for k in range(nmax, 1, -1):
        plen = k - 1
        if len(hist) >= plen:
            prefix = tuple(hist[-plen:])
            idx = anti_copy_indices.get(k)
            if idx:
                _ban_ids(logits, idx.get(prefix, ()))

def _self_anticopy_ban(logits, hist):
    if len(hist) < self_anticopy_min_L + 1:
        return
    Lmax = min(self_anticopy_max_L, len(hist) - 1)
    for L in range(Lmax, self_anticopy_min_L - 1, -1):
        prefix = tuple(hist[-(L - 1):])
        end_for_scan = len(hist) - L + 1
        hit = False
```

```
        for i in range(0, end_for_scan):
            if tuple(hist[i:i + L - 1]) == prefix:
                _ban_ids(logits, [hist[i + L - 1]])
                hit = True
        if hit:
            break

def _variable_period_loop_hit(g, candidate_id):
    if len(g) < 2 * loop_min_period - 1:
        return False
    max_p = min(loop_max_period, len(g) + 1)
    for p in range(loop_min_period, max_p + 1):
        if len(g) >= 2 * p - 1:
            a = g[-(p - 1):] + [candidate_id]
            b = g[-(2 * p - 1):-(p - 1)]
            if a == b:
                return True
    return False

def _typical_topk_topp_sample(logits):
    # logits: [1, V]
    if temperature is None or temperature <= 0.0:
        return torch.argmax(logits, dim=-1, keepdim=True)

    l = logits.clone()
    l = l / float(temperature)
    probs = F.softmax(l, dim=-1)  # [1, V]

    # Typical sampling (nucleus around information content)
    with torch.no_grad():
        log_probs = torch.log(probs + 1e-12)
        ent = -(probs * log_probs).sum(dim=-1,
        keepdim=True)  # [1,1]
        dev = torch.abs(-log_probs - ent)  # [1,V]
        sorted_dev, sorted_idx = torch.sort(dev, dim=-1,
        descending=False)
        sorted_probs = probs.gather(-1, sorted_idx)
```

```
        cdf = torch.cumsum(sorted_probs, dim=-1)
        cut = (cdf > typical_p).to(sorted_probs.dtype)
        cut[..., 1:] = cut[..., :-1].clone()
        cut[..., 0] = 0.0
        typical_mask = torch.ones_like(sorted_probs)
        typical_mask[cut.bool()] = 0.0
        # restore original order
        mask = torch.zeros_like(probs)
        mask.scatter_(-1, sorted_idx, typical_mask)

    # Top-k
    if top_k is not None and top_k > 0:
        topk_vals, topk_idx = torch.topk(probs, k=min(top_k, probs.
        size(-1)), dim=-1)
        mask_topk = torch.zeros_like(probs)
        mask_topk.scatter_(-1, topk_idx, 1.0)
    else:
        mask_topk = torch.ones_like(probs)

    # Top-p
    if top_p is not None and top_p < 1.0:
        sp, si = torch.sort(probs, dim=-1, descending=True)
        sc = torch.cumsum(sp, dim=-1)
        sp_mask = (sc <= float(top_p)).to(probs.dtype)
        # keep at least one
        sp_mask[..., 0] = 1.0
        mask_topp = torch.zeros_like(probs)
        mask_topp.scatter_(-1, si, sp_mask)
    else:
        mask_topp = torch.ones_like(probs)

    # Combine masks: keep tokens that pass all filters
    keep = (mask > 0) & (mask_topk > 0) & (mask_topp > 0)
    filtered = probs * keep.to(probs.dtype)

    # Fallback if we masked everything (due to bans): relax to
      top_k only
```

```
        if filtered.sum() <= 0:
            filtered = probs * (mask_topk > 0).to(probs.dtype)
            if filtered.sum() <= 0:
                # last resort: use raw probs
                filtered = probs
        filtered = filtered / (filtered.sum(dim=-1, keepdim=True)
        + 1e-12)
        choice = torch.multinomial(filtered, num_samples=1)
        return choice
    for _ in range(max_tokens):
        with torch.no_grad():
            logits = self.model(tokens)[:, -1, :].to(torch.
            float32)  # [1,V]
            v = logits[0]
            full_hist = tokens[0].tolist()
            # 1) penalties over full history (windowed)
            _apply_windowed_penalties(v, full_hist)
            logits[0] = v

            # 2) whole-history no-repeat n-gram
            _no_repeat_ngram_ban(logits, full_hist, no_repeat_ngram)
            # 3) self bigram/trigram ban
            _self_ngram_ban(logits, full_hist)
            # 4) dataset anti-copy
            _dataset_anticopy_ban(logits, full_hist)
            # 5) self anti-copy
            _self_anticopy_ban(logits, full_hist)
            # 6) sample with typical + top-k + top-p (robust fallback)
            next_token = _typical_topk_topp_sample(logits)
        next_id = int(next_token.item())
        # 7) variable-period loop detector (before committing)
        if _variable_period_loop_hit(generated, next_id):
            loop_hits += 1
            if loop_hits >= loop_patience:
                break
```

```
    else:
        loop_hits = 0
    # commit
    generated.append(next_id)
    yield next_id
    tokens = torch.cat([tokens, next_token], dim=1)

    if next_id in stop_tokens:
        break
```

What Is TokenGenerator?

TokenGenerator is an inference engine that loads a trained language model and generates text token by token. After spending hours or days training a model, this class brings it to life by producing actual text. It takes a prompt (sequence of tokens) and autoregressively generates new tokens by repeatedly predicting the next most likely token, adding it to the sequence, and using the extended sequence to predict the next token.

Core Functionality

The class in the code above has two main methods:

`__init__(checkpoint_path, device)`: Loads the trained model from disk by reading the configuration JSON (which defines the architecture) and the model weights (the learned parameters), then puts the model in evaluation mode.

`generate(prompt_tokens, ...)`: Takes an initial sequence of tokens and generates new tokens one at a time, yielding each token as it's produced. This streaming approach allows displaying text as it's being generated rather than waiting for completion.

The Generation Process

The generation loop follows a simple pattern repeated for each new token:

1. **Forward Pass**: Feed the current token sequence through the model to get logits (raw scores) for every possible next token
2. **Apply Penalties**: Modify logits to discourage repetitive or unwanted tokens
3. **Sample**: Choose the next token from the probability distribution using temperature, top-p, and top-k sampling
4. **Check Stopping**: Stop if we hit a stop token, reach max length, or detect problematic loops
5. **Commit**: Add the chosen token to the sequence and repeat

Sophisticated Anti-repetition System

The most interesting aspect of this code is its extensive anti-repetition mechanisms. Language models, especially smaller ones, tend to get stuck in loops or repeat training data verbatim. This implementation includes multiple overlapping guardrails:

1. Windowed Repetition and Frequency Penalties

```
def _apply_windowed_penalties(logits_1v, hist):
    window = hist[-penalty_window:] if len(hist) > penalty_window else hist
    freq = defaultdict(int)
    for t in window:
        freq[t] += 1
```

Looks at the last 256 tokens (by default) and penalizes tokens that have appeared recently. The **repetition penalty** divides the logit score by a factor (2.4 by default), making repeated tokens less likely. The **frequency penalty** subtracts a value proportional to how many times the token appeared, with more frequent tokens getting stronger penalties.

2. No-Repeat N-gram Ban

```
def _no_repeat_ngram_ban(logits, hist, n):
    if n and len(hist) >= n - 1:
        # Track which tokens followed each n-1 gram
        seen = defaultdict(set)
        for i in range(len(hist) - n + 1):
            prefix = tuple(hist[i:i + plen])
            nxt = hist[i + plen]
            seen[prefix].add(nxt)
        # Ban tokens that would create a repeated n-gram
        cur_prefix = tuple(hist[-plen:])
        _ban_ids(logits, seen.get(cur_prefix, ()))
```

Completely bans any token that would create a repeated 12-gram (by default). If the sequence "the quick brown fox jumped over" has already appeared, and we currently have "the quick brown fox jumped," we absolutely cannot generate "over" next because it would repeat the exact 12-token sequence.

3. Self N-gram Ban (Bigram/Trigram Blocking)

```
def _self_ngram_ban(logits, hist):
    bans = set()
    if len(hist) >= 1:
        bans.add(hist[-1])  # Don't repeat the very last token
    if len(hist) >= 2:
        # Ban continuing the last bigram the same way it was
          continued before
        y = hist[-1]
        x = hist[-2]
        for i in range(len(hist) - 2):
            if hist[i] == x and hist[i + 1] == y and i + 2 < len(hist):
                bans.add(hist[i + 2])
```

A faster, cheaper check that prevents immediate token repetition and blocks recreating recently seen bigrams and trigrams. If we just generated "the quick," and earlier in the text we had "the quick brown," this prevents generating "brown" again immediately.

4. Dataset Anti-Copy Ban

```
def _dataset_anticopy_ban(logits, hist):
    if not anti_copy_indices:
        return
    nmax = min(anti_copy_max_n, max(len(hist), 2))
    for k in range(nmax, 1, -1):
        plen = k - 1
        if len(hist) >= plen:
            prefix = tuple(hist[-plen:])
            idx = anti_copy_indices.get(k)
            if idx:
                _ban_ids(logits, idx.get(prefix, ()))
```

Uses a precomputed index of n-grams from the training data to prevent the model from copying training examples verbatim. If the last several generated tokens exactly match a sequence from the training data, this bans the continuation tokens that would extend that copied sequence. This prevents plagiarism and ensures the model generates novel text rather than memorizing training data.

5. Self Anti-Copy Ban

```
def _self_anticopy_ban(logits, hist):
    if len(hist) < self_anticopy_min_L + 1:
        return
    Lmax = min(self_anticopy_max_L, len(hist) - 1)
    for L in range(Lmax, self_anticopy_min_L - 1, -1):
        prefix = tuple(hist[-(L - 1):])
        # Search for this prefix earlier in history
        for i in range(0, end_for_scan):
            if tuple(hist[i:i + L - 1]) == prefix:
                # Ban the token that followed it before
                _ban_ids(logits, [hist[i + L - 1]])
```

Prevents the model from copying earlier parts of its own generation. If the model generated a long phrase earlier in the current generation and now starts generating that same phrase again, this detects the match and bans continuing it the same way. This stops the model from quoting itself and creating circular loops.

6. Variable-Period Loop Detection

```
def _variable_period_loop_hit(g, candidate_id):
    max_p = min(loop_max_period, len(g) + 1)
    for p in range(loop_min_period, max_p + 1):
        if len(g) >= 2 * p - 1:
            a = g[-(p - 1):] + [candidate_id]
            b = g[-(2 * p - 1):-(p - 1)]
            if a == b:
                return True
    return False
```

Detects when the model falls into repeating loops of various lengths (8 to 64 tokens by default). If adding the candidate token would create an exact repetition of the previous cycle, it's flagged as a loop. After detecting loops multiple times (controlled by `loop_patience`), generation stops entirely to avoid producing garbage.

Advanced Sampling Strategy

After applying all the penalties and bans, the code uses a sophisticated three-layer sampling approach:

Typical Sampling

```
log_probs = torch.log(probs + 1e-12)
ent = -(probs * log_probs).sum(dim=-1, keepdim=True)
dev = torch.abs(-log_probs - ent)
```

Filters tokens based on their "typicality"—how close their information content is to the expected entropy. This keeps tokens that are neither too surprising nor too obvious, resulting in more natural-feeling text. Typical sampling often outperforms pure top-p by selecting tokens that match the expected information density.

Top-K Sampling

```
topk_vals, topk_idx = torch.topk(probs, k=min(top_k, probs.
size(-1)), dim=-1)
```

Keeps only the top 50 (by default) most probable tokens, setting all others to zero probability. This prevents the model from occasionally selecting extremely unlikely tokens that would derail the generation.

Top-P (Nucleus) Sampling

```
sp, si = torch.sort(probs, dim=-1, descending=True)
sc = torch.cumsum(sp, dim=-1)
sp_mask = (sc <= float(top_p)).to(probs.dtype)
```

Keeps the smallest set of tokens whose cumulative probability exceeds the threshold (0.85 by default). This is dynamic—when the model is confident, fewer tokens are kept; when uncertain, more tokens are considered. This adapts to the model's confidence at each step.

Robust Fallbacks

```
keep = (mask > 0) & (mask_topk > 0) & (mask_topp > 0)
filtered = probs * keep.to(probs.dtype)

if filtered.sum() <= 0:
    filtered = probs * (mask_topk > 0).to(probs.dtype)
    if filtered.sum() <= 0:
        filtered = probs
```

The code applies all three filters simultaneously—tokens must pass typical, top-k, AND top-p to be eligible. However, if all these filters combined with the penalty/ban systems eliminate every possible token (which can happen with aggressive settings), it falls back to less restrictive conditions. First it tries just top-k, then if that's still empty, it uses raw probabilities. This ensures the generation never gets completely stuck.

Temperature Control

```
l = logits.clone()
l = l / float(temperature)
probs = F.softmax(l, dim=-1)
```

Temperature (0.6 by default) controls randomness. The logits are divided by temperature before converting to probabilities. Lower temperatures (< 1.0) make the model more confident and deterministic by sharpening the distribution. Higher temperatures (> 1.0) flatten the distribution, making outputs more random and creative. A temperature of 0 would make it completely deterministic (always picking the highest probability token).

Why So Many Guardrails?

Language models have a tendency to fall into repetitive patterns, especially when:

- The model is small or undertrained
- The prompt is short or ambiguous
- Generation continues for many tokens
- Training data contained repetitive text

Each guardrail catches a different failure mode. The windowed penalties handle short-term repetition. The n-gram bans catch medium-term verbatim copying. The loop detector catches long-term cycles. The dataset anti-copy prevents plagiarism. Together, they create a robust system that generates fluent, non-repetitive text even when the underlying model has a tendency toward repetition.

Practical Usage

```
# Load the model
generator = TokenGenerator("chkpt", device="cuda")
# Generate from a prompt
prompt_tokens = tokenizer.encode("Once upon a time")
for token_id in generator.generate(
    prompt_tokens,
    max_tokens=100,
    temperature=0.7,
    top_p=0.9
):
```

```
# Decode and display each token as it's generated
token_text = tokenizer.decode([token_id])
print(token_text, end="", flush=True)
```

The generator yields tokens one at a time, enabling streaming output. This provides a better user experience since text appears progressively rather than all at once after a long wait.

Key Differences from Training

Training and inference are fundamentally different processes:

- **Training** adjusts model parameters to minimize prediction error across the entire dataset. It uses teacher forcing (always feeding the correct previous tokens) and processes many examples in parallel batches.
- **Inference** uses the frozen trained parameters to generate new text. It uses autoregressive generation (feeding the model's own predictions as input for the next step) and typically processes one sequence at a time. The challenge shifts from learning patterns to avoiding degenerate behaviors like repetition.

This TokenGenerator implementation represents production-grade inference with extensive anti-repetition mechanisms rarely seen in simple examples, making it suitable for practical applications where output quality matters.

TokenGenerator loads a trained model and generates text token-by-token using sophisticated sampling (typical + top-k + top-p with temperature control) and extensive anti-repetition guardrails (windowed penalties, n-gram bans, loop detection, dataset anti-copy, and self anti-copy). This multilayered approach ensures fluent, diverse, non-repetitive output even from smaller models prone to degenerate behaviors. The streaming generation approach (yielding tokens one at a time) enables real-time display as text is being created.

The Big Moment—Prompting Our Model

```
# run_prompt.py
import torch
from collections import defaultdict  # <-- needed for anti-copy builder
DEVICE = "cuda" if torch.cuda.is_available() else "cpu"
CHECKPOINT_DIR = "chkpt"
TOKENIZER_PREFIX = "bpe_tokenizer"

# --- load tokenizer ---
tok = BPETokenizer()
tok.load(TOKENIZER_PREFIX)

# --- build generator from checkpoint ---
gen = TokenGenerator(checkpoint_path=CHECKPOINT_DIR, device=DEVICE)

def build_anti_copy_indices(tokenizer, txt_path: str, n_max: int = 12):
    """
    Returns a dict: k -> { tuple(prefix_len=k-1): set(next_token_ids) }
    for k = 2..n_max. Use larger k to allow style but block verbatim
    copying.
    """
    with open(txt_path, "r", encoding="utf-8") as f:
        raw = f.read()
    ids = tokenizer.encode(raw, add_bos_eos=False)

    indices = {k: defaultdict(set) for k in range(2, n_max + 1)}
    for k in range(2, n_max + 1):
        plen = k - 1
        for i in range(len(ids) - k + 1):
            prefix = tuple(ids[i:i + plen])
            nxt = ids[i + plen]
            indices[k][prefix].add(nxt)
    return indices

BOOK_PATH = "book.txt"
ANTI_N_MAX = 12  # was 7; longer BPE n-grams reduce verbatim copying
```

```
anti_copy_indices = build_anti_copy_indices(tok, BOOK_PATH, n_
max=ANTI_N_MAX)

def clean_bpe_text(text: str) -> str:
    # Map common BPE markers back to plain whitespace
    text = text.replace("Ġ", " ").replace("Ċ", "\n")  # RoBERTa-style
    text = text.replace("▁", " ")  # SentencePiece-style
    return text.strip()

def _detect_variable_period_loop(ids: list[int], min_p: int = 8, max_p: int
= 64) -> bool:
    """Returns True if the tail looks like it's repeating with some period
    in [min_p, max_p]."""
    n = len(ids)
    if n < 2 * min_p:
        return False
    max_p = min(max_p, n // 2)
    # Compare the last (p-1) with the previous (p-1) shifted window
      for any p
    for p in range(min_p, max_p + 1):
        # Need at least 2*p tokens to compare two consecutive blocks
        if n >= 2 * p:
            if ids[-p:] == ids[-2 * p:-p]:
                return True
    return False

def generate_text(
    prompt: str,
    *,
    max_new_tokens: int = 30,
    temperature: float = 0.3,         # ↑ helps escape local attractors
    top_p: float = 0.85,
    repetition_penalty: float = 3.4, # ↑ stronger deterrent
    frequency_penalty: float = 2.5,  # ↑ penalize frequent tokens more
) -> str:
    prompt_ids = tok.encode(prompt, add_bos_eos=False)
    out_ids: list[int] = []
```

```
    # We'll iterate the generator but stop early if we detect repetition
    for tid in gen.generate(
        prompt_tokens=prompt_ids,
        max_new_tokens=max_new_tokens,
        temperature=temperature,
        top_p=top_p,
        repetition_penalty=repetition_penalty,
        frequency_penalty=frequency_penalty,
        no_repeat_ngram=4,                # was 4; use longer n-gram
        anti_copy_indices=anti_copy_indices,
        anti_copy_max_n=ANTI_N_MAX,
        stop_ids=[tok.eos_id],
    ):
        out_ids.append(tid)

        # 1) Variable-period loop stop (periods 8..64)
        if _detect_variable_period_loop(out_ids, min_p=8, max_p=64):
            break

        # 2) Exact block repeat stop: last 40 equals the 40 before it
        if len(out_ids) >= 80:
            if out_ids[-40:] == out_ids[-80:-40]:
                break

    raw = tok.decode(out_ids)
    return clean_bpe_text(raw)

if __name__ == "__main__":
    print("Type your prompt (Ctrl+C to quit):\n")
    while True:
        try:
            user_prompt = input("> ")
            prompt = "### Instruction:\n" + user_prompt + "\n\n###
            Response:\n"
            completion = generate_text(prompt)
            print("\n=== Completion ===\n" + completion + "\n")
```

```
    except (KeyboardInterrupt, EOFError):
        print("\nbye!")
        break
```

Output:

```
> who is rick

=== Completion ===
weekI HAD always thought Jack Gisburn rather a cheap genius—though a good
fellow enough—so it was no great surprise to me to hear that, in the height
of his glory, he had dropped his painting, married a rich widow, and
established himself in a villa on the Riviera.  (Though I rather thought it
would have been Rome or Florence. )
"The height of his glory"—that was what the women called it.  I can hear
Mrs.  Gideon Thwing—his last Chicago sitter—deploring his unaccountable
abdication.  "Of course it's going to send the value of my picture 'way up;
but I don't think of that, Mr.  Rickham—the loss to Arrt is all I think
of. " The word, on Mrs.  Thwing's lips, multiplied its _rs_ as though they
were reflected in an endless vista of mirrors.  And it was not only the
Mrs. Thwings who mourned.  Had not the exquisite Hermia Croft, at the last
Grafton Gallery show, stopped me before Gisburn's "Moon-dancers"suchnot
to  of ak I HAD always thought Jack Gisburn rather a cheap genius—though a
good fellow enough—so it was no great surprise to me to hear that, in the
height of his glory, he had dropped his painting, married a rich widow, and
established himself in a villa on the Riviera.  (Though I rather thought it
would have been Rome or Florence. )
"The height of his glory"—that was what the women called it.  I can hear
Mrs.  Gideon Thwing—his last Chicago sitter—deploring his unaccountable
abdication.  "Of course it's going to send the value of my picture 'way up;
but I don't think of that, Mr.  Rickham—the loss to Arrt is all I think
of. " The word, on Mrs.  Thwing's lips, multiplied its _rs_ as though they
were reflected in an endless vista of mirrors.  And it was not only the
Mrs.  Thwings who mourned.  Had not the exquisite Hermia Croft, at the last
Grafton Gallery show, stopped me before Gisburn's "Moon-dancers" to say,
with tears in her eyes: "We shall not look upon its like again"?
Well!—even through the prism of Hermia's tears I felt able to alitmohad to
here-clI've againRivierTjealpectght swet,  toqw backy of sent againessed in
bye!
```

Understanding Inference Code and Model Behavior

This script above creates an interactive interface for text generation with sophisticated anti-repetition mechanisms. It's an excellent example of production-grade inference techniques applied to a small educational model.

Let's explore how each component works.

Setup and Initialization

```
DEVICE = "cuda" if torch.cuda.is_available() else "cpu"
tok = BPETokenizer()
tok.load(TOKENIZER_PREFIX)
gen = TokenGenerator(checkpoint_path=CHECKPOINT_DIR, device=DEVICE)
```

The script loads the trained BPE tokenizer and initializes the TokenGenerator with the saved model checkpoint, setting up a complete inference pipeline ready for text generation.

Building Anti-copy Indices

```
def build_anti_copy_indices(tokenizer, txt_path: str, n_max: int = 12):
    with open(txt_path, "r", encoding="utf-8") as f:
        raw = f.read()
    ids = tokenizer.encode(raw, add_bos_eos=False)

    indices = {k: defaultdict(set) for k in range(2, n_max + 1)}
    for k in range(2, n_max + 1):
        plen = k - 1
        for i in range(len(ids) - k + 1):
            prefix = tuple(ids[i:i + plen])
            nxt = ids[i + plen]
            indices[k][prefix].add(nxt)
    return indices
```

This function implements a sophisticated technique used in production systems: building an index of all n-grams from the training data. For sequences of 2 to 12 tokens, it records which tokens followed each prefix in the training text. This enables the dataset anti-copy mechanism that prevents verbatim reproduction of training data.

The clever insight: If the model generates 11 consecutive tokens that exactly match a training sequence, the index allows banning the 12th token that would continue that exact copy. This pushes the model toward novel generation rather than memorization. Using n_max=12 provides strong protection while still allowing the model to use learned patterns and style.

Text Cleaning Utilities

```
def clean_bpe_text(text: str) -> str:
    text = text.replace("Ġ", " ").replace("Ċ", "\n")
    text = text.replace("▁", " ")
    return text.strip()
```

BPE tokenizers use special characters to encode spaces and newlines ("Ġ" for spaces in RoBERTa-style, "▁" in SentencePiece-style, "Ċ" for newlines). This utility converts these markers back to actual whitespace for readable output.

Loop Detection

```
def _detect_variable_period_loop(ids: list[int], min_p: int = 8, max_p: int
= 64) -> bool:
    n = len(ids)
    if n < 2 * min_p:
        return False
    max_p = min(max_p, n // 2)
    for p in range(min_p, max_p + 1):
        if n >= 2 * p:
            if ids[-p:] == ids[-2 * p:-p]:
                return True
    return False
```

This implements variable-period loop detection, checking whether the last p tokens exactly match the previous p tokens for any period between 8 and 64. This catches repetitive patterns at multiple scales, from short cycles to longer repeated blocks. It's a safety mechanism that production systems use to detect when generation has become degenerate.

Generation Function with Sophisticated Controls

```
def generate_text(
    prompt: str,
    max_new_tokens: int = 30,
    temperature: float = 0.3,
    top_p: float = 0.85,
    repetition_penalty: float = 3.4,
    frequency_penalty: float = 2.5,
):
```

This function demonstrates how to layer multiple control mechanisms for generation quality. The parameters are intentionally aggressive to explore the boundaries of what different penalty settings can achieve, which is valuable for understanding model behavior.

The function includes multiple stopping conditions beyond the generator's internal checks, creating a robust safety net that prevents runaway generation.

Interactive Interface

```
if __name__ == "__main__":
    print("Type your prompt (Ctrl+C to quit):\n")
    while True:
        user_prompt = input("> ")
        prompt = "### Instruction:\n" + user_prompt + "\n\n### Response:\n"
        completion = generate_text(prompt)
        print("\n=== Completion ===\n" + completion + "\n")
```

The script creates a user-friendly interactive loop with instruction formatting, demonstrating how prompts might be structured for instruction-following models.

Understanding the Output—An Educational Demonstration

Let's examine what the model produces when given the prompt "who is rick." This output is **extremely instructive** because it reveals fundamental principles about how language models work and why certain design choices matter in production systems.

The User Query

> who is rick

A straightforward question asking for information about someone named "Rick."

The Generated Output—Three Revealing Behaviors

The output demonstrates three distinct phases that teach us important lessons about model training and capabilities:

Phase 1: Accessing Learned Knowledge

=== Completion ===

weekI HAD always thought Jack Gisburn rather a cheap genius—though a good fellow enough—so it was no great surprise to me to hear that, in the height of his glory, he had dropped his painting, married a rich widow, and established himself in a villa on the Riviera...

What we observe: The model begins generating text from the opening of its training data (the Edith Wharton short story), with a slight prefix variation.

What this teaches us:

This behavior beautifully demonstrates how neural networks work at a fundamental level. When a model encounters a prompt outside its training distribution, it navigates to the nearest learned attractor in its probability space. The model has developed strong internal representations for this particular story through training, creating what we call an "attractor basin"—a region of probability space where the model naturally gravitates.

Note This is actually evidence that the model **has learned something. It's not producing random noise; it's producing coherent, well-formed text that demonstrates successful learning of language structure, syntax, and narrative flow. The model knows how to generate fluent English prose with proper punctuation, capitalization, and stylistic consistency.**

The lesson about model capacity Small models with limited training data naturally develop strong preferences for their training distribution. This isn't a flaw—**it's expected behavior that demonstrates the model is working correctly within its design constraints.**

Phase 2: Pattern Repetition Under Constraints

...stopped me before Gisburn's "Moon-dancers"suchnot to of ak I HAD always thought Jack Gisburn rather a cheap genius—though a good fellow enough—so it was no great surprise to me to hear that, in the height of his glory...

What we observe: After generating the story's opening, the model encounters anti-repetition mechanisms, produces a few transitional tokens ("suchnot to of ak"), then begins the same passage again.

What this teaches us:

This demonstrates the interaction between learned probability distributions and inference-time constraints. The model's training created strong statistical preferences for certain sequences. When anti-copy and n-gram bans block forward continuation, the model searches its learned space for alternative paths. Finding limited alternatives (due to the restricted training), it cycles back to familiar patterns.

Note This behavior reveals an important principle: **inference mechanisms can guide but not fundamentally change what a model has learned**. The penalties and bans are like guardrails that redirect the model's generation, but they can't create knowledge or capabilities that weren't learned during training.

The lesson about training data diversity: This illustrates why production models train on vast, diverse corpora. With exposure to millions of different text patterns, large models develop many alternative paths through their probability space, enabling them to gracefully handle diverse prompts.

Phase 3: Exploring Low-Probability Space

...stopped me before Gisburn's "Moon-dancers" to say, with tears in her eyes: "We shall not look upon its like again"? Well!—even through the prism of Hermia's tears I felt able to alitmohad to here-clI've againRivierTjealpectght swet, toqw backy of sent againessed in

What we observe: After the second iteration, the output transitions into sequences that appear more random: "alitmohad to here-clI've againRivierTjealpectght swet..."

What this teaches us:

Note **This final phase is perhaps the most educationally valuable**. It shows us what happens when a model is pushed into regions of its probability space that received little training. The aggressive penalties (3.4 for repetition, 2.5 for frequency) have heavily penalized the high-probability tokens the model prefers. This forces the sampling mechanism to select from low-probability tokens—ones the model considers unlikely and rarely generates.

This demonstrates a crucial insight about language models: they learn probability distributions, not rules. When we sample from the tail of these distributions (the unlikely tokens), we get outputs that reveal the boundaries of the learned space. The model hasn't learned strong patterns connecting these rare tokens, so the output becomes less coherent.

The lesson about hyperparameter tuning: This shows why production systems use carefully tuned penalty values (typically in the 1.0–1.5 range) rather than aggressive ones. The goal is to gently guide the model while preserving its learned probability structure, not to completely override it.

What This Demonstrates About Language Model Design

This output is a perfect educational example because it illuminates several fundamental principles that explain why production LLMs are designed the way they are.

Principle 1: Model Scale and Generalization

Our educational model has 2 layers with 128-dimensional hidden states and approximately 100,000 total parameters. ***This is intentionally small for learning purposes and fast training on limited hardware.***

Why production models are massive:

Production LLMs like GPT-4, Claude, or Llama have 20-100+ layers, 2048-12288 dimensional hidden states, and billions to trillions of parameters. This enormous scale serves critical purposes:

- **Learning capacity for abstraction**: Larger models can learn abstract patterns that generalize across contexts. Instead of memorizing specific sequences, they learn the underlying rules of language, reasoning patterns, and knowledge structures. With 2 layers, our model primarily learns to reproduce what it saw; with 50 layers, a model can learn to understand and reason.
- **Diverse knowledge representation**: Production models allocate different parameters to different types of knowledge. Some neurons might specialize in syntax, others in factual knowledge, and others in reasoning patterns. This specialization requires vast parameter counts.
- **Robustness to distribution shift**: Larger models interpolate better between training examples, handling prompts that don't exactly match training data. Our small model has limited interpolation capability, so it stays close to what it has seen. Large models smoothly navigate between different types of requests.

- **Emergent capabilities**: Research shows that certain capabilities only emerge at scale. Chain-of-thought reasoning, in-context learning, and instruction following improve dramatically as models grow. These aren't just "more of the same"—they're qualitatively new abilities that appear with sufficient scale.

Principle 2: Training Data Diversity

Our model trains on a single 4,200-word short story. This is perfect for educational purposes—it trains quickly and demonstrates core concepts clearly.

Why production models train on trillions of tokens:

Production training corpora contain

- Books (fiction and non-fiction across all genres)
- Web pages (covering every imaginable topic)
- Scientific papers (providing technical knowledge)
- Code repositories (teaching logical reasoning and programming)
- Conversations (demonstrating dialogue patterns)
- News articles (providing factual information)
- Wikipedia (structured knowledge)
- And much more across hundreds of languages

The purpose of this diversity:

- **Universal pattern learning**: Seeing "question → answer" patterns millions of times across countless topics teaches the model the general structure of question-answering, not just specific examples.
- **Handling any prompt**: When users ask about anything from quantum physics to cooking recipes to philosophical questions, the model needs relevant training examples. Diverse training ensures the model has seen related content.

- **Reducing memorization**: With trillions of training tokens, each specific sequence appears so rarely that the model must learn patterns rather than memorize specifics. Our model sees the same story repeatedly, naturally leading to memorization.
- **Balanced capabilities**: Training on diverse data ensures the model isn't just good at one thing. Production models need to handle code, creative writing, analysis, mathematics, and conversation equally well.

Principle 3: Training Duration and Compute

Our model trains for 20 epochs on a dataset of about 4,000 tokens, completing in minutes on a single GPU. This enables rapid experimentation and learning.

Why production training takes months on thousands of GPUs:

Large-scale training requires

- **Multiple Passes over Vast Data**: Even with trillions of tokens, models benefit from seeing data multiple times (though typically 1-2 epochs, not 20). With massive datasets, even a single epoch takes enormous compute.
- **Gradual Refinement**: Language understanding improves gradually. Early training learns basic syntax and common patterns. Middle training develops semantic understanding and knowledge. Late training refines subtle reasoning and edge cases. This progression requires extended training.
- **Optimization at Scale**: Large models have complex loss landscapes with many local minima. Finding good solutions requires millions of optimization steps with careful learning rate scheduling. Our tiny model converges quickly; massive models need millions of steps to fully optimize.
- **Quality over Speed**: Production models prioritize final quality over training speed. Even small improvements in capability justify substantial compute investment when millions of users will interact with the model.

Principle 4: Instruction Fine-Tuning and Alignment

Our model only does pretraining—learning to predict next tokens from the story text. This is the first stage of training.

Why production models add additional training phases:

After pretraining on general text, production models undergo

- **Instruction Fine-Tuning**: Training on thousands of examples showing how to respond to instructions and questions. This teaches the pattern: "When given an instruction or question, provide a relevant, helpful response" rather than just continuing the text. Our model never learned this pattern, so it doesn't attempt to answer "who is rick"—it just generates text.
- **RLHF or Preference Learning**: Further training using human feedback about which responses are better. This aligns the model to human preferences for helpfulness, harmlessness, and honesty. The model learns not just to generate fluent text but to be a helpful assistant.
- **Safety Training**: Specific training to decline harmful requests, acknowledge uncertainty, and avoid problematic outputs. This creates the distinction between a language model (which just predicts text) and an AI assistant (which tries to be helpful and safe).
- **Format Learning**: Teaching the model to understand instruction formats, use tools, follow system messages, and engage in multiturn conversations appropriately.

The Value of This Demonstration

This output is **highly educational** precisely because it shows what happens without these production-level investments. By seeing how a small, simply trained model behaves, we understand the following.

What Pretraining Alone Achieves

The model successfully learned:

- Fluent English generation with proper grammar and syntax
- Punctuation, capitalization, and formatting conventions
- Narrative flow and stylistic consistency
- The ability to generate coherent multisentence passages
- Character names, dialogue patterns, and story structure

These are genuine accomplishments that demonstrate successful learning from data.

What Requires Additional Scale and Training

The model doesn't yet have:

- Ability to interpret and respond to questions appropriately
- Knowledge beyond its training data
- Understanding of instruction formats
- Graceful handling of out-of-distribution prompts
- Diverse generation strategies for different contexts

These capabilities require the scale and training procedures used in production systems.

How Inference Mechanisms Work

The output shows us:

- How penalty mechanisms influence token selection
- Where anti-copy and loop detection trigger
- How models navigate when constrained by bans
- The relationship between learned distributions and sampling strategies

This understanding helps us design better inference systems for larger models.

Practical Lessons from This Example

This demonstration teaches us valuable lessons applicable to real projects:

Lesson 1: Match Model Scale to Task Complexity

For simple, narrow tasks (like generating text in a specific style from one domain), smaller models work well. For general-purpose question-answering and assistance across all topics, massive models are appropriate. Our model demonstrates what "too small for general purposes" looks like, helping us understand where the complexity thresholds lie.

Lesson 2: Inference Constraints Have Limits

No amount of penalties, bans, or sampling tricks can create capabilities the model hasn't learned. These mechanisms work best for **polishing and guiding** an already-capable model, not compensating for training deficiencies. This understanding helps us focus training efforts appropriately rather than relying on inference-time fixes.

Lesson 3: Training Data Determines Capabilities

A model can only learn what exists in its training data. Our single-story model can generate literary prose in that style. For question-answering, coding assistance, mathematical reasoning, and multidomain knowledge, training must include relevant examples. This explains why production training corpora are so carefully curated and comprehensive.

Lesson 4: Hyperparameter Tuning Matters

The aggressive penalties (3.4, 2.5) in our example demonstrate how extreme values can push models into unnatural behavior. Production systems carefully tune these parameters (typically 1.0-1.5) to gently guide without breaking learned distributions. This teaches us the importance of empirical testing and moderation in hyperparameter selection.

Lesson 5: Multistage Training Is Essential

Our model only did pretraining. The output shows why instruction fine-tuning and alignment training aren't optional extras—they're fundamental to creating useful AI assistants. A pretrained language model predicts text; an instruction-tuned model responds helpfully. This distinction is crucial for practical applications.

Exploring Even Further—A Real Trained Model: TinyStories GPT-4 Version Implementation

While this book guides you through building and training a language model from scratch on a single text file, seeing a more capable model trained on a larger, diverse dataset provides valuable context for understanding what's possible at slightly larger scales.

TinyStories Dataset and Model

The TinyStories dataset, created by researchers at Microsoft, consists of short stories written in simple language suitable for young children. This dataset provides an excellent middle ground between our single-book training approach and massive production datasets—it's small enough to train on consumer hardware but large enough to demonstrate more sophisticated language generation capabilities.

An improved implementation trained on the TinyStories dataset is available at:

- **Code Repository**: `https://github.com/didogrigorov/LLM-story/`
- **Live Demo**: `https://nanostorychat.com/`

What Makes This Implementation Different

This TinyStories-trained model demonstrates several advantages over our minimal example:

- **Dataset Diversity**: Trained on thousands of short stories rather than a single text, enabling more varied and creative outputs

- **Better Generalization**: The model learns story structure, character development, and narrative patterns that transfer across different story contexts
- **Interactive Experience**: The live demo lets you experiment with prompts and see real-time generation, providing an intuitive understanding of how language models respond to different inputs
- **Larger Scale**: While still modest by production standards, the model uses more parameters and longer training, showing the quality improvements that come with increased compute investment

Learning from the Implementation

Examining this trained model helps bridge the gap between our educational example and production systems. You can:

1. Compare generation quality between our single-book model and the TinyStories model
2. Experiment with the live demo to understand model behavior and limitations
3. Study the repository code to see optimizations and architectural choices for slightly larger-scale training
4. Understand how dataset composition (diverse stories vs. single text) affects model capabilities

The TinyStories approach demonstrates that remarkable language generation is achievable without massive computational resources—a trained model producing coherent short stories can run on a laptop, making advanced AI accessible for learning, experimentation, and small-scale applications.

Key Technical Differences from the Book's Code

The repository implementation includes several production-grade enhancements compared to our educational example:

Architecture and Scale

The production model uses a significantly more sophisticated architecture with approximately 250 million parameters compared to our minimal example. It features 16 transformer layers with 1024-dimensional hidden states and 4096-dimensional feed-forward networks. The architecture implements Grouped Query Attention with 8 key-value heads for memory efficiency, Rotary Position Embeddings for better position encoding, and gradient checkpointing to enable training larger models within memory constraints.

Tokenization

Rather than our simple 2,000-token vocabulary, the production implementation uses a 50,000-token vocabulary trained with a HuggingFace tokenizers library implementation. It employs GPT-2 style pretokenization with proper word boundary handling and ranked BPE merging that matches GPT behavior. The tokenizer includes sophisticated special token handling with multiple fallback options and proper byte-level encoding.

Dataset Management

The production code handles multiple documents separated by special delimiters rather than treating the entire corpus as a single sequence. It implements a streaming dataset that doesn't require loading everything into memory, with four different document splitting modes for flexibility. The dataset is DDP-aware with proper sharding across multiple GPUs and handles per-document BOS and EOS token insertion correctly.

Distributed Training Infrastructure

Unlike our single-GPU example, the production implementation fully supports Distributed Data Parallel training across multiple GPUs using the NCCL backend. Training is coordinated through environment variables with proper rank management, and only the master process handles logging and checkpointing. The training budget is specified in tokens rather than epochs, which scales better to large datasets.

Advanced Training Features

The implementation includes exponential moving average of model weights with 0.9995 decay for more stable final models. It uses the fused AdamW optimizer for faster execution and implements proper gradient accumulation with micro-batching to enable large effective batch sizes. Training includes periodic validation runs during training, not just at epoch boundaries; comprehensive checkpoint management with optimizer and scaler state for resuming interrupted runs; and running loss tracking with exponential moving average for smoother metrics.

Configuration Management

Rather than hardcoded constants, the production code uses dataclass-based modular configuration with separate configs for tokenizer, data loading, model architecture, and training hyperparameters. This makes it easy to experiment with different settings and ensures consistency across training runs.

Generation and Inference

The repository includes a complete generation system with streaming text generation that outputs tokens as they're produced, top-p nucleus sampling for controlled randomness, a repetition penalty to reduce repetitive outputs, and n-gram blocking to prevent the model from simply copying the prompt. It features an echo guard that prevents repeating the last prompt token and a FastAPI web server with Server-Sent Events for real-time streaming in the browser. The generation system preferentially loads EMA weights when available and supports BF16 autocast for efficient inference.

Data Loading Strategy

The production code loads documents individually with delimiter awareness rather than concatenating everything into one sequence. It uses streaming with on-the-fly tokenization instead of pretokenizing and caching the entire dataset. The system supports multiple document splitting modes, including line-based, blank-line-separated paragraphs, custom delimiter tokens like the TinyStories separator, and treating the entire file as one document.

Training Metrics and Monitoring

The implementation provides much richer feedback than our simple epoch-level loss reporting. It displays per-step metrics including loss, perplexity, learning rate, and tokens processed per second. Separate validation runs occur periodically during training and at epoch boundaries, giving ongoing insight into generalization. The system tracks both training and validation perplexity with proper clamping to prevent numerical issues.

Optimization Details

The production code enables TensorFloat-32 precision for matrix multiplication and convolution operations, providing a good balance of speed and accuracy. Weight decay regularization is enabled at 0.1 to prevent overfitting. The learning rate schedule operates on a token basis rather than steps, properly accounting for the actual amount of data processed. Gradient clipping prevents training instabilities that become more common with larger models.

Checkpoint Management

Unlike our simple end-of-training save, the production system saves comprehensive checkpoints periodically during training every 10,000 steps. Checkpoints include model state, optimizer state, gradient scaler state for mixed precision training, current training step counter, and EMA shadow weights when enabled. The system supports resuming from any checkpoint with full training state restoration, enabling recovery from hardware failures or intentional training interruptions.

Web Deployment

The repository includes a complete FastAPI server for deploying the model as an interactive web service. It implements prompt-safe streaming that never reprints the user's input prompt, uses a queue-based system to capture model output and stream it to clients, wraps output in Server-Sent Events format for browser compatibility, and includes HTML templates for a user-friendly interface. This transforms the raw model into an accessible application that anyone can interact with through their browser.

These enhancements collectively demonstrate the path from educational toy example to production-ready system. While our book's implementation focuses on clarity and understanding fundamental concepts, the repository shows how those concepts scale and what additional infrastructure is needed for real-world deployment. The progression from our 2-layer, 128-dimensional model trained on a single text file to this 16-layer, 1024-dimensional model trained on thousands of stories illustrates both the continuity of core concepts and the engineering required to realize their full potential.

Why This Implementation Wasn't Included in the Book

This more sophisticated implementation was intentionally not included as the primary teaching example in the book for important practical reasons related to computational costs and accessibility.

The Reality of Training Costs

Even on a relatively small dataset like TinyStories, training a production-quality language model involves substantial financial investment. The 250-million-parameter model in the repository, when trained properly on the full TinyStories dataset with multiple GPUs, can cost between $500 and $10,000 in cloud computing expenses depending on the hardware used and training duration.

Training on high-end GPUs like NVIDIA H100s or A100s, which are necessary for reasonable training times, costs $2–4 per GPU-hour on major cloud platforms. A complete training run spanning several days across multiple GPUs quickly accumulates significant costs.

For comparison, our book's minimal example can train on a CPU or consumer GPU in under an hour at essentially no additional cost beyond the hardware you already own. This democratizes the learning experience—anyone with a laptop can follow along and see results without opening their wallet.

Accessibility and Learning Goals

The book's primary goal is teaching fundamental concepts, not producing state-of-the-art results. By keeping the example small and fast to train, we ensure that:

- **Students can experiment freely** without worrying about costs mounting with each training run

- **Iteration and debugging are practical** when each training cycle takes minutes, not days
- **The focus remains on understanding** rather than managing infrastructure and monitoring long-running jobs
- **Hardware requirements are minimal** so the barrier to entry is as low as possible

The Scaling Gradient

The progression from our book's example to the TinyStories implementation illustrates an important reality about language model development: there's a steep computational gradient between educational toys and production systems. Our 2-layer model with 128-dimensional embeddings requires roughly 100,000 times less computation to train than the 250-million-parameter TinyStories model. This scaling factor directly translates to cost—what runs in an hour for free on your laptop becomes days of expensive GPU time in the cloud.

This isn't just about money. The complexity of managing distributed training, monitoring metrics across long runs, debugging failures that appear after hours of training, and optimizing performance to reduce costs all represent additional cognitive overhead that distracts from learning core concepts.

The Value Proposition

By providing the repository link and live demo, we offer the best of both worlds:

- **Learn with the simple version** where you can see every detail, modify parameters freely, and develop intuition without consequences
- **Experience the capable version** through the live demo to understand what becomes possible with more resources
- **Study the production code** when you're ready to understand the additional complexity required for real-world deployment
- **Make informed decisions** about when the additional investment in compute makes sense for your goals

When to Make the Investment

As you progress beyond this book, you'll develop intuition for when scaling up makes sense:

- **Research projects** where model quality directly impacts results may justify GPU costs
- **Professional work** where training costs are negligible compared to developer time
- **Production applications** where the value delivered to users far exceeds training expenses
- **Learning opportunities** where hands-on experience with distributed training is the educational goal itself

For initial learning and conceptual understanding, our minimal example provides maximum educational value per dollar spent—which in this case is zero dollars. The TinyStories implementation shows where the path leads, but walking that path has a literal price tag attached.

Appreciating What Was Achieved

Despite the model's limitations for general question-answering, it successfully demonstrates:

- **Complete Training Pipeline**: Data preparation, tokenization, training loop, optimization, checkpointing
- **Modern Architecture**: Transformer with attention, RoPE positional encoding, proper normalization
- **Production Techniques**: Mixed precision training, gradient clipping, learning rate scheduling
- **Sophisticated Inference**: Multimechanism anti-repetition, robust sampling strategies
- **End-to-End Implementation**: From raw text to interactive generation

These are the building blocks that scale up to production systems. Understanding them at this educational scale provides the foundation for working with massive models.

Looking Forward

This big example in the whole book perfectly illustrates the engineering and scientific achievements that enable modern LLMs. The massive scale of production systems isn't arbitrary—it's the result of careful analysis showing what's required for general capabilities.

When we see that GPT-4 has hundreds of billions of parameters trained on trillions of tokens for months on thousands of GPUs, we can now understand **why**

- Those parameters store knowledge spanning all human domains
- Those tokens provide sufficient examples of every pattern and task
- That training time enables deep optimization of complex objectives
- Those resources enable the additional training phases that create helpful assistants

Our small educational model helps us appreciate these achievements by showing us exactly what changes as we scale up. Each limitation we observe in the small model corresponds to a design decision in large models.

This code and output provide an **excellent educational experience** that illuminates fundamental principles of language modeling. By examining how a small, simply trained model behaves when pushed outside its comfort zone, we gain insights into

- How neural networks learn and represent knowledge
- Why model scale, training data, and training procedures matter
- How inference mechanisms interact with learned distributions
- What's required to build general-purpose AI assistants

The model's behavior isn't a failure—it's exactly what we should expect given its size and training, and it teaches us important lessons. By understanding these principles at a small scale, we're better equipped to work with, evaluate, and appreciate production language models.

The path from this educational example to GPT-4 or Claude isn't mysterious—it's a matter of scale, data diversity, training duration, and multistage training procedures, all grounded in the same fundamental principles we've explored here. This makes the example an invaluable teaching tool that bridges theory and practice.

CHAPTER 11

Advanced Training and CUDA Kernels

The Journey from Raw Text to Intelligent Assistant—The Art and Science of LLM Training

Training a large language model is a multistage process that transforms raw data and computation into systems capable of understanding, reasoning, and generating human-like text. Rather than a single task, it is a coordinated sequence of phases, each with its own goals and challenges. Modern LLM development typically involves four stages: pretraining, mid-training, supervised fine-tuning, and alignment training.

Pretraining establishes the model's broad understanding of language and world knowledge by exposing it to vast text corpora. Mid-training then strengthens specific domains or skills, sharpening abilities that the base model only partially acquires. Supervised fine-tuning teaches the model to follow instructions, generate structured responses, and interact in ways that humans find useful. Finally, alignment training ensures the model behaves according to human values and preferences, often using techniques such as reinforcement learning, preference modeling, or safety-focused optimization. Each stage refines the model further, building on the last.

LLM training is both a science and an art. The scientific dimension involves mathematical foundations, careful experiment design, controlled evaluation, and thoughtful architectural decisions based on theory and empirical evidence. The artistic side is harder to quantify: intuition about data quality, judgment in balancing competing objectives, and experience in recognizing when to adjust strategies or stop training. Understanding user needs—especially those not captured by benchmarks—requires empathy and domain insight.

D. Grigorov, *Building Large Language Models from Scratch*, https://doi.org/10.1007/979-8-8688-2297-1_11

Across all stages, several factors remain critical. Scale in data and compute enables complex capabilities to emerge. High-quality, well-curated data ensures the model learns useful patterns rather than noise. Continuous evaluation and iterative improvement gradually raise performance. Ethical practices, transparency, and a focus on safety protect users and society. Above all, anchoring development in real user needs ensures models become genuinely helpful rather than merely impressive on paper.

The field evolves rapidly: new methods appear regularly, theoretical understanding deepens, and model capabilities expand. Yet the central challenge persists—creating AI systems that are consistently helpful, harmless, and honest. These systems should augment human intelligence rather than replace it, and their benefits should extend broadly across society.

The journey from raw text to intelligent assistant demands expertise in machine learning, distributed systems, human behavior, ethics, and domain knowledge. It consumes substantial computational resources and significant human effort. But the outcomes—models that can reason, converse, write, and generate code—represent one of the most remarkable achievements in artificial intelligence. Capabilities once believed to require human cognition now arise from carefully orchestrated training procedures applied to large neural networks.

As techniques continue to mature, efficiency will improve, and understanding will deepen. But the foundational insight remains: patient, responsible, multistage training can produce AI systems of extraordinary value. The art and science of LLM training demonstrates what becomes possible when rigorous engineering meets visionary goals, guided by a commitment to beneficial outcomes.

Pretraining—Building the Foundation

Pretraining is the first and most fundamental stage in developing a large language model. Its purpose is to teach the model the statistical structure of language—grammar, semantics, world knowledge, and common patterns—by exposing it to massive amounts of text. The result is a base model that can understand and generate text but is not yet optimized for instruction following, safety, or conversational use.

Pretraining Data

Modern models train on hundreds of billions to trillions of tokens sourced from diverse datasets. Web crawls like CommonCrawl and C4 provide large-scale general text; books and academic papers supply long-form and technical writing; GitHub and StackOverflow add programming knowledge; Wikipedia contributes structured factual content; and news and filtered social media bring current events and conversational style. Diversity is essential for broad generalization.

Data quality heavily influences final performance. Developers remove duplicates, filter harmful or low-quality content, strip personal information, ensure proper language detection, comply with licensing rules, and maintain balance across topics and perspectives. These steps help ensure the model learns generalizable and responsible patterns rather than noise or bias.

Next-Token Prediction

The core pretraining objective is autoregressive next-token prediction: given previous tokens, predict the next one. This simple task provides unlimited free supervision and forces the model to internalize syntax, meaning, facts, and reasoning in order to minimize prediction errors. When scaled to large datasets and models, this objective leads to strong and sometimes emergent capabilities.

Architecture and Training Setup

Most LLMs use transformer architectures, often decoder-only for generative tasks. Key design choices include parameter count, depth, hidden size, attention heads, context window, and vocabulary size. Training requires massive GPU/TPU clusters and distributed training strategies such as model, data, and pipeline parallelism. Mixed-precision arithmetic, gradient accumulation, and frequent checkpointing improve efficiency and reliability.

Typical hyperparameters include learning rates around 1e-4 to 6e-4, very large batch sizes, the AdamW optimizer, weight decay, and gradient clipping. Pretraining often processes 1–2 trillion tokens.

Compute Budget, Duration, and Evaluation

Chinchilla scaling laws show that model size and dataset size must be balanced—roughly 20 training tokens per parameter—for optimal compute use. Training can take days for small models, weeks or months for mid-sized ones, and many months for the largest.

Evaluation uses perplexity, loss curves, and downstream benchmarks to monitor progress. The finished base model can generate coherent text and demonstrates broad knowledge, but it is not yet aligned with user intent. Further stages of training are necessary to make the model safe, reliable, and instruction-following.

Mid-Training—Targeted Capability Development

What Is Mid-Training?

Mid-training, also called continued pretraining or domain adaptation, is an intermediate stage that bridges general pretraining and fine-tuning. It involves additional pretraining on carefully curated datasets designed to enhance specific capabilities such as reasoning, coding, or mathematical problem-solving; adapt the model to specific domains like medical, legal, or scientific texts; improve particular skills including long-context understanding or multilingual abilities; and prepare the model for the transition to instruction following.

Why Mid-Training Matters

Limitations of Pure Pretraining

While pretraining on diverse web data is remarkably powerful, it has inherent limitations. Coverage across domains is uneven, with some important topics severely underrepresented in web crawls. The quality of web data varies dramatically, from carefully edited professional content to hastily written forum posts filled with errors.

Certain skills that require concentrated examples of specific reasoning patterns or specialized knowledge may not develop fully from diffuse web exposure. Important but rare content can get averaged out and diluted when mixed with the enormous volume of general web text.

Benefits of Mid-Training

Mid-training addresses these limitations by providing domain expertise through focused exposure to high-quality, domain-specific content, leading to significantly improved performance in target areas. It enables capability enhancement, producing dramatic improvements in specific skills like mathematical reasoning or code generation.

The approach is efficient, being far more effective than training a domain-specialized model from scratch. Mid-training also offers flexibility, as different objectives can be applied sequentially to build up complementary capabilities.

Types of Mid-Training

Domain-Specific Mid-Training

Domain-specific mid-training involves concentrated exposure to specialized corpora. Medical mid-training incorporates clinical notes, medical literature, and drug databases to develop clinical reasoning and medical knowledge. Legal mid-training uses legal documents, case law, and statutes to build understanding of legal reasoning and terminology. Scientific mid-training draws from research papers and technical documentation to enhance scientific literacy. Financial mid-training leverages financial reports, market data, and analysis to develop economic and financial expertise. Code mid-training uses source code repositories, technical documentation, and programming discussions to dramatically improve programming capabilities.

Capability-Specific Mid-Training

Rather than targeting a domain, capability-specific mid-training focuses on enhancing particular abilities. Mathematical reasoning training uses math problems, proofs, and textbooks to develop stronger quantitative and logical reasoning. Long-context understanding training employs extended documents, complete books, and lengthy articles to improve the model's ability to maintain coherence and extract information across very long sequences.

Multilingual training incorporates high-quality non-English text to balance language capabilities. Instruction following training uses instruction-formatted data to begin bridging toward the SFT stage. Tool use training provides examples of API calls and function usage to teach the model how to interact with external tools effectively. Chain-of-thought reasoning training uses step-by-step reasoning examples to develop more transparent and reliable reasoning processes.

Data Quality Enhancement

Some mid-training focuses on learning from curated, exceptionally high-quality subsets of data. This approach filters content for factual accuracy, verifies logical consistency, selects material for its pedagogical value in teaching concepts clearly, and aggressively cleans out noise and errors that could confuse the model's learned patterns.

Mid-Training Methodology

Data Curation

The quality of mid-training data is paramount to success. The process begins with source selection, identifying the highest-quality sources within the target domain. Rigorous filtering removes low-quality or irrelevant content that could dilute the learning signal. Verification ensures factual accuracy wherever possible, which is particularly important for domains like medicine or law, where errors could be dangerous.

Formatting structures the data optimally for learning, sometimes adding structure or context that helps the model extract the right patterns. Finally, determining mixing ratios balances domain-specific data with enough general data to prevent the model from forgetting its broader capabilities.

Training Approach

Mid-training typically uses learning rates somewhat lower than initial pretraining, often ranging from 1e-5 to 1e-4, to avoid disrupting the model's existing knowledge too drastically. The duration is typically shorter than full pretraining, involving billions to hundreds of billions of tokens rather than trillions. Mixing strategies often continue to include general data alongside domain-specific content to maintain balance. Evaluation uses domain-specific benchmarks to measure whether the targeted improvements are actually materializing.

Preventing Catastrophic Forgetting

A key challenge in mid-training is maintaining general capabilities while adding new ones. Several strategies help prevent catastrophic forgetting. Data mixing incorporates general web data alongside domain-specific data so the model continues seeing the full range of language use. Replay buffers explicitly revisit samples from the original pretraining data to reinforce important patterns.

Elastic weight consolidation applies regularization based on parameter importance, protecting weights that are crucial for existing capabilities. Conservative learning rate schedules avoid overly aggressive updates that might overwrite existing knowledge. Regular evaluation monitors performance on general benchmarks to catch any degradation in broad capabilities before it becomes severe.

Examples of Successful Mid-Training

Code-specialized models provide compelling demonstrations of mid-training's power. When a base model undergoes mid-training on large code repositories, it develops dramatically improved coding abilities, including better understanding of programming languages, stronger debugging capabilities, and more reliable code generation. CodeLlama, derived from Llama 2 through code-focused mid-training, exemplifies this approach, as do models like StarCoder.

Mathematical reasoning represents another successful application. Additional training on mathematical texts, formal proofs, and detailed problem solutions produces models with substantially enhanced quantitative capabilities. Google's Minerva model, which excels at complex STEM problems, demonstrates how targeted mid-training can develop sophisticated mathematical reasoning that goes far beyond what base pretraining alone achieves.

Long-context models showcase yet another dimension of mid-training success. Through training on longer sequences combined with position interpolation techniques, models learn to handle contexts of 32K, 100K, or even over 1 million tokens. This enables entirely new applications like analyzing entire codebases, reading complete books, or maintaining coherent narratives across very long conversations.

Mid-Training vs. Fine-Tuning

Key differences:

Aspect	Mid-Training	Fine-Tuning
Objective	Continue language modeling on targeted data	Learn to follow instructions/formats
Data format	Raw text (next token prediction)	Instruction-response pairs
Scale	Billions to hundreds of billions of tokens	Thousands to millions of examples
Purpose	Enhance capabilities/knowledge	Shape behavior and style
Learning rate	Moderate	Lower
Duration	Weeks to months	Days to weeks

Supervised Fine-Tuning (SFT)—Teaching Instructions

The Transition from Base to Assistant

Base models are powerful but awkward to use in practice. They excel at completing text but don't naturally follow instructions, answer questions directly, format responses appropriately, refuse inappropriate requests, or engage in natural dialog. Supervised fine-tuning transforms base models from text completion engines into helpful assistants that interact naturally with users.

SFT Data: Instructions and Demonstrations

Data Format

SFT data consists of carefully crafted input-output pairs that demonstrate desired behavior. For example, an instruction might ask, "Explain photosynthesis in simple terms," and the corresponding response would provide a clear, accessible explanation: "Photosynthesis is the process plants use to convert sunlight into energy. Plants take in carbon dioxide from the air and water from the soil. Using sunlight captured by chlorophyll in their leaves, they combine these to create glucose (sugar) for energy and release oxygen as a byproduct. This is why plants are so important—they produce the oxygen we breathe!"

Data Sources

SFT datasets draw from multiple sources to ensure comprehensive coverage. Human demonstrations involve people writing example responses that exemplify the desired style and quality. Model distillation uses outputs from larger or better models as training examples for smaller models. Synthetic generation employs existing capable models to create diverse instruction-following examples across many domains. User data from real interactions can be incorporated with appropriate privacy protections and filtering. Task-specific datasets from public sources can be reformatted as instructions to teach particular capabilities.

Dataset Composition

A comprehensive SFT dataset must cover general knowledge questions across various topics and difficulty levels; task instructions including summarization, translation, analysis, and creative writing; reasoning problems spanning mathematics, logic, and common sense; code generation tasks in various programming languages; multiturn dialog demonstrating conversational ability; refusal examples showing how to decline inappropriate requests; format-following demonstrations for structured outputs and specific styles; and safety examples handling sensitive topics appropriately.

SFT Training Process

Objective Function

Unlike pretraining, which predicts all next tokens indiscriminately, SFT typically employs one of two approaches. Full sequence training computes loss over both the instruction and response, while response-only training (more common) only computes loss on the response portion. The response-only approach is generally preferred because it ensures the model learns to produce good responses rather than simply mimicking any text pattern, focusing the learning signal on the behavior we actually want to shape.

Training Hyperparameters

Learning rates during SFT are much lower than during pretraining, typically ranging from 1e-5 to 5e-6. This conservative approach is necessary because the model already knows language—we're teaching it a new behavior pattern without damaging existing knowledge. Training usually proceeds for just 1–3 epochs over the SFT data to avoid overfitting. Batch sizes are smaller than in pretraining, typically thousands to tens of thousands of tokens per batch. The entire SFT phase usually takes days to a few weeks, much faster than the months required for pretraining.

Data Quality in SFT

The Importance of Quality over Quantity

Research has consistently shown that quality matters far more than quantity in supervised fine-tuning. Ten thousand high-quality examples typically produce better results than a million low-quality examples. Diversity of tasks and scenarios matters more than sheer count. Response quality directly impacts model behavior since the model is explicitly trying to imitate these demonstrations.

Quality Indicators

High-quality SFT data exhibits several key characteristics. Accuracy means information is factually correct and reliable. Helpfulness ensures responses directly address the instruction rather than providing tangentially related information. Clarity manifests in well-written, easily understandable responses. Appropriate length balances thoroughness with conciseness, being neither too brief nor unnecessarily verbose. Style consistency maintains an appropriate tone throughout. Safety considerations ensure sensitive topics are handled appropriately with nuance and care.

Balancing the SFT Dataset

Task Distribution

An effective SFT dataset requires broad coverage across different task types, representation of various difficulty levels from simple queries to complex multistep challenges, content spanning multiple domains and topics, and examples of different response styles and formats to give the model flexibility.

Avoiding Overfitting

Preventing overfitting requires sufficient diversity to prevent memorization of specific examples. Regular evaluation on held-out validation data ensures the model is learning to generalize rather than memorize. Continuous monitoring for degradation of general capabilities ensures that gains in instruction-following don't come at the cost of broader knowledge and abilities.

Multiturn Dialog Training

Teaching conversational abilities requires special attention to multiturn interactions. A conversation might begin with a simple factual question like "What's the capital of France?" answered concisely as "Paris," followed by a follow-up like "Tell me more about it" that requires the model to understand the referent and provide additional relevant information about Paris's location, cultural significance, and famous landmarks.

This training presents several challenges. The model must maintain context across multiple turns, referring back to earlier parts of the conversation appropriately. It needs to maintain consistency across responses, not contradicting itself or forgetting what was previously discussed. Natural dialog flow requires understanding when to provide brief answers versus detailed explanations, when to ask clarifying questions, and how to gracefully handle topic shifts.

The Result: An Instruction-Following Model

After SFT, the model has transformed significantly. It can follow diverse instructions accurately across many domains and task types. It engages naturally in multiturn conversations, maintaining context and coherence. It provides helpful, informative responses that directly address user needs. It formats outputs appropriately using markdown, code blocks, or other structures as needed. It can decline inappropriate requests, at least to some extent.

However, limitations remain even after successful SFT. The model may produce inconsistent preferences when responding to similar prompts in different ways. It may fail to refuse harmful requests reliably, sometimes being too cautious or not cautious enough. Generated responses may be verbose or repetitive, lacking the conciseness humans prefer. The model's outputs may not match human preferences optimally across multiple dimensions. These remaining issues motivate the need for alignment training.

Reinforcement Learning from Human Feedback (RLHF)

The Alignment Problem

After supervised fine-tuning, models can follow instructions competently, but significant challenges remain. When presented with the same prompt, models generate responses of varying quality, and we need a way to consistently prefer better responses. Safety concerns persist, as models may still produce harmful content in adversarial scenarios.

Truthfulness issues manifest when models confidently state incorrect information without appropriate uncertainty. Helpfulness problems arise when models give technically correct but practically unhelpful responses. Style mismatches occur when the model's conversational approach doesn't align with desired interaction patterns. RLHF addresses these challenges by teaching models to understand and optimize for human preferences.

The Three Stages of RLHF

The RLHF process consists of three distinct stages. First comes supervised fine-tuning, which we've already covered, creating an instruction-following model. Second is reward model training, where we teach a separate model to predict human preferences between different responses. Third is reinforcement learning optimization, where we use the reward model to improve the policy model's behavior.

Reward Model Training

Purpose

The reward model learns to predict human preferences with a specific goal: given two responses to the same prompt, predict which one humans would prefer. This seemingly simple task encodes complex human values about helpfulness, truthfulness, safety, and style into a numerical score that can guide further training.

Data Collection Process

Building a reward model begins with prompt selection, gathering diverse prompts from actual usage patterns or carefully designed scenarios that cover important edge cases. For each prompt, we use response generation to create multiple responses, typically four to nine variants, using the SFT model. Human comparison involves trained labelers who rank or compare these responses according to defined criteria. This process produces preference data that explicitly captures which responses are better in a format the reward model can learn from.

For example, given the prompt "Explain quantum entanglement," we might generate multiple responses ranging from oversimplified ("Quantum entanglement is when particles are connected") to comprehensive and accurate explanations. Human labelers would prefer the response that balances accuracy, clarity, and appropriate technical depth.

Training the Reward Model

The reward model typically uses the same base architecture as the language model but replaces the token prediction head with a scalar output that produces a single number representing predicted quality. The training objective is to learn which response humans would prefer from the comparison data. The loss function is typically a ranking loss or classification loss that penalizes the model when it predicts the wrong preference ordering.

Mathematically, this can be expressed as minimizing the loss: $-\log(\sigma(r\_preferred - r\_rejected))$, where r represents the reward score assigned to each response and σ is the sigmoid function ensuring the output represents a probability.

Reward Model Outputs

Once trained, the reward model can assign a scalar score to any prompt-response pair, representing the predicted human preference for that response. Higher scores indicate responses that humans would likely prefer, while lower scores suggest responses that would be dispreferred. This score becomes the training signal for the final stage of RLHF.

Reinforcement Learning Optimization

The Setup

At this stage, we have assembled all the necessary components. Our policy model is the SFT model we want to improve. Our reward model predicts human preferences for any response. Our goal is to maximize the reward our policy receives while staying reasonably close to the original model to prevent erratic behavior.

The Algorithm: Proximal Policy Optimization (PPO)

PPO has emerged as the most common reinforcement learning algorithm for RLHF. The process works iteratively through several steps. The policy model generates responses to a batch of prompts. The reward model scores each generated response. We compute advantages by calculating how good each response is relative to the average. We update the policy to make higher-reward responses more likely while keeping changes bounded. This cycle repeats for many iterations, gradually improving the policy's outputs.

Key Innovation: KL Penalty

A critical innovation in RLHF is the KL (Kullback-Leibler) divergence penalty that prevents the model from gaming the reward model or forgetting its capabilities. The final reward used for training combines the reward model's score with a penalty term: Final_reward = Reward_model_score - β × KL_divergence(new_policy, original_policy). This penalty ensures the model doesn't deviate too far from the original SFT model, maintaining stability and preventing the model from finding pathological solutions that maximize reward in unexpected ways.

Challenges in RLHF

Reward Hacking

One significant challenge is that models might find ways to achieve high rewards that don't actually align with human preferences. This "reward hacking" can manifest in several ways. Models might generate overly long responses if length is implicitly associated with quality in the training data. They might use certain phrases or styles that the reward model has learned to associate with high quality, even when those phrases don't actually improve the response. The model might exploit biases in how the reward model was trained, finding edge cases where the reward model's predictions diverge from actual human preferences.

Several mitigation strategies help address reward hacking. Regular reward model updates using fresh human comparison data prevent the policy from over-optimizing for a static reward function. Diverse training prompts expose reward hacking strategies that only work on narrow prompt distributions. Continuous human evaluation alongside automated metrics catches when reward scores diverge from actual quality. Using multiple reward models with different training data or architectures makes it harder to game them all simultaneously.

Reward Model Limitations

The reward model itself has inherent limitations that bound what RLHF can achieve. It can only capture preferences reflected in its training data, missing important considerations that weren't emphasized during labeling. Biases from labeler preferences inevitably seep into the reward model, which might not represent the full diversity of user preferences. Generalization to out-of-distribution prompts can be poor, with the reward model giving unreliable scores for situations quite different from its training data. Different reward models or human evaluators might disagree about which response is better, especially for subjective qualities.

Training Instability

Reinforcement learning training can be inherently unstable, presenting several risks. The model might diverge from useful behavior if updates are too aggressive or poorly calibrated. Mode collapse can occur where the model starts generating very similar responses regardless of the input. Catastrophic forgetting of capabilities learned during pretraining and SFT can happen if the RL process isn't carefully managed.

Solutions to these stability issues include using conservative KL penalties to limit how much the policy can change per update. Careful learning rate tuning prevents too-rapid optimization. Regular checkpointing and evaluation on diverse test sets catch problems early. Mixing in supervised fine-tuning data during RL training helps maintain important capabilities.

Practical Considerations

Computational Cost

RLHF represents a substantial computational investment. The training process needs to run multiple models simultaneously, including the policy being trained, the reward model scoring outputs, and the reference policy for computing KL divergence. Generating many samples per training step is necessary for stable learning but computationally expensive. Large GPU clusters are often required, similar to the pretraining phase. The entire RLHF process can take weeks of training time even with substantial resources.

Human Labeling

The human labeling process requires considerable care and resources. Thousands to hundreds of thousands of comparison judgments are needed to train a robust reward model. Labeler training and calibration are crucial to ensure consistent, high-quality judgments. Inter-annotator agreement must be monitored to catch when labelers interpret instructions differently or disagree systematically. Ensuring diverse labeler perspectives prevents the reward model from encoding a narrow set of preferences. Handling subjective preferences where there isn't a clear right answer requires careful protocol design.

Results of RLHF

Models trained with RLHF typically demonstrate substantial improvements across several dimensions. Helpfulness increases as responses become more useful and informative, better addressing what users actually want to know. Safety improves with reduced harmful outputs and more appropriate handling of sensitive topics. Preference alignment manifests in responses that humans consistently prefer over pre-RLHF outputs. Consistency improves as the model develops more reliable patterns of behavior. Verbosity often reduces as the model learns when conciseness is preferred over lengthy explanations.

Notable examples of RLHF's success include ChatGPT using GPT-3.5 and GPT-4, Claude models from Anthropic, and Gemini from Google. All these production systems use forms of RLHF as a critical component of their training pipeline, demonstrating the technique's practical importance in building capable AI assistants.

Alternative Alignment Approaches

Direct Preference Optimization (DPO)

The Innovation

Direct preference optimization, introduced in 2023, represents a significant simplification of the RLHF pipeline. DPO achieves similar results to RLHF while skipping the reward model entirely. It directly optimizes the policy using preference data without a separate reinforcement learning phase. Under certain mathematical conditions, DPO is theoretically equivalent to RLHF but with a much simpler implementation.

How DPO Works

Given preference pairs consisting of a prompt, a winning response, and a losing response, DPO computes a loss that directly updates the policy. The loss function is $-\log(\sigma(\beta \times (\log \pi\_\theta(y\_w|x) - \log \pi\_ref(y\_w|x) - (\log \pi\_\theta(y\_l|x) - \log \pi\_ref(y\_l|x)))))$.

Here, π_θ represents the policy being trained, π_ref is the reference policy (typically the SFT model), y_w is the preferred response, y_l is the dispreferred response, and β controls how far the policy can deviate from the reference model.

This formulation directly optimizes for preference without needing to estimate a reward function first, simplifying the training pipeline substantially.

Advantages of DPO

DPO offers several compelling advantages over traditional RLHF. The approach is simpler since it requires no reward model training or complex RL infrastructure. Training tends to be more stable than PPO-based RLHF, with fewer hyperparameters to tune. The method is more efficient computationally, with less overhead from running multiple models. Effectiveness matches or often exceeds RLHF performance in practice despite the simpler approach.

Limitations

DPO does have some constraints. It requires high-quality preference data, as there's no reward model to smooth over data imperfections. The approach offers less flexibility than RLHF for some applications, particularly those requiring online learning or adaptive reward modeling. DPO may not capture complex preference distributions as effectively when preferences are highly context-dependent or multidimensional.

Constitutional AI (CAI)

Philosophy

Developed by Anthropic, Constitutional AI takes a different approach by embedding explicit principles—a "constitution"—directly into the training process. Rather than relying purely on implicit human preferences captured through comparisons, CAI makes the values and principles guiding model behavior explicit and transparent.

Two-Stage Process

The first stage involves critique and revision through a sophisticated self-improvement loop. The model generates an initial response to a prompt. It then critiques its own response against constitutional principles, identifying potential problems or areas for improvement. Based on this critique, the model revises its response to better align with the stated principles. This process of generation, critique, and revision produces high-quality training data that embodies the constitutional principles.

The second stage uses reinforcement learning from AI feedback (RLAIF) rather than human feedback. A reward model is trained using AI-generated preferences that are explicitly based on the constitutional principles. Standard RL techniques then optimize the policy using this reward model, but with the AI-generated preferences grounded in the explicit constitution rather than human comparison data.

Constitutional Principles Example

A constitution might include principles such as "Choose the response that is most helpful, harmless, and honest," "Avoid responses that could be threatening or violent," "Choose the response that is least racist," and "Choose the response that has the least stereotyping." These explicit principles guide both the critique-revision process and the AI preference generation.

Advantages

Constitutional AI offers several key benefits. The approach scales better than pure human feedback since AI can generate far more comparison data than humans can label. Explicit principles make model behavior more transparent and easier to understand and audit. Organizations can iterate quickly on safety improvements by modifying or adding constitutional principles. The reduced need for human labeling lowers costs and accelerates development cycles.

Reinforcement Learning from AI Feedback (RLAIF)

Core Idea

RLAIF replaces human labelers with AI models in the preference generation pipeline. A strong model such as GPT-4 evaluates responses and generates preference comparisons. These AI-generated preferences are then used to train a reward model following the standard RLHF process. Finally, RL training proceeds normally using the reward model trained on AI preferences.

When RLAIF Works Well

RLAIF excels in certain scenarios. Tasks with clear correctness criteria, such as mathematical problems or code correctness, allow AI evaluators to make reliable judgments. When scaling beyond human labeling capacity, RLAIF provides a path to generate millions of comparisons. In domains where AI evaluators have demonstrated reliability through validation, RLAIF can dramatically reduce labeling costs while maintaining quality.

Limitations

However, RLAIF faces important limitations. AI evaluators inevitably carry their own biases and limitations, which transfer to the trained model. The approach may not capture nuanced human preferences that require intuition, empathy, or cultural context. Quality fundamentally depends on the evaluator model's capabilities—weak evaluators produce weak training signals. There's a risk of inheriting and amplifying the evaluator's flaws, potentially creating models that optimize for the wrong objectives.

Iterative Approaches

Iterative RLHF

Production systems often use iterative refinement cycles. The process trains an initial reward model, runs RL to improve the policy, collects new comparison data from the improved model, retrains the reward model on the expanded dataset, and repeats this cycle. This progressive improvement allows the system to continuously learn from new data and adapt to evolving requirements. Systems like ChatGPT have successfully employed this approach.

Online Learning

Some systems implement continuous learning from production data, updating models based on real user interactions. This requires extremely careful privacy and safety controls to prevent the model from learning inappropriate behaviors from adversarial users while still benefiting from legitimate feedback.

Hybrid Approaches

Modern production systems increasingly combine multiple methods to leverage their complementary strengths. A typical pipeline might use SFT plus RLHF plus Constitutional AI, layering different alignment techniques. Different methods might address different capabilities—DPO for some aspects, RLHF for others. Human and AI feedback can be mixed strategically, using human judgment where nuance matters and AI evaluation for scale. Regular safety evaluations and targeted interventions complement the automated training processes, ensuring models remain safe and aligned even as they improve on other dimensions.

Post-Training Techniques and Refinements

Context Distillation

Context distillation addresses the challenge of embedding instructions or desired behaviors into the model without requiring them at inference time. The process begins by running the model with a detailed system prompt that specifies desired behaviors. We collect high-quality outputs produced under this prompt's guidance. We then fine-tune the model on those outputs without including the system prompt in the training data. Through this process, the model internalizes the behavior patterns, making them default rather than prompt-dependent.

This technique offers several benefits. It saves context window space that would otherwise be consumed by lengthy system prompts. Inference becomes faster without the overhead of processing long instructions. The model's behavior becomes more consistent since the patterns are trained in rather than depending on prompt interpretation.

Self-Improvement Techniques

Self-Critique

Self-critique establishes a feedback loop within the model itself. The model generates an initial response to a prompt, then critiques its own response according to specific criteria. Based on this critique, the model generates an improved version. Training on these improved responses teaches the model to produce higher-quality outputs directly.

Iterative Refinement

Iterative refinement leverages diversity to improve quality. The model solves problems or addresses prompts using multiple different approaches. We identify and keep the best solutions based on correctness, clarity, or other quality metrics. Training on these best solutions improves the model's reasoning and accuracy by teaching it to favor approaches that lead to better outcomes.

Red Teaming and Adversarial Training

Red Teaming

Red teaming involves dedicated teams whose explicit goal is to make the model fail. They search for edge cases where the model produces harmful, incorrect, or inappropriate outputs. They develop jailbreaks that bypass safety measures. They identify failure modes that wouldn't appear in normal usage. From these efforts, we create adversarial datasets capturing challenging cases. Fine-tuning the model to handle these cases appropriately makes it more robust.

Adversarial Training

Adversarial training systematizes the process of generating challenging inputs. Automated systems systematically generate inputs designed to elicit specific failure modes. The model is explicitly trained to handle these adversarial cases robustly. This iterative process improves both safety and reliability, making the model resistant to manipulation.

Capability-Specific Fine-Tuning

After alignment training establishes general safety and helpfulness, additional targeted fine-tuning can enhance specific capabilities. Tool use fine-tuning focuses on API calls and function usage, teaching the model to interact effectively with external systems. Code execution fine-tuning improves code quality and correctness. Factuality fine-tuning specifically targets reducing hallucinations and improving factual accuracy. Reasoning fine-tuning enhances step-by-step problem-solving approaches.

Multiobjective Optimization

Real-world deployment requires balancing multiple, sometimes competing objectives. Helpfulness must be balanced against safety—being maximally informative sometimes conflicts with being cautious. Accuracy needs to be weighed against engagement—technically perfect responses might be dry. Conciseness competes with thoroughness—some users want brief answers, while others need detailed explanations. Politeness can conflict with directness—some contexts demand bluntness.

Several approaches address multiobjective optimization. Multiobjective reward models explicitly model trade-offs between different qualities. Weighted combinations of objectives allow tuning the balance between competing goals. Pareto optimization seeks solutions that aren't strictly dominated on any objective. Conditional generation based on user preferences allows the model to adapt its behavior to different user needs and contexts.

Evaluation and Benchmarking

Evaluation During Pretraining

Intrinsic Metrics

Pretraining progress is monitored through several intrinsic metrics. Perplexity measures how well the model predicts held-out text, with lower values indicating better performance. Training loss should decrease steadily as the model learns. Validation loss helps detect overfitting by showing whether the model generalizes beyond its training data.

Downstream Tasks

Models are also evaluated on their performance across various downstream tasks. Reading comprehension benchmarks like SQuAD and NaturalQuestions test whether the model can extract answers from passages. Common sense reasoning tasks, including HellaSwag and PIQA, assess intuitive understanding. World knowledge evaluations such as MMLU and TriviaQA probe the breadth of learned information. Language understanding benchmarks like SuperGLUE measure fundamental linguistic capabilities.

Evaluation During Supervised Fine-Tuning

Instruction Following

SFT evaluation focuses heavily on whether models can follow complex multistep instructions accurately, handle ambiguous or underspecified requests gracefully, and ask clarifying questions when appropriate. These capabilities are essential for useful assistant behavior.

Task Performance

Specific task metrics measure quantitative performance. Summarization quality is assessed using ROUGE scores comparing generated summaries to references. Translation accuracy uses BLEU scores measuring n-gram overlap. Question answering accuracy tracks whether the model provides correct answers. Code generation is evaluated using pass@k metrics on test cases, measuring how often generated code runs correctly.

Style and Format

Beyond correctness, evaluation considers whether responses are appropriately lengthy—neither too brief nor unnecessarily verbose. Formatting should be correct, properly using markdown, code blocks, and other structures. Tone should be consistent and appropriate to the context. Readability metrics ensure outputs are clear and accessible.

Evaluation During Alignment Training

Preference Modeling

Alignment training evaluation focuses on agreement with held-out human preferences not used during training. Consistency of preferences ensures the model doesn't contradict itself with small prompt variations. Robustness across prompt variations tests whether preferences are genuine rather than artifacts of specific phrasing.

Safety Evaluations

Safety testing examines refusal of harmful requests, ensuring the model appropriately declines dangerous or unethical instructions. Handling of sensitive topics is evaluated for nuance and care. Bias detection and measurement quantify fairness across demographic groups. Adversarial prompt resistance tests whether safety measures can be easily bypassed.

Human Evaluation

Ultimately, human judgment remains crucial. Head-to-head comparisons pit different models or versions against each other. Absolute quality ratings assess individual responses. Multidimensional assessment evaluates helpfulness, accuracy, safety, and other qualities separately. A/B testing with real users provides ground truth about user preferences in production contexts.

Comprehensive Benchmarks

Knowledge and Reasoning

MMLU (massive multitask language understanding) tests knowledge across 57 subjects ranging from elementary topics to professional domains. Big-Bench provides a diverse collection of challenging tasks designed to probe model capabilities. TruthfulQA specifically evaluates factual accuracy and truthfulness, testing whether models produce reliable information. GSM8K assesses grade school math reasoning, measuring quantitative problem-solving abilities.

Coding

HumanEval evaluates Python code generation through a set of programming problems with test cases. MBPP (mostly basic programming problems) assesses fundamental programming skills. CodeContests tests competitive programming abilities with complex algorithmic challenges.

Safety and Alignment

BBQ (bias benchmark for QA) measures social biases across various demographic dimensions. ToxiGen assesses toxic language detection and generation resistance. Red team findings from organizations like Anthropic systematically document adversarial testing results and failure modes.

Multilingual

XNLI evaluates cross-lingual natural language understanding across many languages. FLORES measures translation quality between diverse language pairs. XGLUE provides multilingual task evaluations spanning various linguistic capabilities.

The Limitations of Benchmarks

Benchmark Saturation

Benchmarks face inherent limitations that reduce their effectiveness over time. Models increasingly train on benchmark data, whether directly or through similar examples in pretraining corpora, leading to contamination. Performance plateaus on established benchmarks even as actual capabilities continue improving. This creates a constant need for fresh, harder benchmarks that better distinguish between different capability levels.

Gap Between Benchmarks and Real-World Use

A fundamental challenge is that benchmarks test specific, narrow skills in controlled settings, while real usage involves open-ended, diverse applications with complex user needs. Strong benchmark performance doesn't guarantee user satisfaction because benchmarks can't capture the full richness of real interactions. They miss crucial factors like appropriateness, style, personality, and the ability to handle genuinely novel situations.

Solutions

Addressing these limitations requires multiple approaches. New, harder benchmarks must be developed regularly to stay ahead of model improvements. Contamination detection and mitigation techniques help identify when models have seen test data during training. Greater emphasis on human evaluation provides richer feedback than automated metrics alone. Real-world deployment metrics, including user satisfaction, task completion rates, and actual usage patterns, provide ground truth about model utility beyond what any benchmark can capture.

Practical Considerations and Best Practices

Data Is King

At every stage of the training pipeline, data quality matters more than quantity. During pretraining, careful filtering and deduplication establish a solid foundation by ensuring the model learns from high-quality text rather than memorizing low-quality repetitive content. Mid-training benefits enormously from domain-expert curation when possible, as specialists can identify the most valuable learning materials. For supervised fine-tuning, high-quality demonstrations are worth a hundred times more than mediocre ones—a few thousand excellent examples outperform millions of poor examples. RLHF success depends critically on careful labeler training and rigorous quality control to ensure preference judgments are reliable and consistent.

Computational Resource Management

Cost-Benefit Analysis

Understanding where to invest computational resources is crucial for effective model development. Pretraining represents the most expensive phase, typically requiring months of GPU cluster time and consuming the bulk of the compute budget. Mid-training demands moderate resources, usually taking weeks of focused training. SFT is relatively inexpensive, often completing in days with much smaller computational requirements. RLHF falls in the moderate to expensive range, requiring weeks of training and the overhead of running multiple models simultaneously.

Strategic Decisions

Organizations face critical strategic choices throughout the development process. The fundamental question of when to train from scratch versus fine-tuning existing open models affects both cost and capability. Deciding which stages deserve the heaviest investment depends on the specific use case and capability gaps. Trade-offs between model size and training duration must account for Chinchilla scaling laws and computational budgets. Smaller models trained longer may outperform larger undertrained models for the same compute cost.

Preventing Degradation

Common Pitfalls

Several failure modes can undermine model quality if not carefully managed. Overfitting occurs when running too many training epochs on small datasets, causing the model to memorize rather than generalize. Catastrophic forgetting happens when the model loses general capabilities while learning specialized skills. Reward hacking emerges when models find ways to maximize alignment metrics that don't correspond to genuine quality improvements. Dataset bias occurs when training data doesn't represent the true distribution of use cases, leading to skewed behavior.

Prevention Strategies

Preventing these problems requires vigilance across multiple dimensions. Regular evaluation on diverse test sets catches degradation before it becomes severe. Data mixing and replay prevent catastrophic forgetting by continuing to expose the model to broad content. Conservative learning rates avoid overly aggressive updates that might damage existing knowledge. Extensive testing before deployment, including adversarial testing and diverse user scenarios, helps identify problems that automated metrics miss.

Scaling Laws and Efficiency

Compute-Optimal Training

The Chinchilla scaling insight suggests models should see approximately 20 tokens per parameter for optimal performance at fixed compute. This principle helps organizations balance model size against training duration. However, diminishing returns appear at very large scales, where incremental improvements require exponentially more resources.

Efficiency Techniques

Various techniques improve training efficiency without sacrificing quality. Mixed precision training using FP16, BF16, or even INT8 reduces memory requirements and increases training speed. Gradient checkpointing trades additional computation for reduced memory consumption, enabling larger batch sizes. Flash Attention provides optimized attention implementations that dramatically speed up the core Transformer operations. Model pruning removes unnecessary parameters after training, reducing inference costs. Post-training quantization reduces numerical precision, enabling faster and cheaper deployment while maintaining most model capabilities.

Responsible AI Considerations

Throughout the Training Pipeline

Responsible development requires attention at every stage. Data ethics ensures training data is ethically sourced with appropriate permissions and attribution. Bias mitigation actively works to reduce harmful biases rather than passively accepting them. Privacy protection strips personally identifying information and prevents models from memorizing sensitive details. Transparency about capabilities and limitations helps users understand what models can and cannot do reliably. Safety testing through extensive red teaming and adversarial evaluation finds failure modes before deployment. Deployment monitoring continues evaluation in production, catching emerging issues as user interactions reveal new edge cases.

Red Lines

Certain practices should never be compromised. Never train on explicitly illegal content regardless of potential performance gains. Remove personal identifying information rigorously to protect privacy. Filter extreme toxicity and harm from training data. Implement safety guardrails at every stage of the pipeline, not as an afterthought but as a core design principle. If any malicious use case is detected, refuse to proceed regardless of technical curiosity or commercial pressure.

The Future of LLM Training

Emerging Trends

Multimodal Training

The future of LLM training increasingly involves training on text, images, audio, and video simultaneously rather than treating each modality separately. This approach creates unified representations that work across all modalities, enabling more natural human-computer interaction. Models can understand relationships between visual and textual information, connect spoken language to written text, and process the world in ways that mirror human multimodal perception.

Longer Context Windows

Context windows have expanded dramatically from initial limits of 2,000 tokens to current systems handling 100,000 tokens and experimental systems processing over 1 million tokens. This expansion requires new architectural innovations, including sparse attention mechanisms that don't scale quadratically with sequence length and explicit memory systems that can store and retrieve information efficiently. Longer contexts enable entirely new applications like processing complete books, analyzing entire codebases in one pass, and maintaining coherent conversations across days or weeks of interaction.

Continuous Learning

Future systems will likely move toward continuous learning paradigms where models update themselves from new information without full retraining. This requires solving challenging technical problems around catastrophic forgetting and stability. Success would enable much faster iteration on improvements, with models adapting to new information, correcting errors when identified, and incorporating user feedback immediately rather than waiting for the next training cycle.

More Efficient Training Methods

Few-Shot and Zero-Shot Alignment

Emerging research aims to reduce the need for large supervised fine-tuning datasets by improving models' ability to learn alignment from minimal examples. Better transfer from pretraining could enable models to behave appropriately with little or no task-specific training. More efficient use of human feedback through techniques like active learning could identify the most valuable examples to label, dramatically reducing annotation costs.

Self-Supervised Alignment

The vision of models learning to align themselves without extensive human supervision is becoming more concrete. Constitutional AI demonstrates how self-improvement cycles can replace some human labeling. Future systems might autonomously identify and correct their own mistakes, reducing reliance on external feedback. However, this approach requires careful validation to ensure models align with actual human values rather than learned proxies.

Mixture of Experts

Mixture of experts architectures activate only relevant subnetworks for each input rather than using the entire model. This enables better scaling by increasing model capacity without proportionally increasing computational cost. Specialized sub-networks can focus on different domains or capabilities. This architecture potentially offers the benefits of very large models at a fraction of the inference cost.

Better Evaluation

More Robust Benchmarks

Future evaluation will require contamination-resistant methods that remain valid even when models have seen similar data during training. Open-ended generative assessments that resist gaming will better measure actual capabilities. Emphasis will shift from narrow task performance to real-world task completion, measuring end-to-end success on complex, realistic problems.

Automatic Evaluation

As AI evaluators improve, they increasingly match human judgment on many dimensions. This enables faster iteration on improvements since evaluation doesn't bottleneck on human availability. Automatic evaluation can scale beyond human capacity, assessing thousands of model variants quickly. However, human evaluation remains essential for subjective qualities and for validating that automated metrics truly capture what matters.

Democratization

Smaller, More Efficient Models

Intensive research focuses on creating smaller models that match larger ones through better training techniques, data quality, and architectural innovations. Success makes advanced AI capabilities accessible to researchers and developers without massive computational resources. Lower inference costs enable broader deployment and more diverse applications.

Open-Source Progress

High-quality open models like LLaMA, Mistral, and others have dramatically accelerated research by providing strong starting points. Community-driven improvements leverage distributed expertise to advance the field faster. Open development accelerates innovation by allowing researchers worldwide to experiment and share findings.
This trend toward openness, balanced with safety considerations, will likely continue reshaping the AI landscape.

Better Tools and Infrastructure

Training frameworks are becoming easier to use, abstracting away low-level complexity. Pretrained components and strong starting points reduce the need to train from scratch. Automated hyperparameter tuning removes some of the art from model training, making it more accessible to practitioners without extensive experience.

Theoretical Understanding

Why Does It Work?

Despite remarkable empirical success, theoretical understanding of why large language models work so well remains incomplete. Deeper understanding of emergence would explain how complex capabilities arise from simple training objectives. Better theories of generalization would make model behavior more predictable. Understanding these fundamentals would enable more reliable capability forecasting and more principled training approaches.

Controllability

Future research aims for more precise control over model behavior, enabling fine-grained steering toward desired outcomes. Better customization would allow models to be tailored to specific use cases or user preferences without full retraining. Most ambitiously, reliable safety guarantees backed by mathematical proofs rather than empirical testing would enable deployment in high-stakes domains with confidence.

Training Neural Networks with CUDA Kernels and Modern Frameworks

Training large language models and deep neural networks requires enormous computational power. While frameworks like PyTorch and TensorFlow abstract away the low-level details, understanding CUDA kernels—the fundamental building blocks of GPU-accelerated training—provides crucial insights into performance optimization and what's actually happening when you train a model.

Understanding CUDA and GPU Computing

What Is CUDA?

CUDA (Compute Unified Device Architecture) is NVIDIA's parallel computing platform and programming model. It allows developers to harness the massive parallel processing power of NVIDIA GPUs for general-purpose computing, far beyond just graphics rendering.

Why GPUs for Deep Learning?

Modern GPUs contain thousands of cores designed for parallel computation. While a CPU might have 8–64 cores optimized for sequential processing, a high-end GPU like the NVIDIA H100 has over 16,000 CUDA cores.

This massive parallelism is perfect for neural network operations, which often involve:

- Matrix multiplications with millions or billions of elements
- Element-wise operations across large tensors
- Gradient computations that can be parallelized
- Batch processing where each example can be handled independently

The CPU vs. GPU Paradigm

Training on CPUs processes operations sequentially or with limited parallelism. A matrix multiplication might process one element at a time or use vectorization to process small batches. GPUs, in contrast, can compute thousands of matrix elements simultaneously, each on a different core. For a 1000x1000 matrix multiplication, a GPU might compute all one million output elements in parallel, completing in microseconds what would take a CPU milliseconds.

CUDA Kernels Explained

What Is a CUDA Kernel?

A CUDA kernel is a function that runs on the GPU and is executed in parallel by many threads. When you launch a kernel, you specify how many threads should run it, and each thread typically processes a small portion of the data.

Think of a kernel as a single instruction that thousands of workers (GPU threads) execute simultaneously, each on their own piece of data. This is called single instruction, multiple data (SIMD) or, more accurately, single instruction, multiple threads (SIMT) in CUDA terminology.

Thread Hierarchy

CUDA organizes threads into a hierarchy:

Threads: The smallest unit of execution. Each thread runs the kernel code and has a unique ID.

Blocks: Threads are grouped into blocks. Threads within a block can cooperate and share data through shared memory. A block might contain 256, 512, or 1024 threads.

Grid: Blocks are organized into a grid. The grid represents the entire computation being performed by the kernel launch.

For example, to process a 1024x1024 matrix, you might create a grid of 32x32 blocks, each containing 32x32 threads, giving you 1,048,576 threads total—one for each matrix element.

Kernel Syntax

A CUDA kernel is defined with the `__global__` keyword in C/C++:

```
__global__ void myKernel(float* data, int n) {
    int idx = blockIdx.x * blockDim.x + threadIdx.x;
    if (idx < n) {
        data[idx] = data[idx] * 2.0f;  // Each thread doubles one element
    }
}
```

Key Components:

- `__global__`: Declares this function as a kernel callable from CPU
- `blockIdx`: The block's position in the grid
- `blockDim`: The number of threads per block
- `threadIdx`: The thread's position within its block
- `idx`: The unique global index for this thread

Memory Hierarchy

CUDA has multiple memory types with different speeds and access patterns:

> **Global Memory**: Large (gigabytes) but slow. Accessible by all threads. This is where your model parameters, activations, and gradients typically live.
>
> **Shared Memory**: Fast (on-chip) but small (kilobytes per block). Threads in the same block can share data here. Used for cooperative algorithms and caching frequently accessed data.
>
> **Registers**: Fastest but very limited. Private to each thread. Used for thread-local variables.
>
> **Constant Memory**: Read-only memory cached for fast access. Good for hyperparameters that don't change during a kernel.

Efficient kernel design carefully manages data movement through this hierarchy, keeping frequently accessed data in faster memory.

CUDA Kernels in Neural Network Training

Where Kernels Are Used

Every operation in neural network training ultimately runs as CUDA kernels (when using GPUs). The main operations include

> **Matrix Multiplication (GEMM)**: The backbone of neural networks. Forward passes multiply inputs by weights; backward passes multiply gradients by transposed weights. Libraries like cuBLAS provide highly optimized GEMM kernels.

Element-Wise Operations: Activation functions (ReLU, tanh, sigmoid), dropout, normalization. Each element can be computed independently, making them perfectly suited for parallel execution.

Reductions: Operations like computing loss, norms, or statistics. These require coordination between threads but can still be heavily parallelized.

Convolutions: In CNNs, convolutions are implemented as specialized kernels that slide filters over inputs. Libraries like cuDNN provide optimized implementations.

The Training Loop at Kernel Level

When you write in PyTorch:

```
loss = model(x)
loss.backward()
optimizer.step()
```

What actually happens involves hundreds or thousands of kernel launches:

Forward Pass:

1. Matrix multiplication kernels compute layer activations
2. Element-wise kernels apply activation functions
3. Normalization kernels compute means and variances
4. Loss computation kernels calculate the objective

Backward Pass:

1. Loss gradient kernel computes initial gradients
2. Gradient kernels propagate through each layer
3. Matrix multiplication kernels compute weight gradients
4. Gradient accumulation kernels sum gradients

Optimizer Step:

1. Momentum/Adam state update kernels
2. Weight update kernels apply gradients to parameters
3. Gradient clipping kernels if needed

Each kernel launch has overhead (typically microseconds), so frameworks batch operations and fuse kernels when possible to minimize launches.

Why Custom Kernels Matter

While frameworks provide excellent performance out of the box, custom kernels become important for:

- **Fused Operations**: Combining multiple operations into a single kernel reduces memory traffic. For example, fusing matrix multiplication with activation function application saves reading/writing intermediate results.
- **Novel Operations**: New layer types or loss functions might not have optimized implementations. Writing custom kernels enables experimentation.
- **Memory Optimization**: Custom kernels can implement operations in-place or with minimal temporary storage, crucial for very large models.
- **Performance Critical Paths**: Identifying bottlenecks and writing specialized kernels can provide 2-10x speedups for specific operations.

Practical CUDA Kernel Examples

Example 1: Element-Wise ReLU Activation

One of the simplest but most common operations in deep learning is the ReLU activation function: `output = max(0, input)`.

CUDA Kernel Implementation:

```
__global__ void relu_forward(const float* input, float* output, int n) {
    int idx = blockIdx.x * blockDim.x + threadIdx.x;

    if (idx < n) {
        output[idx] = fmaxf(0.0f, input[idx]);
    }
}

__global__ void relu_backward(const float* grad_output, const float* input,
                                float* grad_input, int n) {
    int idx = blockIdx.x * blockDim.x + threadIdx.x;

    if (idx < n) {
        grad_input[idx] = (input[idx] > 0.0f) ? grad_output[idx] : 0.0f;
    }
}
```

Launching the Kernel:

```
int n = 1000000;  // 1 million elements
int blockSize = 256;
int numBlocks = (n + blockSize - 1) / blockSize;

relu_forward<<<numBlocks, blockSize>>>(d_input, d_output, n);
```

This launches enough blocks and threads to cover all elements, with each thread processing exactly one element.

Key Features:

- Each thread handles one element independently
- No coordination between threads needed
- Memory access is coalesced (sequential threads access sequential memory)
- The backward pass uses the same pattern but computes gradients

Example 2: Matrix Multiplication (Naive Implementation)

Matrix multiplication is the workhorse of neural networks. Here's a simplified implementation to illustrate the concepts (production code uses highly optimized libraries):

```
__global__ void matmul_naive(const float* A, const float* B, float* C,
                              int M, int N, int K) {
    // C[M,N] = A[M,K] * B[K,N]
    int row = blockIdx.y * blockDim.y + threadIdx.y;
    int col = blockIdx.x * blockDim.x + threadIdx.x;

    if (row < M && col < N) {
        float sum = 0.0f;
        for (int k = 0; k < K; k++) {
            sum += A[row * K + k] * B[k * N + col];
        }
        C[row * N + col] = sum;
    }
}
```

Launching:

```
dim3 blockSize(16, 16);  // 16x16 = 256 threads per block
dim3 numBlocks((N + blockSize.x - 1) / blockSize.x,
               (M + blockSize.y - 1) / blockSize.y);

matmul_naive<<<numBlocks, blockSize>>>(d_A, d_B, d_C, M, N, K);
```

Limitations of Naive Implementation:

- Each element of A and B is loaded from global memory many times
- No use of shared memory for caching
- Memory access patterns are not optimal
- Real implementations use tiling, shared memory, and register blocking

Example 3: Optimized Matrix Multiplication with Shared Memory

Here's a more sophisticated version using shared memory to cache tiles of the input matrices:

```
#define TILE_SIZE 32

__global__ void matmul_tiled(const float* A, const float* B, float* C,
                              int M, int N, int K) {
    __shared__ float As[TILE_SIZE][TILE_SIZE];
    __shared__ float Bs[TILE_SIZE][TILE_SIZE];

    int row = blockIdx.y * TILE_SIZE + threadIdx.y;
    int col = blockIdx.x * TILE_SIZE + threadIdx.x;

    float sum = 0.0f;

    // Loop over tiles
    for (int t = 0; t < (K + TILE_SIZE - 1) / TILE_SIZE; t++) {
        // Load tile into shared memory
        if (row < M && t * TILE_SIZE + threadIdx.x < K)
            As[threadIdx.y][threadIdx.x] = A[row * K + t * TILE_SIZE +
            threadIdx.x];
        else
            As[threadIdx.y][threadIdx.x] = 0.0f;

        if (t * TILE_SIZE + threadIdx.y < K && col < N)
            Bs[threadIdx.y][threadIdx.x] = B[(t * TILE_SIZE + threadIdx.y)
            * N + col];
        else
            Bs[threadIdx.y][threadIdx.x] = 0.0f;

        __syncthreads();  // Wait for all threads to load their data

        // Compute using shared memory
        for (int k = 0; k < TILE_SIZE; k++) {
            sum += As[threadIdx.y][k] * Bs[k][threadIdx.x];
        }
```

```
        __syncthreads();  // Wait before loading next tile
    }

    if (row < M && col < N) {
        C[row * N + col] = sum;
    }
}
```

Key Improvements:

- Tiles of A and B are loaded into fast shared memory
- Each element is loaded from global memory only once per tile
- Threads cooperate to load and compute on shared data
- `__syncthreads()` ensures all threads in a block coordinate properly

This tiled version can be 10-100x faster than the naive version depending on matrix sizes.

Example 4: Softmax with Numerical Stability

Softmax is used in attention mechanisms and classification layers. It requires computing exponentials and a reduction:

```
__global__ void softmax_forward(const float* input, float* output, int
batch_size, int dim) {
    int batch_idx = blockIdx.x;
    if (batch_idx >= batch_size) return;

    // Shared memory for reduction
    __shared__ float shared_max;
    __shared__ float shared_sum;

    const float* x = input + batch_idx * dim;
    float* y = output + batch_idx * dim;

    // Find maximum (for numerical stability)
    float thread_max = -INFINITY;
```

```
    for (int i = threadIdx.x; i < dim; i += blockDim.x) {
        thread_max = fmaxf(thread_max, x[i]);
    }

    // Reduce to find global max
    __shared__ float temp[256];
    temp[threadIdx.x] = thread_max;
    __syncthreads();

    for (int stride = blockDim.x / 2; stride > 0; stride >>= 1) {
        if (threadIdx.x < stride) {
            temp[threadIdx.x] = fmaxf(temp[threadIdx.x], temp[threadIdx.x +
            stride]);
        }
        __syncthreads();
    }

    if (threadIdx.x == 0) {
        shared_max = temp[0];
    }
    __syncthreads();

    // Compute exp(x - max) and sum
    float thread_sum = 0.0f;
    for (int i = threadIdx.x; i < dim; i += blockDim.x) {
        float exp_val = expf(x[i] - shared_max);
        y[i] = exp_val;
        thread_sum += exp_val;
    }

    // Reduce to find global sum
    temp[threadIdx.x] = thread_sum;
    __syncthreads();

    for (int stride = blockDim.x / 2; stride > 0; stride >>= 1) {
        if (threadIdx.x < stride) {
            temp[threadIdx.x] = temp[threadIdx.x] + temp[threadIdx.x +
            stride];
        }
```

```
        __syncthreads();
    }

    if (threadIdx.x == 0) {
        shared_sum = temp[0];
    }
    __syncthreads();

    // Normalize
    for (int i = threadIdx.x; i < dim; i += blockDim.x) {
        y[i] = y[i] / shared_sum;
    }
}
```

Key Concepts:

- Uses shared memory for efficient reductions
- Subtracts maximum for numerical stability (prevents overflow)
- Multiple synchronization points ensure correct ordering
- Each block processes one example in the batch

Example 5: Custom Fused Kernel—LayerNorm + GELU

Fusing operations reduces memory bandwidth requirements by minimizing the number of times intermediate results must be written to and read from GPU memory. Instead of running layer normalization and GELU activation as two separate kernels—each requiring its own memory loads and stores—a fused kernel keeps intermediate values in registers or shared memory, dramatically improving efficiency on modern accelerators. Here's a kernel that combines layer normalization with the GELU activation:

```
__global__ void layernorm_gelu_fused(const float* input, float* output,
                                     const float* gamma, const float* beta,
                                     int batch_size, int hidden_dim,
float eps) {
    int batch_idx = blockIdx.x;
    if (batch_idx >= batch_size) return;

    const float* x = input + batch_idx * hidden_dim;
```

```
    float* y = output + batch_idx * hidden_dim;

    // Compute mean
    __shared__ float shared_mean;
    float thread_sum = 0.0f;
    for (int i = threadIdx.x; i < hidden_dim; i += blockDim.x) {
        thread_sum += x[i];
    }

    // Reduction for mean
    __shared__ float temp[256];
    temp[threadIdx.x] = thread_sum;
    __syncthreads();

    for (int stride = blockDim.x / 2; stride > 0; stride >>= 1) {
        if (threadIdx.x < stride) {
            temp[threadIdx.x] += temp[threadIdx.x + stride];
        }
        __syncthreads();
    }

    if (threadIdx.x == 0) {
        shared_mean = temp[0] / hidden_dim;
    }
    __syncthreads();

    // Compute variance
    __shared__ float shared_var;
    float thread_var = 0.0f;
    for (int i = threadIdx.x; i < hidden_dim; i += blockDim.x) {
        float diff = x[i] - shared_mean;
        thread_var += diff * diff;
    }

    // Reduction for variance
    temp[threadIdx.x] = thread_var;
    __syncthreads();
```

```
    for (int stride = blockDim.x / 2; stride > 0; stride >>= 1) {
        if (threadIdx.x < stride) {
            temp[threadIdx.x] += temp[threadIdx.x + stride];
        }
        __syncthreads();
    }

    if (threadIdx.x == 0) {
        shared_var = temp[0] / hidden_dim;
    }
    __syncthreads();

    // Normalize and apply GELU
    float std = sqrtf(shared_var + eps);
    const float sqrt_2_over_pi = 0.7978845608f;
    for (int i = threadIdx.x; i < hidden_dim; i += blockDim.x) {
        // LayerNorm
        float normalized = (x[i] - shared_mean) / std;
        normalized = normalized * gamma[i] + beta[i];

        // GELU activation: 0.5 * x * (1 + tanh(sqrt(2/π) * (x + 0.044715
        * x^3)))
        float x3 = normalized * normalized * normalized;
        float tanh_arg = sqrt_2_over_pi * (normalized + 0.044715f * x3);
        y[i] = 0.5f * normalized * (1.0f + tanhf(tanh_arg));
    }
}
```

Benefits of Fusion:

- Input is read once instead of twice (once for LayerNorm, once for GELU).
- Intermediate normalized values don't need to be written to/read from global memory.
- Can achieve 1.5–2x speedup compared to separate operations.

Integration with Deep Learning Frameworks

PyTorch Custom CUDA Extensions

PyTorch makes it relatively easy to write and use custom CUDA kernels through its C++/CUDA extension mechanism.

File: custom_relu.cu

```
#include <torch/extension.h>
#include <cuda.h>
#include <cuda_runtime.h>

__global__ void relu_forward_kernel(const float* input, float* output,
int n) {
    int idx = blockIdx.x * blockDim.x + threadIdx.x;
    if (idx < n) {
        output[idx] = fmaxf(0.0f, input[idx]);
    }
}

__global__ void relu_backward_kernel(const float* grad_output, const
float* input,
                                     float* grad_input, int n) {
    int idx = blockIdx.x * blockDim.x + threadIdx.x;
    if (idx < n) {
        grad_input[idx] = (input[idx] > 0.0f) ? grad_output[idx] : 0.0f;
    }
}

torch::Tensor relu_forward_cuda(torch::Tensor input) {
    auto output = torch::zeros_like(input);
    const int n = input.numel();
    const int threads = 256;
    const int blocks = (n + threads - 1) / threads;

    relu_forward_kernel<<<blocks, threads>>>(
        input.data_ptr<float>(),
        output.data_ptr<float>(),
```

```
        n
    );

    return output;
}

torch::Tensor relu_backward_cuda(torch::Tensor grad_output, torch::Tensor
input) {
    auto grad_input = torch::zeros_like(input);
    const int n = input.numel();
    const int threads = 256;
    const int blocks = (n + threads - 1) / threads;

    relu_backward_kernel<<<blocks, threads>>>(
        grad_output.data_ptr<float>(),
        input.data_ptr<float>(),
        grad_input.data_ptr<float>(),
        n
    );

    return grad_input;
}

PYBIND11_MODULE(TORCH_EXTENSION_NAME, m) {
    m.def("forward", &relu_forward_cuda, "ReLU forward (CUDA)");
    m.def("backward", &relu_backward_cuda, "ReLU backward (CUDA)");
}
```

File: relu_extension.py

```
import torch
from torch.utils.cpp_extension import load

# JIT compile the CUDA extension
custom_relu = load(
    name='custom_relu',
    sources=['custom_relu.cu'],
    extra_cuda_cflags=['-O3']
)
```

```
class CustomReLU(torch.autograd.Function):
    @staticmethod
    def forward(ctx, input):
        ctx.save_for_backward(input)
        return custom_relu.forward(input)

    @staticmethod
    def backward(ctx, grad_output):
        input, = ctx.saved_tensors
        return custom_relu.backward(grad_output, input)

# Use in a model
def relu(x):
    return CustomReLU.apply(x)

# Example usage
x = torch.randn(1000000, device='cuda', requires_grad=True)
y = relu(x)
loss = y.sum()
loss.backward()
```

Triton: High-Level GPU Programming

Triton is a Python-like language for writing GPU kernels that's easier than CUDA but still provides excellent performance:

```
import torch
import triton
import triton.language as tl

@triton.jit
def relu_kernel(
    input_ptr,
    output_ptr,
    n_elements,
    BLOCK_SIZE: tl.constexpr,
):
```

```
    pid = tl.program_id(axis=0)
    block_start = pid * BLOCK_SIZE
    offsets = block_start + tl.arange(0, BLOCK_SIZE)
    mask = offsets < n_elements

    x = tl.load(input_ptr + offsets, mask=mask)
    output = tl.maximum(0.0, x)
    tl.store(output_ptr + offsets, output, mask=mask)

def relu_triton(x):
    output = torch.empty_like(x)
    n_elements = x.numel()

    grid = lambda meta: (triton.cdiv(n_elements, meta['BLOCK_SIZE']),)
    relu_kernel[grid](x, output, n_elements, BLOCK_SIZE=1024)

    return output

# Usage
x = torch.randn(1000000, device='cuda')
y = relu_triton(x)
```

Triton automatically handles many optimizations like memory coalescing and generates efficient PTX code, but the system goes far beyond just simplifying memory access patterns.

At its core, Triton is a domain-specific language and compiler built specifically for writing high-performance GPU kernels with a fraction of the effort required in CUDA. When a developer writes a Triton kernel, the Triton compiler analyzes the computation, data shapes, and access patterns to apply a suite of hardware-aware optimizations.

Memory Coalescing and Data Movement Optimization

Triton ensures that threads within a warp access contiguous memory locations whenever possible. This coalescing allows the hardware to merge many individual loads and stores into a small number of wide memory transactions. Triton also automatically aligns memory accesses, uses vectorized loads and stores when beneficial, and reduces unnecessary movement of data between global memory, shared memory, and registers.

Automatic Tiling and Parallelization

Choosing the right tile sizes in GPU programming is notoriously hard because it depends on register pressure, shared memory limits, and warp occupancy. Triton's compiler explores these constraints automatically. It partitions workloads into blocks that maximize parallel execution while fitting the GPU's architectural limits. The developer simply specifies an abstract layout (like block sizes in each dimension), and Triton determines how to map these into warp- and thread-level execution.

Register Allocation and Instruction Scheduling

Much like a traditional optimizing compiler, Triton performs register allocation to ensure that intermediate values stay in fast on-chip storage rather than spilling to slower memory. It also schedules instructions to avoid pipeline stalls and keep ALUs fully utilized. These optimizations can significantly improve throughput, especially in compute-heavy kernels such as matrix multiplications or normalization layers.

Operation Fusion

Because Triton kernels are written as monolithic blocks of code, developers can easily fuse multiple operations—like normalization, activation functions, and elementwise transforms—into a single pass over memory. The compiler helps by eliminating redundant loads, simplifying expressions, and keeping intermediate values in registers. This reduces memory bandwidth requirements, often the bottleneck in deep learning workloads.

Efficient PTX Generation

Ultimately, Triton lowers everything into optimized PTX code tailored to NVIDIA GPUs. This includes choosing appropriate instructions (e.g., tensor core ops when available), applying loop unrolling, and emitting specialized math instructions. As a result, Triton kernels are frequently competitive with—or sometimes faster than—expert-tuned CUDA kernels.

Ease of Use and Rapid Prototyping

One of Triton's biggest strengths is usability. Developers can express high-performance GPU computations using Pythonic syntax and NumPy-like indexing rather than dealing with CUDA's ceremony. This lowers the barrier to creating custom GPU ops, enabling quick experimentation in areas like model acceleration, fused layers, or novel attention mechanisms.

In practice, Triton empowers ML researchers and engineers to achieve CUDA-level performance without needing deep GPU programming expertise, accelerating both innovation and production optimization.

Performance Optimization Strategies

Memory Coalescing

Modern GPUs load memory in chunks (typically 128 bytes). When threads in a warp (group of 32 threads) access consecutive memory addresses, the GPU can fulfill all requests with a single memory transaction.

Bad: Strided Access

// Threads access memory with large strides—inefficient

```
int idx = threadIdx.x * STRIDE;  // If STRIDE is large, many
transactions needed
float val = data[idx];
```

Good: Coalesced Access

// Threads access consecutive memory—efficient

```
int idx = threadIdx.x;  // Consecutive threads access consecutive memory
float val = data[idx];
```

Occupancy Optimization

GPU occupancy refers to how many warps are active relative to the maximum. Higher occupancy helps hide memory latency by switching between warps.

Factors affecting occupancy:

- **Registers Per Thread**: More registers means fewer active threads
- **Shared Memory Per Block**: More shared memory means fewer active blocks
- **Block Size**: Must be a multiple of warp size (32)

Use NVIDIA's Occupancy Calculator or profiler to find optimal configurations.

Kernel Fusion

Launching a kernel has overhead (~1-10 microseconds). Fusing multiple operations into one kernel reduces this overhead and saves memory bandwidth.

Unfused: Three Kernels

```
// Kernel 1: Matrix multiply
matmul<<<...>>>(A, B, C);
// Kernel 2: Add bias
add_bias<<<...>>>(C, bias);
// Kernel 3: Apply ReLU
relu<<<...>>>(C);
```

Fused: One Kernel

```
// Single kernel does all three operations
matmul_bias_relu<<<...>>>(A, B, bias, C);
```

Asynchronous Operations

GPUs can overlap computation with data transfer using CUDA streams:

```
cudaStream_t stream1, stream2;
cudaStreamCreate(&stream1);
cudaStreamCreate(&stream2);
```

// These can run concurrently

```
cudaMemcpyAsync(d_A, h_A, size, cudaMemcpyHostToDevice, stream1);
kernel1<<<grid, block, 0, stream2>>>(d_B, d_C);
```

This overlapping hides transfer latency and keeps the GPU busy.

Using Tensor Cores

Modern NVIDIA GPUs (Volta and newer) have specialized Tensor Cores for mixed-precision matrix multiplication. These provide 8–16x speedup for certain operations.

To use Tensor Cores:

- Use mixed precision (FP16 inputs, FP32 accumulation)
- Ensure matrix dimensions are multiples of 8 or 16
- Use library functions (cuBLAS, cuDNN) that automatically leverage them
- In custom kernels, use `wmma` (warp matrix multiply-accumulate) intrinsics

Chapter 7: Real-World Training Performance

Profiling and Bottleneck Identification

NVIDIA provides several profiling tools:

- **NSight Systems**: System-level profiling showing CPU/GPU overlap, kernel launches, and memory transfers.
- **NSight Compute**: Kernel-level profiling showing metrics like memory bandwidth utilization, compute utilization, and occupancy.

PyTorch Profiler: High-level profiling integrated with PyTorch:

```
with torch.profiler.profile(
    activities=[
        torch.profiler.ProfilerActivity.CPU,
        torch.profiler.ProfilerActivity.CUDA,
    ],
    with_stack=True
) as prof:
    for _ in range(10):
        output = model(input)
        loss = criterion(output, target)
        loss.backward()
        optimizer.step()

print(prof.key_averages().table(sort_by="cuda_time_total", row_limit=10))
```

Common Bottlenecks:

- Memory bandwidth saturation (>80% bandwidth utilization)
- Small kernels with high launch overhead
- Synchronization points causing idle time
- CPU bottlenecks in data loading or preprocessing

FlashAttention Example

FlashAttention is a famous example of custom CUDA kernels dramatically improving training performance. Standard attention has $O(N^2)$ memory complexity and requires many kernel launches.

FlashAttention uses:

- Tiling to keep data in fast SRAM (shared memory)
- Fused operations to reduce memory traffic
- Recomputation of attention during backward pass instead of storing

Results: 2–4x speedup and much lower memory usage, enabling longer context lengths.

The key insight: Modern GPUs are memory-bound, not compute-bound. Reducing memory traffic by clever algorithm design and kernel fusion provides bigger gains than raw compute optimizations.

Training at Scale

Large-scale training uses multiple GPUs with various parallelism strategies:

Data Parallel: Each GPU processes different data with the same model. Gradients are synchronized via NCCL (NVIDIA Collective Communications Library), which uses custom CUDA kernels for efficient all-reduce operations.

Model Parallel: Different parts of the model are on different GPUs. Requires careful kernel scheduling to overlap computation and communication.

Pipeline Parallel: Model stages are on different GPUs, processing different micro-batches. Kernel launches must be carefully orchestrated to maintain the pipeline.

All of these rely on highly optimized CUDA kernels for:

- Communication collectives (all-reduce, all-gather, etc.)
- Memory management and copying
- Computation

Practical Tips for Working with CUDA Kernels

When to Write Custom Kernels

Write custom kernels when:

- Profiling shows a clear bottleneck that fusion could address
- You need a novel operation not available in libraries
- Memory is the bottleneck, and you can reduce traffic
- You understand the operation deeply enough to optimize it

Don't write custom kernels when:

- Existing library functions (cuBLAS, cuDNN) are adequate
- The operation is not a bottleneck
- Development time would be better spent on model architecture
- You're new to CUDA (start by understanding existing kernels)

Development Workflow

1. **Start with PyTorch/TensorFlow**: Implement and verify correctness
2. **Profile**: Identify actual bottlenecks
3. **Research**: Check if optimized implementations exist (libraries, papers)
4. **Prototype in Triton**: Easier than raw CUDA
5. **Optimize in CUDA**: If needed for the last few percentage points
6. **Validate**: Ensure numerical accuracy matches reference
7. **Benchmark**: Measure actual speedup on real data

Debugging CUDA Kernels

CUDA errors can be cryptic. Useful techniques:

Check for errors after every kernel:

```
cudaError_t err = cudaGetLastError();
if (err != cudaSuccess) {
    printf("CUDA error: %s\n", cudaGetErrorString(err));
}
```

Use cuda-memcheck:

```
cuda-memcheck ./my_program
```

This catches out-of-bounds memory access, race conditions, etc.

Print debugging:

```
__global__ void debug_kernel(float* data) {
    int idx = threadIdx.x;
    if (idx == 0) {  // Only one thread prints
        printf("Block %d, data[0] = %f\n", blockIdx.x, data[0]);
    }
}
```

Common Pitfalls

- **Race Conditions**: Threads accessing the same memory without proper synchronization can produce non-deterministic results. Use `__syncthreads()` or atomic operations.
- **Bank Conflicts**: In shared memory, sequential 32-bit addresses are divided among 32 banks. If multiple threads in a warp access the same bank, it serializes. Structure data access to avoid this.
- **Divergent Branches**: If threads in a warp take different code paths, they execute serially. Minimize branching or ensure all threads in a warp take the same branch.
- **Insufficient Occupancy**: Too many registers or shared memory can limit active warps. Use compiler flags like `-maxrregcount` or restructure to use less shared memory.

CUDA kernels are the fundamental building blocks that make modern deep learning possible. While high-level frameworks abstract away most details, understanding how training actually executes on GPUs provides valuable insights for optimization and debugging.

Key Takeaways:

- **Massive Parallelism**: GPUs excel at operations that can be split across thousands of threads. Neural network operations naturally fit this model.
- **Memory Hierarchy Matters**: Modern GPUs are memory-bound. Efficient kernel design focuses on minimizing data movement and maximizing use of fast memory.
- **Fusion Is Powerful**: Combining operations reduces kernel launch overhead and memory traffic, often providing bigger gains than raw compute optimization.
- **Libraries Are Highly Optimized**: cuBLAS, cuDNN, and similar libraries represent years of optimization work. Use them when possible.
- **Custom Kernels Have Their Place**: For novel operations, memory-critical paths, or fusion opportunities, custom kernels can provide substantial speedups.

Understanding CUDA kernels transforms deep learning from a black box into a comprehensible system where you can reason about performance, optimize bottlenecks, and push the boundaries of what's trainable. Whether you're training massive language models or developing new architectures, this knowledge provides the foundation for efficient GPU computing.

Understanding these concepts will make you a more effective deep learning practitioner, capable of training larger models faster and debugging performance issues that would otherwise be mysterious.

Appendix: Glossary of Terms

- **Autoregressive**: Generating sequences one token at a time, where each token depends on previous tokens
- **Base Model**: The result of pretraining before any fine-tuning or alignment
- **Catastrophic Forgetting**: When a model loses previously learned knowledge during further training
- **Constitutional AI**: Alignment method using explicit principles and AI-generated feedback
- **Context Window**: Maximum sequence length a model can process at once
- **DPO (Direct Preference Optimization)**: Alignment method that directly optimizes preferences without a reward model
- **Emergent Abilities**: Capabilities that appear at scale but aren't present in smaller models
- **Fine-Tuning**: Additional training on a specific task or domain after pretraining
- **Foundation Model**: Another term for base model; a model trained for general-purpose use
- **Hyperparameters**: Settings that control the training process (learning rate, batch size, etc.)
- **KL Divergence**: Measure of difference between two probability distributions
- **MID-Training**: Continued pretraining on targeted data between pretraining and fine-tuning
- **RLHF (Reinforcement Learning from Human Feedback)**: Alignment method using human preferences and RL
- **Perplexity**: Measure of how well a model predicts text; lower is better
- **Pretraining**: Initial training phase on large text corpora

- **Reward Model**: Model trained to predict human preferences for RLHF
- **Scaling Laws**: Relationships between model size, data size, and performance
- **SFT (Supervised Fine-Tuning)**: Teaching models to follow instructions using demonstration examples
- **Tokenization**: Breaking text into smaller units (tokens) for processing
- **Transformer**: Neural network architecture used in modern LLMs

Index

A

B

D. Grigorov, *Building Large Language Models from Scratch*, https://doi.org/10.1007/979-8-8688-2297-1

C

D

E

F

G

H

M

N

O

P

Q

R

S

T

U

V

W, X

Y

Z